Current Topics in Microbiology 96 and Immunology

Editors

W. Henle, Philadelphia · P.H. Hofschneider, Martinsried
H. Koprowski, Philadelphia · F. Melchers, Basle
R. Rott, Gießen · H.G. Schweiger, Ladenburg/Heidelberg
P.K. Vogt, Los Angeles

Gene Cloning in Organisms Other Than E. coli

Edited by P.H. Hofschneider
and W. Goebel

With 63 Figures

Springer-Verlag
Berlin Heidelberg New York 1982

Professor Dr.Dr. P.H. Hofschneider
Max-Planck-Institut für Biochemie
D-8033 Martinsried b. München

Professor Dr. W. Goebel
Institut für Genetik und Mikrobiologie
der Universität
Röntgenring 11
D-8700 Würzburg

ISBN 3-540-11117-4 Springer-Verlag Berlin Heidelberg New York
ISBN 0-387-11117-4 Springer-Verlag New York Heidelberg Berlin

Typesetting: Fotosatz Service Weihrauch, Würzburg
Printing and binding: Universitätsdruckerei H. Stürtz AG, Würzburg
2121/3321-543210

Table of Contents

Indexed in Current Contents

Preface

Gene cloning for the production of recombinant DNA is usually performed with E. coli. There is, however, no doubt that gene cloning in organisms other than E. coli will assume a much more important role in the future: efficient cloning systems are needed for the analysis of gene expression and its regulation in eukaryotic cells, for the elucidation of the genetic mechanisms of transformation, and for the study of the genetic rearrangement during differentiation and embryogenesis. Furthermore, optimal cloning systems will be required for practical applications in the near future: for the optimal production of antibiotics, amino acids, vitamins, enzymes etc., for the expression of gene products under favorable energetic conditions for mass production, for the expression of glycolysated animal proteins, for the genetic manipulation of plants – and in the more distant future, for gene therapy in man.

The editors have therefore invited leading scientists to summarize the present status and the future applicability of gene cloning systems in their fields of research. To ensure a complete coverage of a genetic system, the editors have generally asked two authors to cover one system.

This volume is a first and unique account of our knowledge of cloning systems employing organisms other than E. coli such as B. subtilis, pseudomonas, neurospora, saccharomyces and streptomyces, as well as of vectors and selection procedures for animal and plant systems including the liposome technique.

October 1981
P.H. Hofschneider, Martinsried
W. Goebel, Würzburg

Cloning Vectors Derived from Plasmids and Phage of Bacillus

Jürgen Kreft* and Colin Hughes*

1 Introduction

Bacillus subtilis is a well-characterized, gram-positive, non-pathogenic, spore-forming soil bacterium which produces a wide array of extracellular enzymes (for reviews see *Young* 1980; *Priest* 1977; *Henner* and *Hoch* 1980). The development of molecular cloning systems within this organism will not only greatly assist biochemical and genetic studies but should play a fundamental role in the further development of biotechnological processes based on the *Bacilli*.

As described by *Ehrlich* (this volume), following the realization that antibiotic resistance plasmids from *Staphylococcus aureus* could be transformed into *B. subtilis* (*Ehrlich* 1977), much effort has been devoted to developing *S. aureus* plasmid cloning vectors for the *Bacilli*. In this review we will describe the current status of vectors constructed from plasmids and phage indigenous to the *Bacilli*.

In the wake of the development of recombinant DNA techniques in *E. coli*, studies of *Bacillus* plasmids gave way to searches aimed at isolating potential cloning vectors. Extrachromosomal DNA in *Bacillus* was first demonstrated in *B. megaterium* and since then many reports have been made of plasmids in this species (*Carlton* and *Smith* 1974; *Rostas* et al. 1980), *B. subtilis* (*Lovett* and *Bramucci* 1975; *Tanaka* and *Koshikawa* 1977; *Bernhard* et al. 1978; *Uozumi* et al. 1980), and *B. pumilus* (*Lovett* et al. 1976).

* Institut für Genetik und Mikrobiologie der Universität, Röntgenring 11, D-8700 Würzburg, West Germany

Table 1. *Bacillus* plasmids developed as cloning vehicles

Plasmid	Source	Mol. wt. (x 10^{-6})	Copy number	Restriction sites	Marker	Reference
pLS 28	*B. subtilis* (natto)	4.1	5	*Eco*RI(2), *Bam*HI(1), *Hind*III(5)	–	*Tanaka* and *Koshikawa* 1977
pBS 1	*B. subtilis*	5.5	6	*Eco*RI(1), *Bam*HI(1), *Sal*I(1), *Pst*I(3), *Hind*III(6)	–	*Bernhard* et al. 1978
pBC 16	*B. cereus*	3.0	20	*Eco*RI(2), *Bam*HI(1)	Tc^R	*Bernhard* et al. 1978
pAB 124	*B. stearo-thermophilus*	2.9	?	*Eco*RI(3), *Hind*III(2), single sites for *Bst*EII, *Cau*I, *Hpa*I, *Xba*I	Tc^R	*Bingham* et al. 1979

Numbers in parentheses indicate the number of restriction sites for this enzyme

However, the majority of these plasmids, in most cases isolated from culture collection strains, lack readily identifiable markers and are thus not of immediate use as cloning vehicles. Nevertheless two such cryptic plasmids, *pBS*1 and *pLS*28, have been developed further. Examination of environmental isolates has proved more successful. *Bernhard* et al. (1978) isolated the tetracycline resistance (Tc^R) plasmid *pBC*16 from a strain of *B. cereus* found in soil, and the antibiotic-resistant thermophiles discovered in river sludge and silage yielded the two Tc^R plasmids *pAB*124 and *pAB*128 (*Bingham* et al. 1979).

The following section describes the development of some of these plasmids (Table 1) as cloning vehicles.

2 Development and Use of Vector Plasmids

2.1 Vectors Capable of Replication Only in B. subtilis

The tetracycline resistance plasmid *pBC*16 isolated from *B. cereus* can be transformed into *B. subtilis* (*Bernhard* et al. 1978) in which it replicates quite stably with no detectable segregation after more than 100 generations without selective pressure. It contains two *Eco*RI sites and in order to determine if these sites were within the tetracycline resistance gene(s) *pBC*16 was partially and completely digested with *Eco*RI and ligated with *Eco*RI linearized *pBS*1, a cryptic plasmid isolated from *B. subtilis*.

None of the Tc^R colonies obtained after transformation of competent cells of *B. subtilis* 168 with the ligated mixture contained a complete hybrid of the two parental plasmids but several derivatives were isolated (Fig. 1), *pBC*16-1, *pBS*161, and *pBS*161-1 being of particular interest (*Kreft* et al. 1978). The plasmid *pBC*16-1 is the circularized large *Eco*RI fragment (mol. wt. 1.8 × 10^6) of *pBC*16 which obviously carries both the replication functions and the tetracycline resistance determinant.

*pBS*161 and *pBS*162 have been found together in a large number of tetracycline-resistant colonies, the former plasmid alone carrying a Tc^R determinant.

Recircularization in vitro of the largest *Hind*III fragment of *pBS*161 yielded *pBS*161-1,

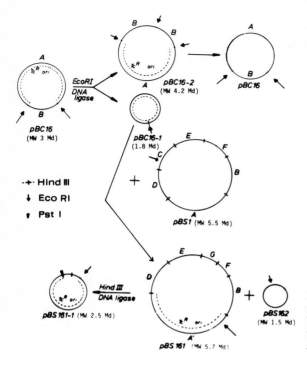

Fig. 1. Genealogy of *Bacillus* plasmids derived from *pBC*16 and *pBS*1. With the exception of *pBS*161-1 only the restriction sites for *Eco*RI and *Hind*III are shown

a small Tc^R plasmid (mol. wt. 2.5×10^6) with a high copy number (ca. 20), and single sites for *Hind*III, *Eco*RI, and *Pst*I, none of which lies within the replication region or the tetracycline resistance determinant.

The recombinant plasmid RSF2124-B. *leu* (*Nagahari* and *Sakaguchi* 1978), contains the leucine A, B, and C genes of *B. subtilis* and can transform not only *leu⁻ E. coli* but also *B. subtilis* to prototrophy. This plasmid is able to replicate only in *E. coli*. After digestion with *Eco*RI, ligation to *Eco*RI-cleaved *B. subtilis* (*natto*) plasmid *pLS*28 (Table 1), and transformation of *B. subtilis* RM125 (leu⁻, recE4), *leu⁺* transformants yielded two recombinant plasmids *pLS*101 (mol. wt. 6.5×10^6) and *pLS*102 (mol. wt. 10.7×10^6). After subcultivation of these clones slowly growing colonies have been observed containing only the plasmid *pLS*103, which is indistinguishable from *pLS*101 (Fig. 2).

Insertion of foreign DNA into the single *Bam*HI site inactivates *leu*A but not *leu*C, which can thus be used as a marker (*Tanaka* and *Sakaguchi* 1978). A derivative of *pLS*103 termed *pLL*10, has only one *Eco*RI site and complements *leu*A and *leu*B but not, in contrast to *pLS*103, *leu*C. In order to see whether DNA insertion into the remaining *Eco*RI site inactivated the *leu* function and also to introduce another marker, an *Eco*RI fragment carrying a *B. subtilis* 168 trimethoprim resistance determinant was recloned from *pBR*322-Tmp^R into *pLL*10 and transformed into *B. subtilis* ML112. *leu⁺* Tmp^R clones yielded *pTL*10 (Fig. 3), a plasmid of mol. wt. 9.4×10^6 giving two fragments of 5.7×10^6 and 3.7×10^6 after *Eco*RI digestion. The latter fragment could convert *B. subtilis* to Tmp^R when inserted in both orientations, indicating its retention of the promoter (*Tanaka* and *Kawano* 1980).

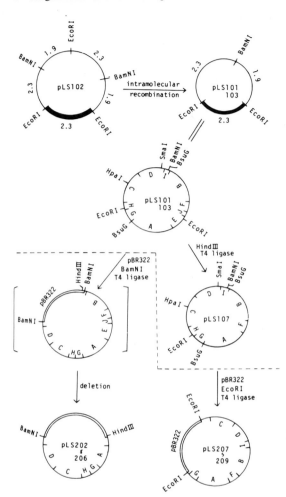

Fig. 2. Structure of constructed plasmids *pLS*101 etc. Numbers denote the molecular weights ($\times 10^{-6}$) of the DNA fragments. The *thick* and *thin lines* indicate the vector and the DNA segments containing the leucine gene respectively. Cleavage sites of *Eco*RI, *Bam*NI, *Sma*I, and *Bsu*G are shown *inside the circles* and those of *Hind*III outside. Courtesy of *T. Tanaka* and *K. Sakaguchi*

In order to reduce both the size and number of *Eco*RI and *Bgl*II sites on *pTL*10, the derivative *pTL*12 has been constructed (Fig. 3); a *leu*⁺ Tmp^R plasmid of mol. wt. 6.4 × 10⁶ carrying single sites for *Eco*RI, *Bgl*II, *Bam*HI, and *Xma*I. *leu* inactivation occurs following insertion at the *Bam*HI and *Xma*I sites, and *Bam*HI cleavage leaves a cohesive end (GATC), making possible the use of *Bgl*II, *Bcl*I, and *Mbo*I, which also leave this sequence. The presence of Tmp^R as a marker allows direct selection of transformants.

The Tc^R plasmid *pAB*124 isolated from *B. stearothermophilus* has three *Eco*RI sites (Table 1). The circularized *Eco*RI-A fragment (pAB224) (*Bingham* et al. 1980) is capable of autonomous replication and carries the tetracycline resistance determinant. It contains single sites for seven restriction enzymes, three of which produce cohesive termini. *pAB*524 has only one *Eco*RI fragment of *pAB*124 deleted (Fig. 4). Table 2 summarizes the properties of the plasmids described in this section. With the exception of *pTL*10 and *pTL*12, all these plasmids carry only one easily detectable genetic marker and do not allow

Table 2. Vectors derived from plasmids listed in Table 1

Vector	Source	Mol. wt. (x 10^{-6})	Single restriction sites	Markers
pBC 16-1	*pBC* 16	1.8	*Eco*RI	Tc^R
pBS 161-1	*pBC* 16/*pBS* 1	2.5	*Eco*RI, *Hind*III, *Pst*I	Tc^R
pLS 103	*pLS* 28/ RSF 2124-B. leu	6.5	*Bam*HI, *Sma*I, *Hpa*I, *Xma*I	leu
pLL 10	*pLS* 103	5.7	*Eco*RI, *Bam*HI, *Bgl*I, *Xma*I	leu
pTL 10	*pLL* 10	9.4	*Bam*HI, *Bgl*I, *Xma*I	leu, Tmp^R
pTL 12	*pTL* 10	6.4	*Bam*HI, *Eco*RI, *Bgl*II, *Xma*I	leu, Tmp^R
pAB 224	*pAB* 124	1.95	*Eco*RI, *Hpa*I, *Hpa*II, *Hha*I, *Tha*I, *Cau*II, *Bst*EII	Tc^R
pAB 524	*pAB* 124	2.3	*Hpa*I, *Hpa*II, *Cau*II, *Bst*EII	Tc^R

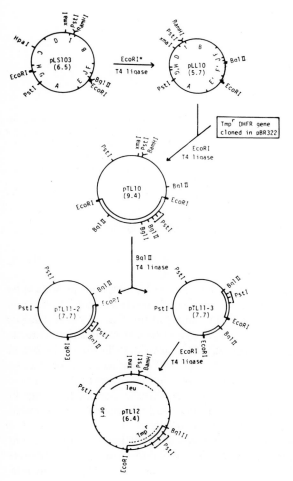

Fig. 3. Structure of plasmids *pLS*103 to *pTL*12. Numbers in parentheses are molecular weights. *Hind*III sites of *pLS*103/*pLL*10 are shown inside the circles and these were preserved in *pTL*10. Courtesy of *T. Tanaka* and *N. Kawano*

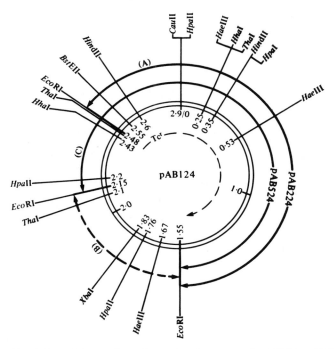

Fig. 4. Restriction endonuclease cleavage maps of *pAB*124, *pAB*224, and *pAB*524. Courtesy of *A.H.A. Bingham*, *C.J. Bruton*, and *T. Atkinson*

identification of recombinant molecules by insertional inactivation. They are, therefore, of limited value as vectors for molecular cloning.

2.2 Hybrid Vectors Capable of Replication in E. coli and B. subtilis

The construction of hybrid vectors attempts to combine the advantages of the well-defined *E. coli* cloning systems with those of the *Bacillus* host.

Hybrid replicons comprising *E. coli* vector plasmids and antibiotic resistance plasmids from *Staphylococcus aureus* have been described by *Ehrlich* (this volume). We have also constructed several *S. aureus/E. coli* hybrid plasmids of this type (*pJK*310, *pJK*312, *pJK*321, *pJK*521, and *pJK*523) (*Goebel* et al. 1979; *Kreft* and *Goebel*, manuscript in preparation). Two of them, *pJK*310 (*pUB*110 + *pBR*325) (*Gryczan* et al. 1978; *Bolivar* 1978) and *pJK*523 (*pC*221 + *pBR*322) (*Novick* 1976; *Bolivar* et al. 1977), express resistance to two antibiotics in *B. subtilis* and carry single restriction sites in these markers.

In addition we have developed hybrid replicons consisting of the *E. coli* vectors *pBR*322 (*Bolivar* et al. 1977), *pACYC*184 (*Chang* and *Cohen* 1978) and the *Bacillus* plasmids *pBS*161-1 and *pBS*1 (*Kreft* et al. 1978; *Goebel* et al. 1979).

*pJK*3 and *pJK*3-1 have been constructed by ligation of *Hind*III-cleaved *pBR*322 and *pBS*161-1. From the resulting complete hybrid *pJK*3 several duplex restriction sites have been removed by religation of *Pst*I cleaved *pJK*3, thus yielding *pJK*3-1 (Fig. 5). This plas-

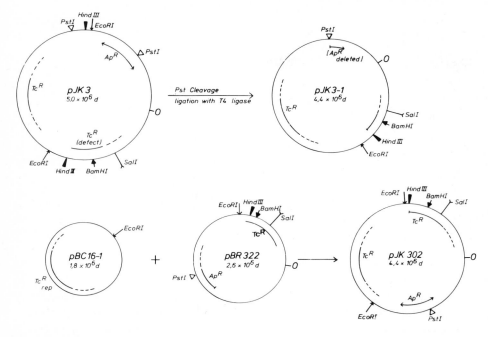

Fig. 5. Construction and restriction endonuclease cleavage maps of *pJK*3-1 and *pJK*302. Construction of *pJK*3 is described in the text

mid has retained only the tetracycline resistance determinant, but has single sites for five restriction enzymes and is capable of replication in both *E. coli* and *B. subtilis*. After transformation of *E. coli* with a ligation mixture from *Hind*III-cleaved *pACYC*184 and *pBS*161-1, the largely deleted hybrid plasmid *pJK*201, which carries Cm^R and Tc^R determinants, has been obtained (*Goebel* et al. 1979).

*pJK*302 is a hybrid consisting of *Eco*RI cleaved *pBR*322 and *pBC*16-1. It has single sites for four restriction enzymes, the *Pst*I site being situated in the Ap^R gene. Cleavage with *Eco*RI of both parental plasmids does not inactivate the Tc^R determinants on these; the hybrid *pJK*302 expresses a high level (more than 100 μg/ml) in both *E. coli* and *B. subtilis* (Fig. 5).

Ligation of the *Eco*RI-cleaved plasmids *pBR*322 and *pBS*l yields in *E. coli* the expected complete hybrid *pJK*501. After transformation of competent cells or protoplasts of *B. subtilis* tetracycline-resistant colonies yield numerous derivatives of *pJK*501 which have deleted different parts of the original plasmid (*Kreft* and *Parrisius,* unpublished observations). One of those derivatives which do not undergo further rearrangements, *pJK*502, has single restriction sites for *Hind*III, *Bam*HI, *Sal*I (in the Tc^R determinant), and *Pst*I (in the Ap^R gene).

In order to convert such a bifunctional plasmid into a cosmid system, we have introduced the *cos* site from *pHC*79 (*Hohn* and *Collins* 1980) into *pJK*3, yielding *pLK*103 (Fig. 6). But for unknown reasons all attempts to package this plasmid in vitro into lambda heads have failed so far (*G. Luibrand,* unpublished observations).

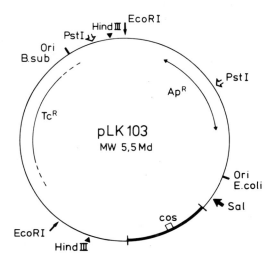

Fig. 6. Restriction endonuclease cleavage map of *pLK*103. The *thick line* indicates the *Bgl*II fragment containing the *cos* site which has been inserted into the *Bam*HI site of *pJK*3

A third type of bifunctional plasmid has been obtained by ligation of *Eco*RI-cleaved *pBS*161-1 and M13mp2 phage (*Gronenborn* and *Messing* 1978) RF double-stranded DNA. The resulting hybrid molecule *pKJB*200 (mol. wt. 7.4×10^6) replicates and expresses tetracycline resistance in both *E. coli* and *B. subtilis,* and gives rise to single-stranded DNA and phage in *E. coli* (*K.J. Burger,* unpublished observations). This plasmid has single restriction sites for *Hind*III and *Pst*I and should facilitate DNA sequencing of cloned fragments.

The plasmids *pJK*3, *pJK*302, *pKJ*502 and *pLK*103 carry two antibiotic resistance markers (Ap^R and Tc^R), both spanning single restriction sites, thus allowing the detection of recombinant plasmids by insertional inactivation. However, due to the nonexpression of the Ap^R gene from *pBR*322 in *B. subtilis* (see Sect. 4) its inactivation can only be detected in *E. coli*. Nevertheless recombinant plasmids can subsequently be transformed into *B. subtilis,* using Tc^R as a selective marker.

As will be discussed later, nonexpression and instability of cloned DNA fragments is an important problem in *B. subtilis*. It is, therefore, worthwhile to clone foreign DNA first in *E. coli* and to use the "bridge" character of the described hybrid replicons to introduce the cloned genes into *B. subtilis*.

2.3 Transformation

Bacillus subtilis cells can be transformed by DNA either at the stage of natural competence (*Spizizen* 1958; *Bott* and *Wilson* 1976) or after creation of protoplasts by lysozyme treatment (*Chang* and *Cohen* 1979).

In competent cells the transformability follows the same time course for chromosomal and plasmid DNA (*Contente* and *Dubnau* 1979). Competent cells are efficiently transformed only by oligomers of plasmid DNA (*Mottes* et al. 1979), whereas protoplasts can be transformed with equal efficiency by monomeric or oligomeric plasmid DNA (*Kreft,* unpublished observations).

It has been reported that a restriction-deficient mutant strain of *B. subtilis* can be more efficiently transformed by plasmid DNA than the restriction-proficient parental strain (*Tanaka* 1979). It has also been claimed that strains carrying the *recE* mutation transform poorly (*Dubnau* et al. 1980). However, in our hands the protoplast system shows no difference between the transformation rates of *B. subtilis* BR151 and BR151 *recE4* or between *B. subtilis* MT120 r^-_M m^-_M *recE4* and MT128 r^+_M m^+_M *recE4* (*Tanaka* 1979). In all cases plasmids *pBS*161-1 (a *Bacillus* plasmid) and *pJK3* (a *Bacillus/E. coli* hybrid plasmid) have been tested and, interestingly, no difference in transformation rate was seen regardless of whether the *pJK3* plasmid DNA used for transformation had been isolated from *E. coli* or from *B. subtilis* (*Kreft,* unpublished observations). Clearly the influence of restriction and/or recombination systems on the transformation rate depends upon the particular plasmid used. The transformation efficiency with competent cells or protoplasts is highest with small plasmids, but nevertheless we have been able to transform *B. subtilis* protoplasts with a plasmid of mol. wt. 17×10^6.

2.4 Stability of Vector Plasmids

One prerequisite for the application of a host-vector system to the molecular cloning of DNA is the stable replication of vector plasmids and recombinant molecules in the host. It seems, however, that recombinant plasmids show a remarkable tendency to undergo alterations (mainly deletions) in *B. subtilis.*

Several different mechanisms may be responsible for the phenomena observed. Plasmids can of course recombine with each other or with the chromosome if they contain homologous segments (*Keggins* et al. 1978; *Tanaka* and *Sakaguchi* 1978), the *recE4* mutation (*Dubnau* and *Cirigliano* 1974) preventing this recombination process. Intramolecular recombination, on the other hand, can occur without involvement of the *recE4* function (*Tanaka* 1979b). In one case it has been shown that such an event can occur in a site-specific way, giving rise to two daughter plasmids containing the entire DNA sequence of the parental plasmid (*Fujii* and *Sakaguchi* 1980). The plasmids *pBS*161 and *pBS*162 (see. Sect. 2.1) may have been generated by the same process, as suggested by the distribution of restriction sites on these and the parental plasmids and the fact that the sum of the molecular weights of *pBS*161 and *pBS*162 is roughly the same as the sum of *pBS*1 and the large *Eco*RI fragment of *pBC*16.

All the hybrid plasmids described in Sect. 2.2 replicate stably in *E. coli* without detectable segregation. In *B. subtilis* the segregation rate after ten generations without selective pressure is 4%–65%, depending on the plasmid examined. They show no deletions or rearrangements while replicating in *E. coli,* but frequently display extensive deletions when isolated from *B. subtilis* transformants.

For example, from *B. subtilis* transformed by *pJK*501 a large variety of deleted plasmids may be obtained, the deletions affecting both parts (*pBS*1 and *pBR*322) of the hybrid plasmid and ranging from $3.6–6.1 \times 10^6$ in size. In some cases if plasmid DNA from *B. subtilis* cells transformed with this plasmid is isolated immediately after transformation, plasmids indistinguishable in size from *pJK*501 are found, while after subcultivation large deletions are again observed. This clearly demonstrates that in this case the deletion event occurs after the uptake of the plasmid into the cell, as has also been proposed by others (*Gryczan* and *Dubnau* 1978).

The occurrence of deletions, at least in the case of *pJK*501, is independent of the *recE* function and the restriction/modification system of the recipient. Also the type of *E. coli* modification of *pJK*501 plasmid DNA isolated from *E. coli* and used for transformation of *B. subtilis* had no influence on the deletion phenomenon (*Kreft* and *Parrisius*, unpublished observations).

Insertion of different foreign DNA segments into the same vector plasmid showed that not the vector itself but rather the particular combination of vector with another segment of DNA determines whether this new structure is stable in *B. subtilis* (see. Sect. 4).

2.5 Clone Analysis

The screening for recombinant plasmids is difficult in cases where no marker inactivation and no primary selection for a cloned fragment is possible. To screen for plasmids with inserts (or deletions) the use of a rapid lysis procedure is of great advantage, the method of *Birnboim* and *Doly* (1979) giving in our hands satisfactory results for *B. subtilis*.

It should be kept in mind, however, that due to the remarkable tendency for deletions to occur in *B. subtilis* the mere size of a plasmid isolated from transformants is not a reliable criterion in assessing its structure.

Immunologic (*Broome* and *Gilbert* 1978) and colony hybridization methods (*Grunstein* and *Hogness* 1975) can help to identify particular recombinant plasmids, but have yet to be adapted to *B. subtilis*.

For studies on the expression of cloned DNA fragments a minicell system is available for *B. subtilis* (*Reeve* et al. 1973).

3 Use of Bacteriophage Vectors

Several phage systems of *B. subtilis* have been well characterized (*Graham* et al. 1979; *Cregg* and *Ito* 1979; *Mizukami* et al. 1980). In particular, early work has involved the phages Φ3T and ρll which may be termed specialized transducing phages as they carry the thymidilate synthetase gene *thy*P3 (*Dean* et al. 1976). *Kawamura* et al. described in 1979 a method to construct specialized transducers of *B. subtilis* based on the phage ρll. This involved cleavage by *Eco*RI of DNA from ρll and the defective phage PBSX induced from *B. subtilis* 168. The latter phage contains only host chromosomal DNA (*Okamoto* et al. 1968), thus limiting the method to the cloning of homologous DNA. It has been extended by *Yoneda* et al. (1979) to permit the cloning into *B. subtilis* of foreign DNA for which no primary selection exists. They chose to construct a specialized transducing phage containing the α-amylase gene(s) from *B. amyloliquefaciens*. Chromosomal DNA from this strain and Φ3T DNA were digested with *Bgl*II, mixed, and ligated. This ligated mixture was then added to a preparation of chromosomal DNA from *B. subtilis* RUB200, a strain prototrophic for threonine and defective in α-amylase synthesis. This mixture was in turn incubated with *B. subtilis* RUB201, a threonine auxotroph lysogenic for Φ3T. *thr*+ transformants were selected and tested for α-amylase production. Competent cells may take up more than one fragment of DNA (congression) so that by selecting, in this case, *thr*+ transformants one effectively enriches for cells carrying foreign DNA (such selection gave a 10^4-fold enrichment for *amy*+ clones). Seven of 10^5 *thr*+ transformants of RUB201

acquired α-amylase activity and from five of these Φ3T could be induced. Infection of the amy⁻ strain RUB200 with these phages showed a 100% correlation between establishment of lysogeny and the amy⁺ phenotype. Transformation and selection of amy⁺ clones showed cotransformation with the phage specific *thy*P3 gene.

The technique has been utilized for the cloning of *spo* 077 (quoted in *Kawamura* et al. 1980) and *amy*E (*Nomura* et al. 1979) into ρ11, but this phage, having a genome with mol. wt. 80×10^6, generates a large number of fragments after routine digestions. *Iijima* et al. (1980) have, therefore, adapted the procedure to the temperate phage Φ105, which has a genome size of 26×10^6. Chromosomal DNA of *B. subtilis* 168 (*trp*C2) was prepared from phage PBSX and after *Eco*RI digestion ligated with *Eco*RI digested DNA from Φ105C. Ligated DNA was used to transform *B. subtilis* (*trp*C2 *lys* 3 *met* B10) lysogenic for Φ105C. Selection for auxotrophic markers, subsequent mitomycin C induction, and transduction of the resulting lysate into *B. subtilis* (*trp*C2 *lys* 3 *met* B10) allowed isolation of *met* B⁺ transducing particles. While Φ105C DNA is insensitive to *Bam*HI, incorporation of the new *met* B fragment introduced a single site for this enzyme. This seems possible with other phages, e.g., the virulent Φ1 (*Kawamura* et al. 1980), which also have no *Bam*HI site. A deletion mutant of Φ1, Φ1E2Δ1, has been isolated with increased cloning capacity and this has been used to clone ρ11 fragments which introduced *Bam*HI and *Hae*III restriction sites.

To summarize, initial bacteriophage systems have been shown to be very efficient especially in shotgun cloning of heterologous DNA. Due to the selection marker thymidilate synthetase the phage Φ3T is particularly useful. Major limitations of the method are that by lysogenization normally only single gene copies can be introduced into recipient cells and that induction of lysogens leads to lysis, which might cause containment problems.

4 Molecular Cloning with Plasmid Vectors

In a strict sense the construction of vector plasmids like *pJK*3 (*Kreft* et al. 1978) or *pTL*12 (*Tanaka* and *Kawano* 1980) has already involved the cloning of either heterologous or homologous DNA. This section describes the cloning and expression of isolated genomic fragments and the expression of genetic markers on hybrid replicons.

It has been shown that, at least in certain cases, even DNA cloned in an *E. coli* vector plasmid can transform *B. subtilis* without replication of the recombinant plasmid in the *Bacillus* host (see also Sect. 2.1). The thymidilate synthetase gene from *B. subtilis* bacteriophage Φ3T, which has extensive homology to the chromosomal gene, can transform *thy*⁻ *B. subtilis* to *thy*⁺ when cloned in *pSC*101 or *pMB*9 (*Ehrlich et al.* 1976). On the other hand, the nonhomologous thymidilate synthetase gene from phage β22 can only transform *thy*⁻*B. subtilis* when cloned into an *E. coli* plasmid carrying a small fragment of DNA homologous to the *B. subtilis* chromosome (*Duncan* et al. 1978; *Young* 1980). In this case the whole recombinant plasmid becomes integrated into the chromosome. Recently it has been shown that the thymidilate synthetase gene from *E. coli* can also transform *B. subtilis* when cloned into *pBR*322 or *pMB*9 (*Rubin* et al. 1980). It is not yet clear, however, if there exists sequence homology between the cloned gene and the *B. subtilis* chromosome.

Most of the hybrid plasmids described in Sect. 2.2 carry more than one antibiotic

Table 3. Expression of antibiotic resistance markers on hybrid replicons in *E. coli* and *B. subtilis*

Plasmid	Marker	Source	Expression in	
			E. coli	*B. subtilis*
pJK 3	Ap^R	*pBR* 322	+	−
	Tc^R	*pBS* 161-1	+	+
pJK 201	Cm^R	*pACYC* 184	+	−
	Tc^R	*pBS* 161-1	+	+
pJK 502	Ap^R	*pBR* 322	+	−
	Tc^R	*pBR* 322	+	+

resistance marker. Transformation studies have revealed, however, that these markers are not always expressed in both host bacteria, as is shown in Table 3.

Hybridization of in vivo radioactively labeled RNA from *B. subtilis* carrying *pJK*3 or *pJK*201 (*Goebel* et al. 1979; *Kreft* et al., manuscript in preparation) to restriction fragments of these plasmids showed that the nonexpression of the *E. coli* Ap^R (β-lactamase) and the Cm^R (chloramphenicol acetyl transferase) genes in *B. subtilis* is due to a transcriptional block. That no structural rearrangement in the DNA sequence of these genes is responsible for this lack of expression was shown following successful retransformation of *E. coli* with plasmid DNA isolated from *B. subtilis* (*Goebel* et al. 1979).

The tetracycline resistance determinant of *pBR*322 (combined with *pBS*1 to give *pJK*501) can be expressed in *B. subtilis*. It is not, however, clear if this resistance determinant indeed originates from *E. coli*. In addition, since the Tc^R determinant in *pJK*501 is very close to the *Bacillus* part of the hybrid plasmid, it is possible that its transcription starts at a *Bacillus* promoter.

The tetracycline resistance specified by *pBC*16 is expressed in *E. coli*, although at a reduced level (*Kreft* et al. 1978). The differences observed in expression of *E. coli* and *Bacillus* genes in the nonhomologous host might be explained by the promoter specificity of the RNA polymerase. In vitro studies with RNA polymerase from *B. subtilis* have shown that this enzyme preferentially binds to and transcribes from *Bacillus* promoters (*Williamson* and *Doi* 1978; *Lee* et al. 1980) in contrast to *E. coli* RNA polymerase, which transcribes nonhomologous genes quite efficiently (*Davison* et al. 1979).

It has been shown by *Ehrlich* (1977) that plasmids from *S. aureus* can express Cm^R or Tc^R in *B. subtilis*. In order to see whether the β-lactamase (E.C. 3.5.2.6.) mediating Ap^R in *S. aureus* can also be expressed in *B. subtilis* we recloned an *Eco*RI fragment, containing the β-lactamase gene (from *pSC*122, *Timmis* et al. 1975), into *pJK*3-1. Ap^R/Tc^R transformants of *E. coli* contained the recombinant plasmid *pJK*401 and restriction analysis showed that deletions had occurred in both the vector and in the fragment originating from *S. aureus*. *B. subtilis* can be transformed to Tc^R with *pJK*401 and the plasmid replicates stably in this host. Ap^R is not expressed nor can β-lactamase activity be detected intracellularly (*Kreft*, unpublished observations). It remains to be seen whether this nonexpression in *B. subtilis* is due to the removal of regulatory DNA sequences by the observed deletion.

Shotgun cloning experiments in *B. subtilis* seem, from the experience of ourselves and several others, to be rather difficult to perform. The main problem, in addition to

nonexpression of heterologous genes, seems to be the difficulty, seen particularly in shotgun experiments, in cloning large fragments of DNA. In one case it was shown that the mean size (mol. wt. 1×10^6) of DNA inserts found in recombinant plasmids is only one-third of the mean size of the fragments in the *Eco*RI or *Hin*dIII digested donor DNA (*Michel* et al. 1980). It is not yet clear if this phenomenon reflects a preferential transformation of recombinant plasmids with small inserts or is due to posttransformational deletions.

A similar experience has been made during attempts to clone the sporulation gene *spo*OF from *B. subtilis* 60015. *Eco*RI digested DNA from this strain, enriched 870-fold for the *spo*OF gene, was ligated to *Eco*RI linearized plasmid *pBS*161-1, and the asporogenous mutant *B. subtilis* strain JH756b was transformed with the ligated mixture. One of the tetracycline-resistant transformants contained *pBS*161-1 with a small insert of DNA, but this plasmid could not complement the sporulation deficiency (*Rhaese* et al. 1979).

Several methods have been proposed to circumvent these difficulties (*Dubnau* et al. 1980). In order to examine the usefulness of cloning first into *E. coli* as an intermediate host we have tried to study the stability and expression in *B. subtilis* of DNA fragments cloned in *E. coli* following their recloning into *B. subtilis/E. coli* hybrid vectors.

We have recloned the *E. coli pho*A gene encoding the periplasmic enzyme alkaline phosphatase (E.C. 3.1.3.1) which is located on a *Hin*dIII/*Bam*HI fragment (mol. wt. 3×10^6) of *pSB*53 (*W. Boidol* and *G. Siewert*, manuscript in preparation) into *pJK*3-1. The resulting plasmid *pJK*353 can transform *E. coli* SB44 (*pho*A) to *pho*$^+$ but not *B. subtilis* GSY172 (*pho*P8 *arg*A11) (*Le Hégarat* and *Anagnostopoulos* 1973). All the Tc^R *B. subtilis* transformants tested contain a derivative of *pJK*353 with a large deletion of mol. wt. 5.6×10^6.

This deletion affects the *pho*A gene thereby precluding until now study of its expression in *B. subtilis*. Nevertheless, further investigations should indicate how *Bacilli*, which transport extracellular enzymes to the external medium (for a review see *Priest* 1977), transport *E. coli* enzymes which are normally periplasmic, i.e., are only carried through the inner (cytoplasmic) membrane.

Also using *pJK*3-1 as a vector the *pen* gene of *B. licheniformis*, specifying β-lactamase, has been cloned in *B. subtilis*. Either the gene already cloned into a λ vector (*Brammar* et al. 1980) or cloned during a shotgun experiment into phage fd was used as a DNA source. The *Eco*RI fragment of mol. wt. 2.8×10^6 carrying the β-lactamase genes was inserted into *pJK*3-1 in both orientations and in both cases gave Tc^R/Ap^R transformants in *E. coli* and *B. subtilis*. The recombinant plasmid containing the fragment from λ *pen* was rather stably maintained in *B. subtilis*, whereas the plasmid with the fragment from the shotgun cloning became deleted after a few generations (*Neugebauer* 1980). The expression of the β-lactamase was dependent upon the particular recombinant plasmid and the recipient strain. In *B. subtilis* SB202 only about 1% of the activity found in the donor strain of *B. licheniformis* was expressed. In *B. subtilis* BD170 the expression of β-lactamase from the fragment obtained by shotgun cloning was very good. With the fragment from lambda *pen* only 10%–30% of this activity was seen in BD170 (*Sprengel*, personal communication).

A similar cloning experiment has recently been reported (*Gray* and *Chang* 1981) using a bifunctional replicon from *E. coli* and *S. aureus* as a vector. The *B. licheniformis* β-lactamase was efficiently expressed and processed in both *E. coli* and *B. subtilis* BD 224 and secreted into the medium by the *Bacillus* host.

Recombinant plasmids containing genes for exoproteins like β-lactamase may be

very useful as exportation vectors facilitating the secretion of other proteins, the genes for which being fused to the essential parts of the exoenzyme gene.

It has been shown in several cases that genes from the yeast *S. cerevisiae* can be expressed after cloning and introduction into *E. coli* and can complement auxotrophic mutations (*Struhl* et al. 1976; *Ratzkin* and *Carbon* 1977). A small *Hind*III fragment of mol. wt. 2×10^6 containing the *arg*4 gene of *S. cerevisiae* (*Clarke* and *Carbon* 1978) has been recloned from *pYe* (*arg*4) 402-11 into *pJK*3-1 and the resulting recombinant plasmid *pJK*3-1 (*arg*4) transforms *E. coli* JA228 *arg*H to *arg*+, but not *B. subtilis* GSY172 *pho*P8 *arg*A11. In addition, no argininosuccinate lyase (E.C. 4.3.2.1.) activity can be detected in the cells (*Kreft*, unpublished observations). The recombinant plasmid is stable in *B. subtilis* and plasmid DNA isolated from this host can retransform *E. coli arg*H to *arg*+. It has been demonstrated that the *Hind*III fragment containing the *arg*4 gene carries a promoter which functions in *E. coli* (*Clarke* and *Carbon* 1978) but which from preliminary studies seems to allow no transcription in *B. subtilis*.

5 Conclusions

Compared to the very sophisticated vectors and recombinant DNA techniques available in *E. coli*, the application of this methodology to *B. subtilis* is still in its infancy. However, the rapidly increasing amount of research in this field may soon allow exploitation of the particular advantages of *B. subtilis* as a host for cloned DNA.

These include (i) nonpathogenicity, (ii) lack of endotoxin, (iii) direct selection of cloned genes which are specific for this species e.g., genes encoding for sporulation and exoenzymes, the latter being of additional interest in the development of "export vectors", and (iv) the possibility of using the transformability of competent cells of *B. subtilis* by homologous chromosomal DNA for the "scaffolding technique" (*Young* 1980) and also for the enrichment of specific markers in DNA samples prior to cloning.

Until now the use of indigenous bacteriophage systems and of *E. coli* as an intermediate host for *E. coli/Bacillus* hybrid plasmid vectors has proved most promising in shotgun cloning, but large problems remain, particularly the nonexpression of heterologous genes and the instability of cloned DNA fragments, which clearly are much more important in this host than in *E. coli*.

Further developments which seem particularly necessary to achieve successful application of the recombinant DNA technique in *Bacillus* are (i) host mutants which allow more stable maintenance of cloned DNA and (ii) vectors which can express heterologous genes regardless of the presence of suitable transcription and translation signals on the cloned fragment.

An in vitro packaging system into bacteriophage heads, comparable to the cosmid system of *E. coli*, should increase the cloning capacity of plasmid vectors.

For the practical application of the recombinant DNA technique in *B. subtilis* to the microbial production of commercially and medically important compounds, as well as the study of regulation processes in this bacterium, vectors with variable copy number or inducible expression of cloned genes will be of great importance.

Acknowledgements. The authors wish to thank *Dr. W. Goebel* for initiating this work, critical reading of the manuscript and for helpful suggestions. *Drs. K.J. Burger, G. Luibrand, K. Neugebauer, G.*

Siewert, and *R. Sprengel* are thanked for communicating some of their unpublished results and *Dr. H.J. Rhaese* for sending reprints of his work. We are grateful to *Drs. T. Atkinson, A.H.A. Bingham, C.J. Bruton, N. Kawano, K. Sakaguchi*, and *T. Tanaka*, Elsevier/North-Holland Biomedical Press, the Society for General Microbiology, and Springer-Verlag for the permission to reproduce figures and to *Mrs. E. Appel* for typing the manuscript. This work was supported by the Deutsche Forschungsgemeinschaft (SFB 105 A 12).

References

Bernhard K, Schrempf H, Goebel W (1978) Bacteriocin and antibiotic resistance plasmids in *Bacillus cereus* and *Bacillus subtilis*. J Bact 133:897–903

Bingham AHA, Bruton CJ, Atkinson T (1979) Isolation and partial characterization of four plasmids from antibiotic-resistant thermophilic bacilli. J Gen Microbiol 114:401–408

Bingham AHA, Bruton CJ, Atkinson T (1980) Characterization of *Bacillus stearothermophilus* plasmid pAB124 and construction of deletion variants. J Gen Microbiol 119:109–115

Birnboim HC, Doly J (1979) A rapid alkaline extraction procedure for screening recombinant plasmid DNA. Nucleic Acids Res 7:1513–1523

Bolivar F (1978) Construction and characterization of new cloning vehicles. III. Derivatives of plasmid pBR322 carrying unique *Eco*RI sites for selection of *Eco*RI generated recombinant molecules. Gene 4:121–136

Bolivar F, Rodriguez RL, Greene PJ, Betlach MC, Heyneker HL, Boyer HW, Crosa JH, Falkow S (1977) Construction and characterization of new cloning vehicles. II. A multipurpose cloning system. Gene 2:95–113

Bott KF, Wilson GA (1967) Development of competence in the *Bacillus subtilis* transformation system. J Bacteriol 94:562–570

Brammar WJ, Muir S, McMorris A (1980) Molecular cloning of the gene for the β-lactamase of *Bacillus licheniformis* and its expression in *Escherichia coli*. MGG 178:217–224

Broome S, Gilbert W (1978) Immunological screening method to defect specific translation products. Proc Natl Acad Sci USA 75:2746–2749

Carlton BC, Smith MPW (1974) Size distribution of the closed circular deoxyribonucleic acid molecules of *Bacillus megaterium*. Sedimentation velocity and electron microscope measurements. J Bacteriol 117:1201–1209

Chang ACY, Cohen SN (1978) Construction and characterization of amplifiable multicopy DNA cloning vehicles derived from the P15A cryptic miniplasmid. J Bacteriol 134:1141–1156

Chang S, Cohen SN (1979) High frequency transformation of *Bacillus subtilis* protoplasts by plasmid DNA. MGG 168:111–115

Clarke L, Carbon J (1978) Functional expression of cloned yeast DNA in *Escherichia coli*: Specific complementation of argininosuccinate lyase (argH) mutations. J Mol Biol 120:517–532

Contente S, Dubnau D (1979) Characterization of plasmid transformation in *Bacillus subtilis*: kinetic properties and the effect of DNA conformation. MGG 167:251–258

Cregg JM, Ito J (1979) A physical map of the genome of temperate phage Φ 3T. Gene 6:199–219

Davison BL, Leighton T, Rabinowitz JC (1979) Purification of *Bacillus subtilis* RNA polymerase with heparinagarose. In vitro transcription of Φ29 DNA. J Biol Chem 254:9220–9226

Dean DH, Orego JC, Hutchinson KW, Halvorson HO (1976) A new temperate phage for *Bacillus subtilis*, ρ11. J Virol 20:509–519

Dubnau D, Cirigliano C (1974) Genetic characterization of recombination-deficient mutants of *Bacillus subtilis*. J Bacteriol 117:488–493

Dubnau D, Gryczan T, Contente S, Shivakumar AG (1980) Molecular cloning in *Bacillus subtilis*. In: Setlow JK, Hollaender A (eds) Genetic engineering, vol 2. Plenum Press, New York, pp 115–131

Duncan CH, Wilson GA, Young FE (1978) Mechanism of integrating foreign DNA during transformation of *Bacillus subtilis*. Proc Natl Acad Sci USA 75:3664–3668

Ehrlich SD (1977) Replication and expression of plasmids from *Staphylococcus aureus* in *Bacillus subtilis*. Proc Natl Acad Sci USA 74:1680–1682

Ehrlich SD, Bursztyn-Pettegrew H, Stroynowski I, Lederberg J (1976) Expression of the thymidilate sythetase gene of the *Bacillus subtilis* bacteriophage phi-3-T in *Escherichia coli*. Proc Natl Acad Sci USA 73:4145–4149

Fujii M, Sakaguchi K (1980) A site-specific recE4-independent intramolecular recombination between *Bacillus subtilis* and *Staphylococcus aureus* DNAs in hybrid plasmids. Gene 12:95–102

Goebel W, Kreft J, Burger KJ (1979) Molecular cloning in *Bacillus subtilis*. In: Timmis KN, Pühler A, (eds) Plasmids of medical, environmental and commercial importance. Elsevier/North-Holland Biomedical Press, Amsterdam, pp 471–480

Graham S, Yoneda Y, Young FE (1979) Isolation and characterization of viable deletion mutants of *Bacillus subtilis* bacteriophage SP02. Gene 7:69–77

Gray O, Chang S (1981) Molecular cloning and expression of *Bacillus licheniformis* β-lactamase gene in *Escherichia coli* and *Bacillus subtilis*. J Bacteriol 145:422–428

Gronenborn B, Messing J (1978) Methylation of single-stranded DNA in vitro introduces new restriction endonuclease cleavage sites. Nature 272:375–377

Grunstein M, Hogness DS (1975) Colony hybridization: A method for the isolation of cloned DNAs that contain a specific gene. Proc Natl Acad Sci USA 72:3961–3965

Gryczan TJ, Dubnau D (1978) Construction and properties of chimeric plasmids in *Bacillus subtilis*. Proc Natl Acad Sci USA 75:1428–1432

Henner DJ, Hoch JA (1980) The *Bacillus subtilis* chromosome. Microbiol Rev. 44:57–82

Hohn B, Collins J (1980) A small cosmid for efficient cloning of large DNA fragments. Gene 11:291–298

Iijima T, Kawamura F, Saito H, Ikeda Y (1980) A specialized transducing phage constructed from *Bacillus subtilis* phage Φ105. Gene 9:115–126

Kawamura F, Saito H, Ikeda Y (1979) A method for construction of spezialized transducing phage ρ11 of *Bacillus subtilis*. Gene 5:87–91

Kawamura F, Saito H, Ikeda Y (1980) Bacteriophage Φ1 as a gene cloning vector in *Bacillus subtilis*. MGG 180:259–266

Keggins KM, Duvall EJ, Lovett PS (1978) Recombination between compatible plasmids containing homologous segments requires the *Bacillus subtilis* recE4 gene product. J Bacteriol 134:514–520

Kreft J, Bernhard K, Goebel W (1978) Recombinant plasmids capable of replication in *B. subtilis* and *E. coli*. MGG 162:59–67

Lee G, Talkington C, Pero J (1980) Nucleotide sequence of a promoter recognized by *Bacillus subtilis* RNA polymerase. M G G 180:57–65

Le Hégarat J-C, Anagnostopoulos C (1973) Purification, subunit structure and properties of two repressible phosphohydrolases of *Bacillus subtilis*. Eur J Biochem 39:525–539

Lovett PS, Bramucci MG (1975) Plasmid deoxyribonucleic acid in *Bacillus subtilis* and *Bacillus pumilus*. J Bacteriol 124:484–490

Lovett PS, Duvall EJ, Keggins KM (1976) *Bacillus pumilus* plasmid pPL10: properties and insertion into *Bacillus subtilis* 168 by transformation. J Bacteriol 127:817–828

Michel B, Palla E, Niaudet B, Ehrlich SD (1980) DNA cloning in *Bacillus subtilis*. III. Efficiency of random-segment cloning and insertional inactivation vectors. Gene 12:147–154

Mizukami T, Kawamura F, Takahashi H, Saito H (1980) A physical map of the genome of the *Bacillus subtilis* temperate phage ρ11. Gene 11:157–162

Mottes M, Grandi G, Sgaramella V, Canosi U, Morelli G, Trautner TA (1979) Different specific activities of the monomeric and oligomeric forms of plasmid DNA in transformation of *B. subtilis* and *E. coli*. MGG 174:281–286

Nagahari K, Sakaguchi K (1978) Clining of *Bacillus subtilis* leucine A, B and C genes with *Escherichia coli* plasmids and expression of the *leuC* gene in *E. coli*. MGG 158:263–270

Neugebauer K (1980) Untersuchungen zur Konstruktion von Einzelstrangvektoren in *E. coli* und eines Exportvektors für *B. subtilis*. Ph. D. thesis, Heidelberg

Nomura S, Yamane K, Masuda T, Kawamura F, Mizukami T, Saito H, Takatuki M, Tamura G, Maruo B (1979) Construction of transducing phage ρ11 containing α-amylase structural gene of *Bacillus subtilis*. Agric Biol Chem 43:2637–2638

Novick R (1976) Plasmid-protein relaxation complexes in *Staphylococcus aureus*. J Bacteriol 127:1177–1187

Okamoto K, Mudd JA, Mangan J, Huang WM, Subbaiah TV, Marmur J (1968) Properties of the defective phage of *Bacillus subtilis*. J Mol Biol 34:413–428

Priest FG (1977) Extracellular enzyme synthesis in the genus *Bacillus*. Bacteriol Rev 41:711–753

Ratzkin B, Carbon J (1977) Functional expression of cloned yeast DNA in *E. coli*. Proc Natl Acad Sci USA 74:487–491

Reeve JN, Mendelson NH, Coyne SI, Hallock LL, Cole RM (1973) Minicells of *Bacillus subtilis*. J Bacteriol 114:860–873

Rhaese HJ, Groscurth R, Vetter R, Gilbert H (1979) Regulation of sporulation by highly phosphorylated nucleotides in *Bacillus subtilis*. In: Koch G, Richter D (eds) Regulation of macromolecular synthesis by low molecular weight mediators. Academic Press, New York, pp 145–159

Rostas K, Dobritsa SV, Dobritsa AP, Koncz C, Alfödi L (1980) Megacinogenic plasmid from *Bacillus megaterium* 216. MGG 180:323–329

Rubin EM, Wilson GA, Young FE (1980) Expression of thymidylate synthetase activity in *Bacillus subtilis* upon integration of a cloned gene from *Escherichia coli*. Gene 10:227–235

Spizizen J (1958) Transformation of biochemically deficient strains of *Bacillus subtilis* by deoxyribonucleate. Proc Natl Acad Sci USA 44:1072–1078

Struhl K, Cameron JR, Davis RW (1976) Functional genetic expression of eukaryotic DNA in *Escherichia coli*. Proc Natl Acad Sci USA 73: 1471–1475

Tanaka T, Koshikawa T (1977) Isolation and characterization of four types of plasmids from *Bacillus subtilis (natto)*. J Bacteriol 131:699–701

Tanaka T, Sakaguchi K (1978) Construction of a recombinant plasmid composed of *B. subtilis* leucine genes and a *B. subtilis (natto)* plasmid: its use as cloning vehicle in *B. subtilis* 168. MGG 165:269–276

Tanaka T (1979a) Restriction of plasmid-mediated transformation in *Bacillus subtilis* 168. MGG 175:235–237

Tanaka (1979b) *rec*E4-independent recombination between homologous deoxyribonucleic acid segments of *Bacillus subtilis* plasmids. J Bacteriol 139:775–782

Tanaka T, Kawano N (1980) Cloning vehicles for the homologous *Bacillus subtilis* host-vector system. Gene 10:131–136

Timmis K, Cabello F, Cohen SN (1975) Cloning, isolation and characterization of replication regions of complex plasmid genomes. Proc Natl Acad Sci USA 72:2242–2246

Uozumi T, Ozaki A, Beppu T, Arima K (1980) New cryptic plasmid of *Bacillus subtilis* and restriction analysis of other plasmids found by general screening. J Bacteriol 142:315–318

Williamson VM, Doi RH (1978) Delta factor can displace sigma factor from *Bacillus subtilis* RNA polymerase holoenzyme and regulate its initiation activity. MGG 161:135–141

Yoneda Y, Scott G, Young FE (1979) Cloning of a foreign gene coding for α-amylase in *Bacillus subtilis*. Biochem Biophys Res Commun 91:1556–1564

Young FE (1980) Impact of cloning in *Bacillus subtilis* on fundamental and industrial microbiology. J Gen Microbiol 119:1–15

Use of Plasmids from Staphylococcus aureus for Cloning of DNA in Bacillus subtilis

S.D. EHRLICH*, B. NIAUDET*, AND B. MICHEL*

1 Introduction

Suitable cloning vectors are a prerequisite for DNA cloning in any organism. Small, high copy number plasmids, carrying easily selectable genetic markers, are excellent potential vectors for any transformable bacterial host. The common laboratory strains of *Bacillus subtilis* are devoid of such plasmids (*Lovett* and *Bramucci* 1975). The search for vectors which could be used in this host was therefore extended to other organisms. Two main lines of investigation were followed. The first involved screening of various *Bacilli* for the presence of plasmids which were a priori likely to replicate in *B. subtilis* since they would be issued from phylogenetically close bacteria. Such plasmids were indeed found, and after initial difficulties due to the fact that they were in general cryptic, could be used for DNA cloning in *B. subtilis* as reviewed by *Kreft* (this volume). The second line of research involved testing the capacity of well-characterized plasmids present in organisms phylogenetically distant from *B. subtilis* to replicate and express the genetic markers in this host. The attempts to introduce plasmids used for cloning in *E. coli* into *B. subtilis* failed (unpublished data). The failure can be explained a posteriori by the lack of heterospecific expression of both the genetic markers (*Ehrlich* 1978a) and the replication functions tested (*Niaudet* and *Ehrlich* 1979). More success was obtained with small, high copy number plasmids carrying antibiotic genetic markers which were issued from *Staphylococcus aureus*. Most of the plasmids of this class could be introduced by transfor-

* Groupe d'Etude des Acides Nucléiques, Département de Microbiologie, Institut de Recherche en Biologie Moléculaire, 2, Place Jussieu, 75005 Paris

mation into *B. subtilis*, where they were maintained stably and expressed the resistance markers they carried (*Ehrlich* 1977). The vectors developed from these plasmids and their use for DNA cloning are discussed in this review.

2 Transformation Procedures

The transformability of an organism which is to be used as a host for DNA cloning is of utmost importance. Transformation of *B. subtilis* was described over 20 years ago (*Spizizen* 1958). Competence for transformation is induced by growing this bacterium first in a relatively rich medium and then exposing it to an essentially minimal medium (*Anagnostopoulos* and *Spizizen* 1961). The original two-step procedure is still in general use, although a number of minor modifications have been introduced by various researchers (cf. *Niaudet* and *Ehrlich* 1979). It yields cultures containing usually about 0.1% of cells which can be transformed with saturating amounts of DNA. This competence level is largely sufficient for DNA cloning experiments. For particular purposes, cells competent for transformation, smaller than the ones which are not, can be enriched further by centrifugation in renografin gradients (*Cahn* and *Fox* 1968; *Hadden* and *Nester* 1968). Cultures containing about 10% of cells transformable at saturating DNA concentrations can be obtained in this way.

More recently, transformation of *B. subtilis* protoplasts by plasmid DNA in the presence of polyethylene glycol was described (*Chang* and *Cohen* 1979). This method, which allows transformation of up to 80% of regenerated protoplasts but is somewhat less easy to reproduce than that of *Anagnostopoulos* and *Spizizen*, appears very appropriate for the selection of transformants resistant to certain antibiotics. It is not suitable for the selection of nutritional genetic markers, due to the complexity of the medium used to regenerate the protoplasts.

3 Plasmid-Mediated Transformation

To use plasmids efficiently for cloning in *B. subtilis* it is important to understand the mechanism by which they transform this host. Transformation of *B. subtilis* competent cells by plasmids, first reported several years ago (*Lovett* et al. 1976), could be studied more extensively only when plasmids carrying easily selectable genetic markers became available (*Ehrlich* 1977). It was rapidly found that in many respects plasmid-mediated and chromosome-mediated transformation resemble each other: the same growth regimen induces competence for both types of transformation, chromosomal and plasmid DNAs compete during transformation, a single plasmid DNA molecule is sufficient to produce a transforming event (*Ehrlich* et al. 1978; *Contente* and *Dubnau* 1979a). Several differences were, nevertheless, found. Plasmid transformation is affected by the host restriction system (*Tanaka* 1979; *Canosi* et al. 1981), in the same way as phage DNA-mediated transfection, whereas chromosomal transformation is insensitive to restriction (*Trautner* et al. 1974; *Bron* et al. 1980). More surprisingly, plasmid oligomers, but not monomers, are the molecular species active in transformation (*Canosi* et al. 1978; *Mottes* et al. 1979). The entire molecule, however, need not be repeated, partial internal repetition is sufficient to endow it with transforming activity. Furthermore, circular structure of internally re-

Table 1. *S. aureus* plasmids replicating in *B. subtilis*

Plasmid	Genetic markers	Size (kb)	Unique restriction sites	Reference
pC194	Cm^R	2.8	*Bgl*I, *Hae*III, *Hind*III, *Hpa*I, *Hpa*II	a, b, c
pC221	Cm^R	4.5	*Bst*EII, *Eco*RI, *Hind*III	a, d
pC223	Cm^R	4.5	*Bgl*II, *Hind*III, *Hpa*II	a, b, c
pUB112	Cm^R	4.1	*Hind*III, *Hpa*II	a, b, c
pT127	Tc^R	4.5	*Hinf*I, *Hpa*II, *Kpn*I, *Tac*I, *Xba*I	a, c, e
pUB110	Km^R	4.5	*Bam*HI, *Bgl*II, *Eco*RI, *Tac*I, *Xba*I	f, g, h
pE194	Em^R	3.4	*Bcl*I, *Hae*III, *Hpa*I, *Pst*I, *Xba*I	f, g, i
pS501 (pS194)	Sm^R	4.1	*Eco*RI, *Hind*II, *Hind*III, *Xba*I	g, j
pSA2100 (pSC194)	Cm^R Sm^R	6.7	*Eco*RI, *Hae*III, *Hind*II, *Xba*I	g, j

[a] *Ehrlich* 1977; [b] *Grandi* et al. 1981; [c] *A. Goze,* unpublished data; [d] *Wilson* and *Baldwin* 1978; [e] *Iordanescu* et al. 1978; [f] *Gryczan* and *Dubnau* 1978; [g] *Gryczan* et al. 1978; [h] *Keggins* et al. 1978; [i] *Weisblum* et al. 1979; [j] *Löfdahl* et al. 1978

peated plasmids is not essential for their activity (*Michel* et al. 1980a). These data fit a model postulating that plasmids are endonucleolytically cleaved in contact with *B. subtilis* competent cells, similarly to chromosomal DNA (*Davidoff-Abelson* and *Dubnau* 1973). From such a linearized plasmid molecule a viable circular genome can be reconstituted within the cell only by an intramolecular recombination between repeated DNA sequences (*Michel* et al. 1980a). This intramolecular recombination is independent of the *recE* gene which is essential for chromosomal transformation or transfection where intermolecular recombination takes place (*Dubnau* et al. 1973).

The efficiency with which plasmids transform competent cells depends on the fraction of oligomers present in the DNA preparation. Over 10^7 transformants/µg of DNA are regularly obtained in our laboratory with plasmids such as pHV33 (Table 4), extracted from an *E. coli* host. With plasmids prepared from *B. subtilis,* where the degree of oligomerization is lower, 10–100 times lower values are commonly observed, (*Ehrlich* 1978a; *Gryczan* et al. 1978).

Polyethylene glycol-induced protoplast transformation by plasmids has been less studied than the transformation of competent cells. Two important differences have nevertheless been observed: 1. plasmid monomers transform protoplasts (*Dubnau* et al. 1980), 2. a very high level of cotransformation occurs (*Michel* 1979) similar to that observed with eukaryote cells transformed in the presence of polyethylene glycol (*Wigler* et al. 1979). This type of transformation is quite efficient, reportedly yielding over 10^7 transformants/µg of plasmid DNA (*Chang* and *Cohen* 1979).

4 S. aureus Plasmids that Replicate in B. subtilis

Cloning vectors can easily be developed in any transformable host for which small plasmids, endowed with easily selectable genetic markers, are available. A number of such plasmids which were observed originally in strains of *S. aureus* could be introduced by DNA transformation in *B. subtilis* (Table 1).

Table 2. Genes inactivated by insertion

Gene	Parental genome	Inactivating sites	Reference
Km^R	pUB110	BglII	a, b
Sm^R	pSA501	EcoRI, HindIII	a, c
Em^R	pE194	BclI, HpaI	a
Tc^R	pT127	KpnI	b
Ap^R	B. licheniformis chromosome	BglII, PstI	d
trpC	B. pumilus chromosome	HindIII	e
trpC	B. licheniformis chromosome	HindIII	e
trpC	B. subtilis chromosome	HindIII	e

[a] Gryczan et al. 1980b; [b] Michel et al. 1980; [c] Löfdahl et al. 1978; [d] Gray and Chang 1981; [e] Keggins et al. 1979

Several of these plasmids were used for cloning without further improvements by in vitro procedures, similarly to first E. coli plasmid vectors (i.e., pSC101, ColE1, Cohen et al. 1973; Hershfield et al. 1974). Foreign DNA could be inserted into the HindIII site of pC194 (Ehrlich 1978a) and the EcoRI or BamHI site of pUB110 (Gryczan and Dubnau 1978; Keggins et al. 1978) with no interference with plasmid replication or the expression of the resistance gene. On the other hand, the plasmid pSA2100, which is an in vivo isolated cointegrate between pC194 and pSA501 (Iordanescu 1975), carries an EcoRI site situated within its streptomycin resistance gene and could be used as an insertional inactivation vector (Löfdahl et al. 1978; Gryczan et al. 1980b).

5 Constructed Plasmid Vectors

In vitro joining of different S. aureus plasmids, or their segments, to each other, to E. coli plasmids, and to chromosomal segments of various Bacilli gave rise to a panoply of cloning vectors. Two main kinds can be distinguished, the insertional inactivation vectors and the so-called bridge plasmids, which can be used for cloning both in B. subtilis and in E. coli.

Several genes expressed in B. subtilis, and carrying a site which can conveniently be used for cloning are shown in Table 2. They have been used to construct insertional inactivation cloning vectors, such as those shown in Table 3. In each case at least one inactivation site is unique in the vector. In addition a number of other sites available for cloning are present on most of the vectors.

There is no fundamental difference among the vectors shown in Table 3, and a preference for one over the others in a particular cloning experiment may well be determined by considerations such as convenience of the restriction sites. Size difference (3.2–6 Md) for plasmids shown in Table 3 may, however, be important in some cases, the smaller plasmids offering a more favorable insert to vector DNA ratio. A significant consideration may also be that of a copy number of a particular cloning vector. The values reported for plasmids which serve as replicons of vectors shown in Table 3 vary from 8 for pSA501 to 35 or 40 for pC194 and pUB110. These may be amplified 4–8 times in the presence of hydroxyurea (Shivakumar and Dubnau 1978b). In addition, in certain bac-

Table 3. *B. subtilis* insertional inactivation vectors

Plasmid	Parental genomes	Mol wt. (Md)	Genetic markers	Unique sites	Reference
pBD6	pUB110, pSA501	5.8	Km^R Sm^R	*Bam*HI, *Bgl*II, *Hind*III, *Tac*I	a, b
pBD8	pUB110, pSA2100	6.0	Km^R Sm^R Cm^R	*Bam*HI, *Bgl*II, *Eco*RI, *Hind*III, *Xba*I	a, b
pBD9	pUB110, pE194	5.4	Km^R Em^R	*Bam*HI, *Bcl*I, *Bgl*II, *Eco*RI, *Hpa*I, *Pst*I, *Tac*I	a, b
pBD10	pBD8, pE194	4.4	Km^R Cm^R Em^R	*Bam*HI, *Bcl*I, *Bgl*II, *Hpa*I, *Xba*I	a, b
pBD12	pUB110, pC194	4.5	Km^R Cm^R	*Bam*HI, *Bgl*II, *Eco*RI, *Hind*III, *Tac*I, *Xba*I	a, b
pBD64	pUB110, pC194	3.2	Km^R Cm^R	*Bam*HI, *Bgl*II, *Eco*RI, *Tac*I, *Xba*I	b
pHV11	pC194, pT127	3.3	Tc^R Cm^R	*Hpa*I, *Kpn*I, *Xba*I	c, d
pHV41	pC194, pUB110, pBR322	4.5	Km^R Cm^R	*Bam*HI, *Bgl*II, *Eco*RI, *Xba*I	d
pSL103	pUB110, *B. pumilus* chromosome	5.0	Km^R trpEDCF	*Hind*III	e
pSL105	pUB110, *B. licheniformis* chromosome	5.4	Km^R trpDCF	*Hind*III	e, f

[a] *Gryczan* and *Dubnau* 1978; [b] *Gryczan* et al. 1980b; [c] *Ehrlich* 1978; [d] *Michel* et al. 1980; [e] *Keggins* et al. 1978; [f] *Keggins* et al. 1979

Table 4. Bzidge vectors for cloning in *B. subtilis* and *E. coli*

Plasmid	Parental genomes	Mol. wt. (Md)	Genetic markers		Unique sites	Reference
			E. coli	*B. subtilis*		
pHV14	pC194, pBR322	4.6	Ap Cm	Cm	*Ava*I, *Bam*HI, *Eco*RI, *Pst*I, *Pvu*II, *Sal*I	a
pHV33	pC194, pBR322	4.6	Ap Tc Cm	Cm	*Ava*I, *Bam*HI, *Eco*RI, *Pst*I, *Pvu*II, *Sal*I	b
pHV23	pC194, pT127, pBR322	6.1	Ap Cm	Tc Cm	*Ava*I, *Bam*HI, *Eco*RI, *Hpa*I, *Kpn*I, *Pst*I, *Pvu*II, *Sal*I, *Xba*I	c
pHV428	pC194, pUB110, pBR322	6.1	Ap Cm Km	Cm Km	*Ava*I, *Bgl*II, *Eco*RI, *Pst*I, *Pvu*II, *Sal*I, *Xba*I	d
pOG2165	pC194, pUB110, pOPΔ6, *B. licheniformis* chromosome	5.0	Ap Cm	Ap Cm	*Bgl*II, *Hind*III, *Pst*I, *Sst*I	e

[a] *Ehrlich* 1978; [b] *Primrose* and *Ehrlich* 1981; [c] *Michel* et al. 1980; [d] Unpublished data; [e] *Gray* and *Chang* 1981

terial mutants thermosensitive for DNA replication (*dnaC30, dnaII02)* the copy number of pUB110 was found to increase at the restrictive temperature to about 1000 (*Shivakumar and Dubnau* 1978a).

Bridge cloning vectors active in *E. coli* and *B. subtilis* are shown in Table 4. In several cases *S. aureus* plasmids were joined to pBR322 (*Bolivar* et al. 1978). In this way vectors carrying numerous cloning sites and easy to prepare in large amounts from an *E. coli* host were obtained. All of the plasmids described in Table 4 can be used as insertional inactivation vectors, pHV14 and pHV33 in *E. coli*; pHV23, pHV428, and pOG2165 in both *E. coli* and *B. subtilis*. Bridge cloning vectors are mainly useful for the construction of desired hybrids in the host which may be more suitable for a particular experiment and followed by examination of the behavior of inserted genes in either *E. coli* or *B. subtilis*.

Vectors which are likely to be developed in the near future include those which would allow selection of hybrid genomes (perhaps by inactivation of a lethal gene) and those facilitating expression of foreign genes in *B. subtilis*, for which the insertional inactivation cloning vectors are a good starting material.

6 Cloning of Nonreplicating DNA Segments

Most of the plasmid vectors described in the preceding section were constructed by cloning in *B. subtilis*. No particular problems, besides instability of some of the constructs (see below) were encountered. These experiments illustrate that cloning segments from relatively simple mixtures can easily be achieved in this host. A very high proportion of hybrids (over 80%) can be obtained in such experiments, when a relatively high DNA concentration (150 µg/ml) and a low ratio of vector to segment being cloned (1 : 20) is used during ligation reaction (unpublished data). The reason is that under these conditions molecules composed of several segments being cloned, joined to a single vector genome, are preferentially synthesized. Such molecules transform very efficiently *B. subtilis* competent cells, due to their internal repetitions, and by intramolecular recombination yield only hybrid plasmids. Monomeric circular vector molecules, which are also synthesized during ligation, are devoid of transforming activity (see above) and oligomeric vector molecules which are active in transformation are formed only rarely under the given ligation conditions.

Cloning of DNA segments from more complex mixtures ("shotgun cloning") in *B. subtilis* has been attempted less often. One of the first experiments of this type was cloning of genes of the tryptophan operon from several *Bacilli* (*Keggins* et al. 1978). A systematic study of shotgun cloning revealed a relatively high proportion (10%–20%) and yield (10^3–10^4/µg) of hybrid clones in experiments where selection for a vector-carried genetic marker was used (*Michel* et al. 1980). The average size of the cloned segments was, however, substantially lower than that of the segments used for cloning (1 and 3 Md respectively). The predominance of the small inserts in hybrid plasmids which does not have a fully satisfactory explanation, may have been the main reason for the reported difficulties in the isolation of active genes by shotgun cloning in *B. subtilis* (*Gryczan* et al. 1980a). Genes carried on relatively short DNA segments are likely to be efficiently cloned (*Michel* et al. 1980).

A procedure which allows larger DNA segments to be cloned in a shotgun type of experiment has been developed. It is based on the observation that heterologous DNA seg-

ments can integrate by recombination into a replicon present in *B. subtilis* competent cells (i.e., chromosome, resident plasmid), if they are carried on circular molecules partially homologous with the resident replicon (*Duncan* et al. 1978) or if they are inserted in linear molecules in such a way as to be flanked by sequences homologous to the resident replicon (*Harris-Warrick* and *Lederberg* 1978; *Contente* and *Dubnau* 1979b). Molecules of this type were constructed by in vitro joining of segments being cloned to a cloning vector and used to transform *B. subtilis* competent cells harboring a plasmid partially homologous to the cloning vector. By in vivo recombination between such molecules and the resident plasmid, hybrids were obtained which contained relatively large DNA inserts (up to 4.7 Md). Several *B. licheniformis* genes were cloned quite efficiently in this way since 7–210 hybrids carrying a given gene were obtained per µg of DNA (*Gryczan* et al. 1980a).

7 Cloning of Replication Regions

Cloning of replication regions active in a given host (*Timmis* et al. 1975) depends on the availability of DNA segments which are unable to replicate autonomously, but carry genetic markers easy to select in that host. Two such segments deriving from *S. aureus* plasmids have been used to clone replication regions in *B. subtilis*.

The plasmid pT127, which confers tetracycline resistance upon *B. subtilis* (Table 1) is cleaved by *Hind*III endonuclease into three segments. The largest segment (1.5 Md), which carries the Tc^R gene, cannot replicate autonomously. It was used to label genetically and to isolate the replication region of the cryptic *B. subtilis* plasmid pHV400 (*Niaudet* and *Ehrlich* 1979).

A somewhat different approach was used to set up an alternative system for cloning replication regions active in *B. subtilis*. A nonreplicative segment was incorporated into a plasmid maintained in *E. coli*, but unable to replicate in *B. subtilis*. This was achieved by joining the *S. aureus* plasmid pC194 to *E. coli* plasmid pBR322 and by deleting subsequently from the hybrid the replication region of pC194 (*Primrose* and *Ehrlich* 1981). The main advantage which the resulting plasmid, named pHV32, has over the Tc^R segment of pT127 is that it can be prepared in large amounts from the *E. coli* host absolutely free of any contaminating replicon active in *B. subtilis*. In addition, pHV32 has all of the unique restriction sites of pBR322 available for cloning (with the exception of *Cla*I, which is present twice). It has been used for cloning the replication region of pHV400 (*Niaudet* and *Ehrlich* 1979 and pT127 (unpublished data; *Goze* and *Ehrlich* 1980). Similar vectors unable to replicate in *B. subtilis*, which contain Km^R, Tc^R or *hisH* genes, deriving from pUB110, pT127 or *B. amyloliquefaciens* chromosome, respectively, inserted in pBR322 were also constructed (unpublished data).

8 Stability of Hybrid Plasmids

A problem often encountered in DNA cloning experiments, but less often reported in the literature is that newly constructed hybrid plasmids are not stable (*Ehrlich* et al. 1976; *Cohen* et al. 1978). Two types of events are observed, the loss of the entire hybrid from the cell and the rearrangements, most often loss via deletions, of plasmid sequences. The former may be termed segregational, the latter structural instability.

Segregational instability was observed in some cloning experiments using *B. subtilis* as a host. Insertion of DNA sequences into the *Hind*III site of pC194 rendered the hybrids less stable than the parental plasmid. 10%–80% of cells grown for 20 generations without selective pressure were devoid of hybrid plasmids, while none (<0.5%) lost pC194 (*Goze* and *Ehrlich* 1980; unpublished data). It is unlikely that the instability of hybrids was due to insertional inactivation of a function or a region conferring stability on pC194, since a plasmid obtained by in vitro deletion of some 600 nucleotides surrounding the *Hind*III site of pC194 was maintained in *B. subtilis* as stably as pC194 itself (*A. Goze,* unpublished observations). No satisfactory explanation for the segregational instability of the hybrids studied is presently available.

Structural instability of hybrid plasmids in *B. subtilis* has been reported in several instances (*Goebel* et al. 1978; *Gryczan* and *Dubnau* 1978; *Gryczan* et al. 1980b; *Grandi* et al. 1981; *Gray* and *Chang* 1981). One of the most extensively studied examples is the attempt to construct in *B. subtilis* a hybrid between pSA2100 (Table 1) and *E. coli* plasmid pHisG (*Grandi* et al. 1981). The two plasmids were joined in vitro at their unique *Eco*RI site and used to transform either *E. coli* or *B. subtilis*. Expected hybrids were obtained in the former host, but not in the latter where either of the two following types of plasmid were found. The first had lost by deletion all of pSA501 and pHisG sequences, and was very similar to pC194; the second had lost most of the pSA501 and pHisG sequences — some, however, still remained. The rearrangements were not a consequence of transforming *B. subtilis* with molecules joined in vitro, possibly into a particular combination unstable in vivo, since the hybrid isolated in *E. coli* invariably underwent deletions yielding pC194-like plasmids when introduced in *B. subtilis*.

Joining of foreign sequences to pSA2100 was clearly the cause of instability since 1. the parental plasmid could be propagated for many generations without rearrangements; 2. cleavage with *Eco*RI and joining linearized pSA2100 to itself prior to transformation gave rise to no deletion.

Instability of hybrids was observed also with pHV41 cloning vector (Table 3). Some DNA segments were deleted upon insertion in its *Bgl*II site, yielding plasmids very similar to pHV41, but lacking the *Bgl*II site. On the other hand, some segments were carried stably in this site (*Michel* 1979; *Michel* et al. 1980).

Hybrid plasmid instability in *B. subtilis* may possibly be explained by postulating the existence, in this organism, of an efficient recombination machinery active on sequences with a limited amount of homology. Deletions would then be introduced at a relatively high rate in plasmids harbored in this host, and the variants would in some instances rapidly become predominant when competing with larger, in vitro constructed genomes. The hypothesis that recombination is more efficient in *B. subtilis* than in *E. coli* is supported by the fact that plasmids containing internal homologies 200–2000 bp long could be maintained quite stably in the latter host, while they undergo deletions at a very high frequency (90% within 20 generations) in the former (unpublished data). Systematic study of dependence of recombination frequency on the extent of homology may give us more insight into the phenomenon of instability of newly constructed hybrid plasmids.

9 Conclusions

Cloning in *B. subtilis* was described for the first time in 1978 (*Dubnau* and *Gryczan* 1978; *Keggins* et al. 1978; *Ehrlich* et al. 1978; *Ehrlich* 1978a). Great progress has been accom-

plished in the development of various cloning vectors deriving from *S. aureus* plasmids during the past three years. It should be stressed that such vectors could very likely be used in other *Bacilli*, and perhaps in numerous other bacteria, since the host range of plasmids from *S. aureus* appears to extend not only to *B. subtilis* but also to organisms as distant as *E. coli* (*Goze* and *Ehrlich* 1980). Further development of vectors, possibly in the direction of facilitating selection of hybrid clones and expression of foreign genes in this host, is certainly to be expected in the near future.

Use of the cloning technology in *B. subtilis* (which is mostly outside the scope of the present review) has already been fruitful, in a number of research fields. Foreign genes were introduced in this host and found to be functional in some cases, but silent in others (*Ehrlich* and *Sgaramella* 1978). Transformation of *B. subtilis* with plasmids has been studied in some detail (*Ehrlich* et al. 1978; *Canosi* et al. 1978; *Contente* and *Dubnau* 1979a; *Michel* et al. 1980), replicons were analyzed (*Niaudet* and *Ehrlich* 1979; unpublished data), and the synthesis in *Bacilli* of new products, useful for humans, is just round the corner. Much greater progress in the near future is to be anticipated.

A counterpoint which deserves to be mentioned is that the question of stability of hybrid genomes, most pertinent to the successful use of cloning not only in *B. subtilis* but also in other hosts, remains to be solved. A systematic study of recombination between sequences with a low degree of homology may be a way of providing a satisfactory answer.

Acknowledgments. This work was supported by grants ATP 72-79-104 from the Institut National de la Santé et de la Recherche Médicale and ATP "Microbiologie" from the Centre National de la Recherche Scientifique.

References

Anagnostopoulos C, Spizizen J (1961) Requirements for transformation in *Bacillus subtilis*. J Bacteriol 81:741–746

Bolivar F, Rodriguez RL, Greene PJ, Betlach MC, Heyneker HL, Boyer HW, Crosa JH, Falkow S (1978) Construction and characterization of cloning vehicles. II. A multiple cloning system. Gene 2:95–113

Bron S, Luxen E, Venema G, Trautner TA (1980) Restriction and modification in *B. subtilis*. Effects on transformation and transfection with native and single-stranded DNA. MGG 179:103–110

Cahn FH, Fox MS (1968) Fractionation of transformable bacteria from competent cultures of *Bacillus subtilis* on renografin gradients. J Bacteriol 95:867–875

Canosi U, Morelli G, Trautner TA (1978) The relationship between molecular structure and transformation efficiency of some *S. aureus* plasmids isolated from *B. subtilis*. MGG 166:259–267

Canosi U, Iglesias A, Trautner TA (1981) General aspects of plasmid transformation. In: Polsinelli M (ed) Transformation – 80

Chang S, Cohen SN (1979) High frequency transformation of *Bacillus subtilis* protoplasts by plasmid DNA. MGG 168:111–115

Cohen SN, Chang ACY, Boyer HW, Helling RB (1973) Construction of biologically functional bacterial plasmids in vitro. Proc Natl Acad Sci USA 70:3240–3244

Cohen SN, Brevet J, Cabello F, Chang ACY, Chou J, Kopecko DJ, Kretschmer PJ, Nisen P, Timmis K (1978) Macro- and microevolution of bacterial plasmids. In: Schlessinger D (ed) Microbiology. American Society for Microbiology, Washington, pp 217–220

Contente S, Dubnau D (1979a) Characterization of plasmid transformation in *Bacillus subtilis*: kinetic properties and the effect of DNA conformation. MGG 167:251–258

Contente S, Dubnau D (1979b) Marker rescue transformation by linear plasmid DNA in *Bacillus subtilis*. Plasmid 2:555–571

Davidoff-Abelson R, Dubnau D (1973) Kinetic analysis of the products of donor deoxyribonucleate in transformed cells of *B. subtilis*. J Bacteriol 116:154–162

Dubnau D, Davidoff-Abelson R, Scher B, Cirigliano C (1973) Fate of transforming DNA after uptake by competent *Bacillus subtilis:* phenotype characterization of radiation-sensitive recombination-deficient mutants. J Bacteriol 114:273–286

Dubnau D, Gryczan TJ, Contente S, Shivakumar AG (1980) Molecular cloning in *Bacillus subtilis*. In: Setlow JK, Hollaender A (eds) Genetic engineering, principles and methods, vol 2. Plenum Press, New York London, pp 115–131

Duncan CH, Wilson GA, Young FE (1978) Mechanism of integrating foreign DNA during transformation of *Bacillus subtilis*. Proc Natl Acad Sci USA 75:3664–3668

Ehrlich SD (1977) Replication and expression of plasmids from *Staphylococcus aureus* in *Bacillus subtilis*. Proc Natl Acad Sci USA 74:1680–1682

Ehrlich SD (1978a) DNA cloning in *Bacillus subtilis*. Proc Natl Acad Sci USA 75:1433–1436

Ehrlich SD (1978b) *Bacillus subtilis* and the cloning of DNA. Trends Biochem Sci 3:184–186

Ehrlich SD, Sgaramella V (1978) Barriers to the heterospecific gene expression among procaryotes. Trends Biochem Sci 3:259–261

Ehrlich SD, Bursztyn-Pettegrew H, Stroynowski I, Lederberg J (1976) Expression of the thymidylate synthetase gene of the *Bacillus subtilis* bacteriophage Ø3T in *Escherichia coli*. Proc Natl Acad Sci USA 74:4145–4149

Ehrlich SD, Niaudet B, Goze A, Primrose SB (1978) Transformation of *Bacillus subtilis* with plasmid DNAs. In: Glover SW, Butler LO (eds) Transformation 1978. Cotswold Press, Oxford, pp 269–277

Goze A, Ehrlich SD (1980) Replication of plasmids from *Staphylococcus aureus* in *Escherichia coli*. Proc Natl Acad Sci 77:7333–7337

Goebel W, Kreft J, Bernhard K, Schrempf H, Weidinger G (1978) Replication and gene expression of extrachromosomal replicons in unnatural bacterial hosts. In: Boyer HW, Nicosia S (eds) Genetic engineering. Elsevier/North Holland Biomedical Press, New York Amsterdam, pp 47–58

Grandi G, Mottes M, Sgaramella V (1981) Instability of *Escherichia coli* HisG gene cloned in *Bacillus subtilis:* an example of convergent macro and micro evolution in plasmids. Plasmid 6:99–111

Gray O, Chang S (1981) Molecular cloning and expression of *Bacillus licheniformis* β-lactamase gene in *Escherichia coli* and *Bacillus subtilis*. J Bacteriol 145:422–428

Gryczan TJ, Dubnau D (1978) Construction and properties of chimeric plasmids in *Bacillus subtilis*. Proc Natl Acad Sci USA 75:1428–1432

Gryczan TJ, Contente S, Dubnau D (1978) Characterization of *Staphylococcus aureus* plasmids introduced by transformation into *Bacillus subtilis*. J Bacteriol 134:316–329

Gryczan TJ, Contente S, Dubnau D (1980a) Molecular cloning of heterologous chromosomal DNA by recombination between a plasmid vector and a homologous resident plasmid in *Bacillus subtilis*. MGG 177:459–467

Gryczan TJ, Shivakumar AG, Dubnau D (1980b) Characterization of chimeric plasmid cloning vehicles in *Bacillus subtilis*. J Bacteriol 141:246–253

Hadden C, Nester EW (1968) Purification of competent cells in the *Bacillus subtilis* transformation system. J Bacteriol 95:876–885

Harris-Warrick RM, Lederberg J (1978) Interspecies transformation in *Bacillus*: mechanism of heterologous intergenote transformation. J Bacteriol 133:1246–1253

Hershfield V, Boyer HW, Yanofsky C, Lovett MA, Helinski DR (1974) Plasmid ColE1 as a molecular vehicle for cloning and amplification of DNA. Proc Natl Acad Sci USA 71:3455–3459

Iordanescu S (1975) Recombinant plasmid obtained from two different compatible staphylococcal plasmids. J Bacteriol 124:597–601

Iordanescu S, Surdeanu M, Della Latta P, Novick R (1978) Incompatibility and molecular relationships between small staphylococcal plasmids carrying the same resistance marker. Plasmid 1: 468–479

Keggins KM, Lovett PS, Duvall EJ (1978) Molecular cloning of genetically active fragments of *Bacillus* DNA in *Bacillus subtilis* and properties of the vector plasmid pUB110. Proc Natl Acad Sci USA 75:1423–1427

Keggins KM, Lovett PS, Marrero R, Hoch SO (1979) Insertional inactivation of *trpC* in cloned

Bacillus trp segments: evidence for a polar effect on *trpF*. J Bacteriol 139:1001–1006

Löfdahl S, Sjöström JE, Philipson L (1978) A vector for recombinant DNA in *Staphylococcus aureus*. Gene 3:161–172

Lovett PS, Bramucci MG (1975) Plasmid deoxyribonucleic acid in *Bacillus subtilis* and *Bacillus pumilus*. J Bacteriol 124:484–490

Lovett PS, Duvall EJ, Keggins KM (1976) *Bacillus pumilus* plasmid pPL10: properties and insertion into *Bacillus subtilis* 168 by transformation. J Bacteriol 127:817–828

Lovett PS, Keggins KM (1980) *Bacillus subtilis* as a host for molecular cloning. Methods Enzymol 68:342–357

Michel B (1979) Clonage de DNA chez *Bacillus subtilis*. Rapport de DEA, Université Paris VI

Michel B (1981) Recombinaison génétique homoloque chez *Bacillus subtilis*. Thèse de 3e cycle, Université Paris VI

Michel B, Palla E, Niaudet B, Ehrlich SD (1980) DNA cloning in *B. subtilis*. III. Efficiency of random-segment cloning and insertional inactivation vectors. Gene 12:147–154

Michel B, Palla E, Ehrlich SD (1981) Internally repeated monomeric plasmids transform *Bacillus subtilis* competent cells. In: Polsinelli M (ed) Transformation 1980. Cotswald Press, Oxford

Mottes M, Grandi G, Sgaramella V, Canosi U, Morelli G, Trautner TA (1979) Different specific activities of the monomeric and oligomeric forms of plasmid DNA in transformation of *B. subtilis* and *E. coli*. MGG 174:281–286

Niaudet B, Ehrlich SD (1979) In vitro genetic labeling of *Bacillus subtilis* cryptic plasmid pHV400. Plamid 2:48–58

Primrose SB, Ehrlich SD (1981) Instability associated with deletion formation in a hybrid plasmid. Plasmid 6

Shivakumar AG, Dubnau D (1978a) Plasmid replication in *dna*ts mutants of *Bacillus subtilis*. Plasmid 1:405–416

Shivakumar AG, Dubnau D (1978b) Differential effect of hydroxyurea on the replication of plasmid and chromosomal DNA in *Bacillus subtilis*. J Bacteriol 136:1205–1207

Spizizen J (1958) Transformation of biochemically deficient strains of *Bacillus subtilis* by deoxyribonucleate. Proc Natl Acad Sci USA 44:1072–1078

Tanaka T (1979) Restriction of plasmid-mediated transformation in *Bacillus subtilis*. MGG 175:235–237

Timmis K, Cabello F, Cohen SN (1975) Cloning, isolation and characterisation of replication regions of complex plasmid genomes. Proc Natl Acad Sci USA 72:2242–2246

Trautner TA, Pawlek B, Bron S, Anagnostopoulos C (1974) Restriction and modification in *B. subtilis*. Biological aspects. MGG 131:181–191

Weisblum B, Graham MY, Gryczan T, Dubnau D (1979) Plasmid copy number control: isolation and characterization of high-copy-number mutants of plasmid pE194. J Bacteriol 137:635–643

Wigler M, Sweet R, Sim GK, Wold B, Pellicer A, Lacy E, Maniatis T, Silverstein S, Axel R (1979) Transformation of mammalian cells with genes from procaryotes and eucaryotes. Cell 16:777–785

Wilson CR, Baldwin JN (1978) Characterization and construction of molecular cloning vehicles within *Staphylococcus aureus*. J Bacteriol 136:1205–1207

Vectors for Gene Cloning in Pseudomonas and Their Applications

KENJI SAKAGUCHI*

1 Introduction

The bacteria belonging to the genus *Pseudomonas* consist of various phenotypically different species, but homology studies of their DNA by *Palleroni* et al. (1972) have revealed that many of them, including *P. fluorescens, P. putida, P. aeruginosa, P. cichorii, P. syringe, P. stutzeri, P. mendonica, P. alcaligenes,* and *P. pseudoalcaligenes,* belong to an almost homogeneous group with 80%–90% of DNA homology and can be differentiated from other groups that have a "similarity index" of 30%–50%. The DNA homology of *Pseudomonas* species with their related bacterial genera was studied by *DeLey* and others (*DeLey* and *Friedman* 1965; *DeLey* and *Park* 1966; *DeLey* et al. 1966), who indicated relative DNA homology of almost 100% with *Xanthomonas* species, of 40%–50% with *Azotobacter vinelandii,* of 35%–46% with *Azomonas macrocytogenes,* of 28%–29% with *Beijerinckia derxii,* and of 19%–22% with *Azotococcus agilis.* The genera *Serratia, Gluconobacter, Acetobacter,* and *Escherichia* share some 20%–30% DNA with *Pseudomonas,* but *Bacillus* DNA was almost

---

* Mitsubishi-Kasei Institute of Life Sciences, 11, Minamiooya, Machida-Shi-Tokio, Japan

entirely different. These studies suggest that the bacteria belonging to or related to *Pseudomonas* species have very divergent phenotypic characters that lead to classification into many different species, even though their genetic bases or their DNA sequence arrangements may be very similar.

This notion is verified by the existence of the transmissible plasmids among these wide varieties of gram-negative microorganisms. RP1, RP4, RK2, and other drug resistance plasmids which belong to the P1 incompatibility group and specify resistance to carbenicillin, tetracycline, kanamycin and sometimes streptomycin, sulfonamides, chloramphenicol, and trimethoprim, can be readily transferred among various species of *Pseudomonas*, as well as among various genera of bacteria including *Rhizobium*, *Serratia*, *Escherichia*, *Vibrio*, *Azotobacter*, and others (*Chakrabarty* 1976). The degradative plasmids which confer abilities to degrade the aromatic compounds such as toluene also have a wide range o host organisms among similar species (*Yano* 1980).

2 Introducing Escherichia coli Genes into Pseudomonas and Pseudomonas Genes into Escherichia coli

2.1 Construction of E. coli Plasmids Containing E. coli Tryptophan Operon

The gene-dosage effect of tryptophan synthesizing enzymes which are under an operon control was investigated by constructing four composite plasmids of various copy numbers: pSC101-trp, RSF1010-trp, RSF2124-trp, and RP4-trp were constructed in vitro by ligation of *Eco*RI-cleaved vehicle plasmid molecules and fragments obtained from bacteriophage λtrpE-A$_{60-3}$ and transformed into *E. coli* C600r⁻m⁻trp⁻ cells. The tryptophan synthetase activity in crude extracts of cells containing each composite plasmids, which were grown under repressed conditions with the addition of 100 µg/ml tryptophan in medium or under derepressed conditions with the addition of 10 µg/ml indolylacrylic acid, were assayed.

The data in Table 1 show that the repression system of host cells affects more strongly the expression of genes than the effect of gene dosage; however, the cells containing higher gene dosage consistently exhibit more enzyme formation.

Table 1. Trytophan synthetase activity in plasmid-containing strains

	No addition[a]	Tryptophan[b] (100 µg/ml)	Indolyl-acrylic acid[b] (100 µg/ml)	Approximate number of plasmids
pSC101-trp	21	14	101	4.2
RSF1010-trp	60	63	163	11.2
RSF2124-trp	50	18	153	11.9
RP4-trp	25	3	123	1.6

[a] Original strain 10 units
[b] Unit/mg protein; 1 unit = 0.1 µmol substrate conversion for 20 min at 37 °C

2.2 Derepression of E. coli Tryptophan Operon in Pseudomonas Cells and Weaker Expression of Pseudomonas Genes in E. coli

A transmissible composite plasmid RP4-trp comprising whole *E. coli* tryptophan operon was transferred into *Pseudomonas aeruginosa* trp⁻ cells through cell contact. The cloned cells grew on minimal medium. The activities of anthranilate synthetase (ASase) and tryptophan synthetase (TSase) were assayed on the crude extract of the *Pseudomonas* cells grown in the presence or absence of 100 μg tryptophan/ml of the culture medium. Both enzymes belonging to the tryptophan operon were fully expressed under both sets of conditions, showing the impotency of *Pseudomonas* tryptophan repression system to the *E. coli* tryptophan operon (*Nagahari* et al. 1977).

The tenfold increase of *E. coli* TSase in *Pseudomonas* cells even in the presence of tryptophan in the culture media implies that the *trp* repressor protein of *P. aeruginosa* cannot bind to the *trp* operator region of *E. coli*, and the *trp* gene products are expressed constitutively since the copy number of RP4-trp plasmid in both cells does not differ. By contrast, the introduction of *Pseudomonas* tryptophan genes into *E. coli* cells gave extremely low tryptophan synthetase activity (*Hedges* et al. 1977).

Such observations are fairly universal when the genes of both bacterial genera, including toluene oxidizing genes of *Pseudomonas putida* are mixed. *Nakazawa* et al. (1978) and other groups (*Chakrabarty* et al. 1978; *Jacoby* et al. 1978) isolated recombinants of RP4 and TOL plasmid which propagate in *E. coli* and in *P. putida*. The construction of TOL and its regulatory mechanism was extensively studied (*Nakazawa* et al. 1980). The expression of catechol 2,3-oxygenase, which is specified by the TOL segment in *E. coli* cells, was as low as the one-hundredth of the level produced in *P. putida* cells. *Chakrabarty* et al. (1978) transferred the TOL segment into *E. coli*, *Salmonella typhimurium*, *Agrobacterium tumefaciens*, *Azotobacter vinelandii* and *A. chroococcum*, but could not find the phenotypic expression of the TOL segment. *Inouye* et al. (1981) recently constructed by DNA cleaving and ligation a noninducible, highly expressing TOL-pBR322 plasmid in *E. coli*. In this case, the gene for catechol 2,3-oxygenase (*xylE*) were inserted in the opposite directions to usual, and the genes were under the control of the *E. coli tet* promoter in pBR322 plasmid.

These results suggest that *Pseudomonas* promoter sites are recognized poorly by the RNA polymerases in other gram-negative bacteria, including *E. coli*, and that on the contrary, the expression of the *E. coli* gene in *Pseudomonas* is very efficient. The implication may be that the *E. coli* promoter binds to *Pseudomonas* RNA polymerase efficiently, but

Table 2. Specific activities of tryptophan-synthesising enzymes in *P. aeruginosa* PA01 wild type and M12 trp⁻ carrying RP4-*trp*

Strain	Trp	ASase	TSase β
P. aeruginosa wild strain PA01	+	1.41	0.38
	−	1.53	2.17
P. aeruginosa trp⁻ strain M12 (RP4-*trp*)	+	6.59	19.0
	−	6.11	21.2

Cells were grown in glutamate minimal medium with or without tryptophan (100 μg/ml) and harvested in late exponential phase

that the *Pseudomonas* promoter does not bind well to *E. coli* RNA polymerase. The existence of species barriers is worth stressing (for a review see *Sakaguchi* 1980).

3 Introducing E. coli Plasmids into Pseudomonas

The inefficient or phenotypically indiscernible expression of *Pseudomonas* genes in *E. coli* and in other gram-negative bacteria necessitates the construction of a *Pseudomonas* host-vector system for the cloning of *Pseudomonas* genes. In introducing *E. coli* plasmids into *Pseudomonas* cells, RP4 remained at its original copy number, and RSF2124, a derivative of colicinogenic factor El, did not propagate in *Pseudomonas* cells; however, RSF1010, a derivative from streptomycin and sulfonamide resistance factor R6, multiplied efficiently in *Pseudomonas* species and was stably maintained. The host range of RSF1010 is very broad. Thus, RSF1010 is expected to be a strong candidate for a cloning vector in *Pseudomonas* cells.

3.1 Multicopy and Stability of RSF1010 Plasmid in Pseudomonas

The isolated RSF1010 DNA from *E. coli* J5 strain was transformed into *P. putida* ATCC 12633 cells and into *P. aeruginosa* M12 (PAO trp⁻) cells (*Nagahari* and *Sakaguchi* 1978) through the calcium salt procedure employed for *E. coli* (*Sano* and *Kageyama* 1977). The plasmid replicated more efficiently in *Pseudomonas* cells than in *E. coli* cells. As indicated in Fig. 1 and Table 3, the analysis by CsCl-ethidium bromide density gradient and by the density scan of electrophoretic pattern of the so-called chromosomal fraction of the CsCl-ethidium bromide centrifugation revealed that three times more plasmid DNA was contaminated in the chromosomal fraction. The total plasmid number per chromosome

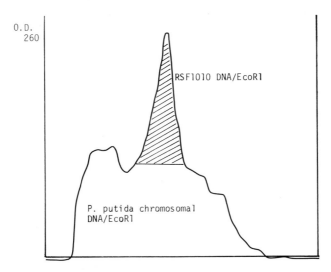

Fig. 1. Electrophoretic pattern of RSF1010 in *Pseudomonas putida* chromosomal fraction

Table 3. Copy number of plasmid RSF1010 in *P. putida*

	% Total DNA as plasmid	Copies per chromosome
CsCl-EtBr plasmid fraction	5.9	43
Chromosomal DNA fraction *Eco*R1	16.9	123
Total	22.8	166

reached 160 or more, 22% of the total DNA. The replication machinery of *Pseudomonas* cells worked better for the RSF1010 plasmid replication unit, raising the resistance ability against streptomycin 100-fold through greater production of streptomycin inactivating enzyme in the cell. However, this ability soon decreased through mutation and selection in the natural medium, which did not contain antibiotics, because the production of such a large amount of a specific enzyme was a heavy burden for the cell.

3.2 High Level Enzyme Production from E. coli Genes in Pseudomonas Cells

RSF1010-*trp* hybrid plasmid DNA containing the tryptophan operon of *E. coli* was intro-duced into *P. aeruginosa* trp⁻ cells by transformation (*Nagahari* 1978).

The cleavage map of RSF1010 is shown in Fig. 2 (*Bagdasarian* et al. 1979; *Gantier* and *Bonewald* 1980; *Nagahari* 1981a). Restriction enzymes available for the unique cleavage are: *Eco*RI, *Eca*I, *Bst*EII, *Sst*I, *Pvu*II, and *Hpa*I. Insertional inactivation may be possible on *Pst*I, *Hpa*I, *Eca*I, *Pru*II, *Bst*EII, and *Hinc*II sites. This plasmid was not cleaved by *Bam*-HI, *Sal*I, *Hind*III, *Kpn*I, *Cla*I, *Bgl*II, *Xho*I, *Sma*I or *Xba*I.

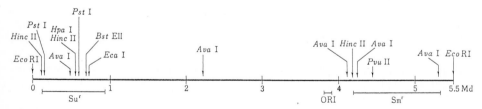

Fig. 2. Cleavage map of RSF1010

Table 4. Stability of RSF1010 in different strains

Host strain	Stability[a] %
E. coli J5	98.9
P. putida ATCC 12633	99.8
P. aeruginosa M12	92.6

[a] Stability is indicated as the percentage of cells resistant to streptomycin after one generation in an antibiotic-free L broth. The concentration of streptomycin in the plates was 10 μg/ml for *E. coli* and 100 μg/ml for *P. putida* and *P. aeruginosa*

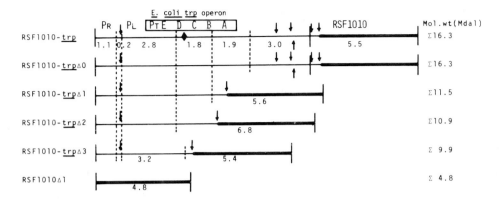

Fig. 3. Restriction maps of plasmid DNAs. Each plasmid is represented as a linear form whose left and right termini form an *Eco*RI site. *Bold lines* represent the 5.0-Md portion of RSF1010 plasmid DNA yielded by double digestion with *Eco*RI and *Pst*I. All parts of the RSF1010Δ1 are also represented by a *bold line. Dotted line* depicts the cleavage site with *Hind*III, (↓) with *Pst*I, (↑) with *Bam*NI. ✦, position of the internal promoter of the *E. coli* tryptophan operon Masses (Md) of DNA fragments generated by the double digestion with *Eco*RI and *Hind*III are indicated below the lines. *E, D, C, B,* and *A,* structural genes of the *E. coli* tryptophan operon; P_T, trp operon promoter; P_R, λ rightward promoter; P_L λ leftward promoter

This plasmid can serve as a stable vector for various *Pseudomonas* species and possibly other gram-negative bacteria. RSF1010 was more stably maintained in *P. putida* cells than in *E. coli* cells (Table 4). *Gantotti* successfully transformed RSF1010 and pBR322 into *P. phaseolicola* (*Gantotti* et al. 1979), but in many cases pBR322 and other *Col*E1 derivatives failed to propagate in *Pseudomonas* species, and the hybrid plasmid with RSF1010 was necessary. From the Trp⁺ transformants, various deletion plasmids were easily obtained as their physical maps with restriction endonucleases show in Fig. 3. This result suggests an efficient method for obtaining deletion mutants on a specific gene through cloning. *P. aeruginosa* cells harboring four kinds of hybrid plasmids produced strikingly high levels of *E. coli* tryptophan synthetase. The level of the enzyme produced from RSF1010-trpΔ2 plasmid reached 270 times higher than that of parent cells (Table 5). The abundant production of other *E. coli* enzymes may be possible by the same procedure of introduction into *Pseudomonas*. However, the high activity decreased very quickly. Even overnight culture was enough to reduce activity by 90%, and after 7 days of pedigree culturing every day, the specific activity of tryptophan synthetase per milligram protein settled at the level of 3–6 times the parent *Pseudomonas* strain (Table 5). It was noteworthy that the molecular weight of the plasmid did not change after the lowering of the activity. Therefore, some single mutations(s) probably at a promoter site(s), caused the drastic decrease in the enzyme activity. Also, through experiments transferring the plasmids into other *Pseudomonas* hosts, the decrease was found to be due to alteration of the plasmids, not the host cells. The tryptophan synthetase produced in *P. aeruginosa* cells was proved to be the *E. coli* type through immunologic testing.

The *E. coli* tryptophan operon was introduced into cells of another pseudomonad, *P. putida*, using the same RSF1010-trp hybrid plasmid and its derivatives (*Nagahari* 1981b).

Table 5. Decrease in the specific activity of TSase β after 1 week of successive daily transfer of culture

P. aeruginosa strain	Sp. act.of TSase β[a]	
	Initially	After 1 week of culture
M12 (RSF1010-trpΔ0)	238	11.8
M12 (RSF1010-trpΔ1)	349	7.8
M12 (RSF1010-trpΔ2)	548	13.3
M12 (RSF1010-trpΔ3)	8.1	3.6
PAO1 (parent)	2.2	ND[b]
M12[c]	0.5	ND

[a] A unit of enzyme activity is defined as the utilization of 1 nmol of substrate per min at 37 °C. Specific activity is given here as units per milligram of protein
[b] ND, not determined
[c] P. aeruginosa trp M12 cells were cultured in glutamate minimal medium containing 50 μg tryptophan/ml

Table 6 and Fig. 4 indicate that the synthesis of E. coli tryptophan synthetase in P. putida cells was 10 times amplified by the introduction of RSF1010-trp, whereas the synthesis of anthranilate synthetase (ASase) was decreased to half. The reason for less expression of the E. coli gene in P. putida than in P. aeruginosa is not clear, but the data in Table 6 and ·Fig. 4 indicate that the internal promoter present on the top of the E. coli trpC gene is working, since the strains pKNT3 and pKNT4 delete the main promoter portion of inserted trp operon but still produce the same level of TSase with pKNT1 and RSF1010-trp. Crawford's hypothesis on the autogenous regulation seems to elucidate the repressed

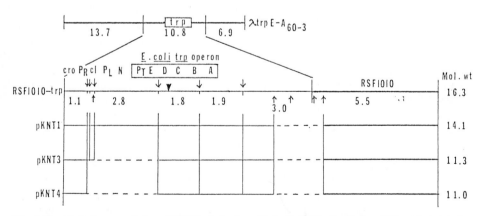

Fig. 4. Restriction maps of plasmid DNAs introduced into P. putida. Each plasmid is represented as a linear form whose left and right termini form an EcoRI site. Bold lines represent the RSF1010 plasmid DNA. (↓) depicts the cleavage site with HindIII, (↑) with PstI. ▼, position of the internal promoter of the E. coli tryptophan operon. Size (Md) of DNA fragments generated by the double digestion of RSF1010-trp plasmid with EcoRI and HindIII are indicated below the line. E, D, C, B, and A, structural genes of the E. coli tryptophan operon; P_T, trp operon promoter; P_R, rightward promoter; P_L, λ leftward promoter

Table 6. Specific activities of tryptophan-synthesizing enzymes in *P. putida trp*B cells carrying derivatives of RSF1010-*trp* hybrid plasmid

Strain	Anthranilate synthetase	Tryptophan synthetase β
*P. putida trp*B		
RSF1010-*trp*	0.23	14.6
pKNT1	0.15	14.6
pKNT3	0.03	11.8
pKNT4	0.05	12.9
P. putida (wild type)	0.40	1.4

production of both enzymes since he used *P. putida*, and he claims that the TSase α-chain
forms at least part of the repressor protein (*Proctor* and *Crawford* 1975). pKNT plasmids
described in Table 6 and Fig. 4 are the artificially cleaved and ligated plasmids. pKNT3
and pKNT4 delete the ASase gene and tryptophan operon promoter gene.

3.3 Construction of a RSF1010-pBR322 Hybrid Plasmid

In order to obtain cloning vectors susceptible to various useful restriction enzymes, a hybrid plasmid between RSF1010 and pBR322 was constructed (*Nagahari* 1980, 1981a) (Fig.
5). The two plasmid DNA were digested with restriction endonuclease *Pst*I, ligated, and
transformed into *E. coli*. Colonies appearing on a plate containing streptomycin 20 μg/ml
and tetracycline 10 μg/ml harbored a plasmid consisting of the sum of the two parent plasmids, but were deprived of the short segment between the two *Pst*I sites on RSF1010. The

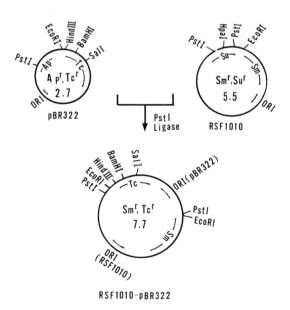

Fig. 5. Construction of plasmid RSF
1010-pBR322. Restriction sites are
drawn to scale on a circular map. *Ap*,
Tc, *Su*, and *Sm* represent the ampi-
cillin-resistant, tetracycline-resistant,
sulfonamide-resistant, and streptomy-
cin-resistant genes respectively. *ORI*
represents replication origin

resulting plasmid was maintained stably in *E. coli*, and confered on its host resistance to streptomycin and tetracycline, but no resistance to ampicillin and sulfonamide. This plasmid allows the use of *Hind*III, *Bam*HI, *Sal*I, and *Pvu*II for unique insertional inactivation sites.

4 Cloning of the P. putida Leucine Gene in P. putida

Using plasmid RSF1010 and *Eco*RI endonuclease, the *P. putida* leucine gene was cloned in *P. putida*. Chromosomal DNA was isolated from *P. putida* cells, cleaved with *Eco*RI, and ligated with plasmid RSF1010 with the usual procedure employed for *E. coli*. The ligated DNA mixture was transformed into the *P. putida* leu⁻ strain and two colonies appeared on minimal medium. The autotrophic strain contained the plasmid depicted at the top of Fig. 6. A *P. putida* chromosomal fragment of 6.6 Md which contained an *Eco*RI cleavage site was inserted into the unique *Eco*RI site of RSF1010. The plasmid was again cleaved by *Eco*RI, ligated, and transformed into *P. putida*. Ten autotrophic colonies appeared. Among them, six colonies contained B-type hybrid plasmid (Fig. 6), three colonies contained both A-type and B-type plasmids, and one colony contained both A-type and C-type plasmids. Therefore, the 3.0-Md DNA segment of A- and B-type plasmids is responsible for the synthesis of isopropylmalate dehydrogenase in *P. putida* cells.

The enzyme activity of *P. putida* cells containing cloned leucine genes was assayed as shown in Table 7. The increase in β-isopropylmalate dehydrogenase activity was only twice that found when the culture medium contained no leucine. This is a rather low val-

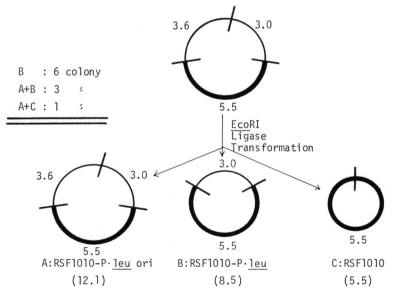

Fig. 6. Cloning of *P. putida* chromosomal leucine gene on RSF1010. *Bold line,* RSF1010 vehicle DNA; *thin line,* inserted *P. putida* chromosomal DNA; *crossed line,* the *Eco*RI cleavage site. Mol. w. of each DNA fragment is indicated in Md. For further explanation see text

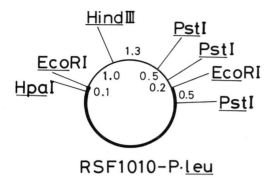

RSF1010-P·leu

Fig. 7. Cleavage map of RSF1010-P · *leu*

ue when the high copy number of RSF1010 plasmid in *Pseudomonas* is considered. Either a repression system is operative in the cell, or the high production of the enzyme caused the rapid selection against such a producer since too much production of a specific protein is unnecessary and inconvenient for growth. The existence of the repression system was proved by the lowering of enzymic activity to one-fifth in the presence of external leucine 100 µg/ml.

It is noteworthy that *P. putida* is classified on the basis of its optimal growth at 30 °C and poor or no growth at 37 °C. This accounts for the biologic safety of this bacteria. It is desirable to select and use as host bacteria *Pseudomonas* strains which do not grow at 37 °C.

The cleavage map of B-type plasmid designated as RSF1010-P. *leu* is shown in Fig. 7.

5 Plasmids Possibly Useful as Cloning Vectors and for Applied Use

Chakrabarty described in his review article various plasmids of drug resistance, of sex factor, of degradation of aromatic compounds, and of mercury resistance (*Chakrabarty* 1976a). Various kinds of drug resistance plasmids were classified according to their compatibility and their transmissibility to the bacteria of other genera. The nature of degradative plasmids of CAM, SAL, NAH, OCT, XYL, and TOL was described along with the plasmids for detoxifying organic and inorganic mercury compounds. Further, the possi-

Table 7. β-IPM dehydrogenase activity in *P. putida* crude extract

Strain	No addition	Leucine added (100 µg/ml)
P. putida wild type	1	0.17
P. putida leu⁻	–	ND[a]
P. putida leu⁻ (RSF1010-P · *leu* ori)	2.1	0.63
P. putida leu⁻ (RSF1010-P · *leu*)	2.3	0.54

Activities relative to the activity of *P. putida* wild-type cells grown in the absence of leucine are represented; [a] ND, not determined

bility for and examples of molecular cloning systems using these plasmids were reported (*Chakrabarty* 1976b). *Yano* (1980) reviewed recent features in the same field in relation to the microbial evolution. Here I describe the more recent results suggesting uses for the *Pseudomonas* molecular cloning system.

5.1 pKJ1

A naturally occurring conjugative plasmid pKJ1 coding for toluene degradation via the meta cleavage pathway and resistance to streptomycin and sulfonamides was isolated from a *Pseudomonas* sp. obtained from an enrichment culture to assimilate *m*-toluate (*yano* and *Nishi* 1980). The plasmid DNA, 150 Md, was transferable to various gram-negative bacteria, including *P. putida*, *P. aeruginosa*, *P. ovalis*, *Alcaligenes faecalis*, *Azotobacter vinelandii*, *Azomonas agilis*, and *E. coli*. Resistance to streptomycin was well expressed in many genera of bacteria, but toluene degrading activity was scarcely expressed in *E. coli*. The plasmid did not belong to the *Pseudomonas* incompatibility group P-1, P-2, P-3, or P-9.

5.2 Introduction and Expression of TOL Gene in E. coli

Nakazawa et al. constructed two recombinant plasmids, one *Pseudomonas* TOL plasmid with RP4, and the other TOL with pBR322. They were introduced into *E. coli*, and the genetic and cleavage maps were studied in relation to the degradative pathway of toluene and its regulation system (*Inouye* et al. 1981; *Nakazawa* et al. 1978, 1980). The expression of the TOL gene was poor in *E. coli* cells, but became potent in direct conjunction with an *E. coli* promoter.

5.3 Mercury and Organomercurial Resistance

Tonomura et al. studied *Pseudomonas* strains possessing enzymes catalyzing the splitting of carbon-mercury linkages (*Tezuka* and *Tonomura* 1976; *Tonomura* and *Kanzaki* 1969). Metallic mercury releasing enzyme in mercury-resistant *Pseudomonas* was studied by the same group (*Furukawa* and *Tonomura* 1972). The involvement of plasmids in the mercury and organomercurial resistances was reported by several groups (*Clark* et al. 1977; *Olson* et al. 1979; *Stanisch* et al. 1977).

5.4 Dehalogenation of Haloacetates

Fluoroacetate is a well-known inhibitor of aconitase, and has been used for the control of pests such as rat, rabbit and opposum. *Tonomura* and his group found a strain of *Moraxella* species which assimilates fluoro- and chloroacetate. The two abilities were attributed to the presence of two halidohydrolases I and II. Halidohydrolase I was active on fluoroacetate, chloroacetate, bromoacetate and produced inducibly, while halidohydrolase II was active on chloroacetate, bromoacetate, iodoacetate and produced constitutively.

The enzymes were purified, characterized, and compared with similar enzymes from different sources. These two enzymes as well as mercury reductase were specified by a plasmid pVO1 of 41 Md. The sizes of DNA fragments generated by the cleavage with restriction enzymes were determined. This plasmid is transmissible into *Pseudomonas acidovorans* through conjugation (*Kawasaki* et al. 1981a, b). Therefore, the natural and artificial introduction of this plasmid into *Pseudomonas* is an interesting topic.

5.5 Degradation of Nylon Oligomers

Okada and his group isolated *Flavobacterium* sp. KI72 strain from soil which could grow on a medium containing waste from factories producing nylon, 6-aminohexanoic acid cyclic dimer was the sole carbon and nitrogen source. Two new enzymes, 6-aminohexanoic acid cyclic dimer hydrolase (EI) (EC 3.5.2) and 6-aminohexanoic acid linear oligomer hydrolase (EII) (EC 3.5.1), were responsible for the hydrolysis (*Kinoshita* et al. 1977).

$$\begin{array}{c} CO\text{-}(CH_2)_5\text{-}NH \\ | \qquad\qquad | \\ NH\text{-}(CH_2)_5\text{-}CO \end{array} \xrightarrow{EI} NH_2\text{-}(CH_2)_5\text{-}CONH\text{-}(CH_2)_5COOH \xrightarrow{EII} 2\ NH_2(CH_2)_5COOH$$

| 6-aminohexanoic | 6-aminohexanoic acid | 6-aminohexanoic |
| acid cyclic dimer | linear dimer | acid |

Both enzymes were purified to homogeneity. EI was only active on its original substrates. More than 100 kinds of natural linear and cyclic amide compounds were tested in vain. EII was active on 6-aminohexanoic acid oligomers from dimer to hexamer and icosamer but not on the hectamer. The activity decreased with the increase of the polymerization number of the oligomer. The enzyme hydrolyzed so as to successively remove the 6-aminohexanoic acid residue from the amino terminal (*Kinoshita* et al. 1981).

The KI72 strain harbored three kinds of plasmid, pOAD1 (26.2 Md), pOAD2 (28.8 Md), and pOAD3 (37.2 Md). Among them, the two hydrolyzing enzymes were borne by pOAD2, as proved by curing experiments with mitomycin C (*Negoro* 1980). Sites of restriction endonucleases on this plasmid were determined and mapped. Hybrid plasmids with pBR322 were constructed with *Hind*III and ligation, and introduced into *E. coli*. Each of the two enzymes was separately cloned in *E. coli* and produced enzymes in *E. coli* cells. EI enzyme produced in *E. coli* was immunologically identical with the enzyme in *Flavobacterium*, but its specific activity was around 2% of its original enzymic level in *Flavobacterium*.

5.6 Degradation of Phenanthrene and Other Polycyclic Aromatic Hydrocarbons

Kiyohara, Nagao, and *Yano* found an unique method to detect bacteria assimilating nonvolatile polycyclic hydrocarbons. To the nutrient agar plate in which colonies of bacteria had formed, hydrocarbons dissolved in ether (10% w/v) were sprayed and incubated for 2-3 days. Bacteria assimilating hydrocarbons formed distinctive clear zones around their colonies (Fig. 8) (*Kiyohara* H, unpublished results). Using this method together with enrichment culture, many strains able to assimilate phenanthrene as a sole carbon source were isolated. They belonged to *Aeromonas, Vibrio, Pseudomonas,* etc. Among them,

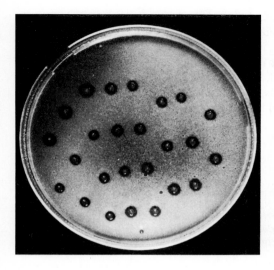

Fig. 8. Growth of *A. faecalis* AFK2 forming clear zones on a mineral salts agar covered with the thin layer of phenanthrene. AFK2 was transferred onto a mineral salts agar plate, which was then sprayed with the solution of phenanthrene in ether and incubated for 2 days at 30 °C. AFK2 was able to grow utilizing solid phenanthrene, forming clear zones around its colonies

Alcaligenes faecalis AFK2 harbored two plasmids. One of them, pHK1, had a molecular weight of 44×10^6, and the other, pHK2, of 22×10^6. The cured strain (AFK211, Tc^r) lost both plasmids. Through mating of AFK2 (Tc^s, Phn+) strain with AFK211 (Tc^r, Phn⁻) strain the phenanthrene assimilating activity was transferred to the cured strain. All the transconjugants carried both of the plasmids. However, the attempts to transfer to other species of bacteria, e.g., *Pseudomonas putida* through mating, were unsuccessful (*Kiyohara* et al. 1980).

This method is applicable to detecting bacteria assimilating other hydrocarbons. An *Aeromonas* sp. S45P1 which assimilates both phenanthrene and naphthalene was isolated.

6 Conclusions

The importance of genetic engineering in *Pseudomonas* cells is to be stressed for the following reasons:

1. *Pseudomonas* and related bacteria have abilities to decompose various strange compounds which are not usually metabolized by other microorganisms, for instance, phenol, toluene, hydrocarbons, cyanides, haloacetates or even new manufactured substrates such as nylon oligomers. These capabilities provide their strong efficacy against pollution and also provide the basis for their future use as biocatalysts in chemical industries.

2. *Cloning genes* of *Pseudomonas* or related bacteria in *E. coli* results in weak or very poor expression of enzymes. Their expression is expected and realized in the *P. putida* system. The *Pseudomonas* gene engineering system is more potent than the *E. coli* system.

3. *Striking overproduction* of *E. coli* enzymes is possible by introducing the *E. coli* gene into *Pseudomonas*.

4. *RSF1010 and its derivatives* exhibit convenient characteristics as carriers in *Pseudomonas* cells, such as their stability, their high copy number, selective drug resistance, ability of insertional inactivation etc.

The *E. coli* gene engineering system is neither the best nor universal. The host vector systems of each bacterium, yeast, mold, Streptomyces, and even plant have advantages, especially for developing their original specificities. They should be and will be investigated further for their applications and for the construction of safer systems.

References

Bagdasarinan M, Bagdasarinan MM, Coleman S, Timmis KN (1979) New vector plasmids for gene cloning in *Pseudomonas*. In: Timmis KN, Pühler A (eds) Plasmids of medical, environmental and commercial importance. Elsevier/North-Holland, Biochemical Press, pp 411–422

Chakrabarty AM (1976a) Plasmids in *Pseudomonas*. Annu Rev Genet 10: 7–30

Chakrabarty AM (1976b) Molecular cloning in *Pseudomonas*. Microbiology 1976, American Society for Microbiology, pp 579–582

Chakrabarty AM, Friello DA, Bopp LH (1978) Transposition of plasmid DNA segments specifying hydrocarbon degradation and their expression in various microorganisms. Proc Natl Acad Sci USA 75: 3109–3112

Clark DL, Weiss AA, Silver S (1977) Mercury and organomercurial resistances determined by plasmids in *Pseudomonas*. J Bacteriol 132:186–196

DeLey J, Friedman S (1965). Similarity of *Xanthomonas* and *Pseudomonas* deoxyribonucleic acid J Bacteriol 89: 1306–1309

DeLey J, Park IW (1966) Molecular biological taxonomy of some free living nitrogen-fixing bacteria. Antonie van Leeuwenhoek, 32: 6–16

DeLey J, Park IW, Tijtgat R, van Ermengeni J (1966) DNA homology and taxonomy of *Pseudomonas* and *Xanthomonas*. J Gen Microbiol 42: 43–56

Furukawa K, Tonomura K (1972) Induction of metallic mercury-releasing enzyme in mercury-resistant *Pseudomonas*. Agric Biol Chem 36: 2441–2448

Gantier F, Bonewald R (1980) The use of plasmid R1162 (RSF1010) and derivatives for gene cloning in the methanol utilizing *Pseudomonas* AM1. MGG 178: 375–380

Gantotti BP, Patil SS, Mandel M (1979) Genetic transformation of *Pseudomonas phaseolicola* by R-factor DNA. MGG 174: 101–103

Hedges RW, Jacob AE, Crawford IP (1977) Wide ranging plasmid bearing the *Pseudomonas aeruginosa* tryptophan synthase genes. Nature 267: 283–284

Inouye S, Nakazawa A, Nakazawa T (1981) Molecular cloning of TOL genes and *xyl*B and *xyl*E in *Escherichia coli*. J Bacteriol 145:1137–1143

Jacoby GA, Rogers JE, Jacob AE, Hedges RW (1978) Transposition of *Pseudomonas* toluene-degrading genes and expression in Escherichia *coli*. Nature 274: 179–180

Kawasaki H, Tone N, Tonomura K (1981a) Plasmid determined dehalogenation of haloacetates in *Moraxella species*. Agric Biol Chem 45: 29–34

Kawasaki H, Tone N, Tonomura K (1981b) Purification and properties of haloacetate halidohydrolase specified by plasmid from *Moraxella* sp. strain B. Agric Biol Chem 45:35–42

Kinoshita S, Negoro S, Muramatsu M, Bisaria VS, Sawada S, Okada H (1977) 6-Aminohexanoic acid cyclic dimer hydrolase, a new cyclic amide hydrolase produced by *Achromobacter guttatus* KI72. Eur J Biochem 80:498–495

Kinoshita S, Terada T, Taniguchi T, Takene Y, Masuda S, Matsunaga N, Okada H (1981) Purification and characterization of 6-aminohexanoic acid oligomer hydrolase of *Flavobacterium* sp. KI72. Eur J Biochem (in press)

Kiyohara H, Nagao K, Yano K (1980) Plasmid involvement in bacterial degradation of phenanthrene. Proc Inter Fermentation Symposium, London, Canada

Nagahari K, Tanaka T, Hishinuma F, Kuroda M, Sakaguchi K (1977) Derepression of *E. coli* trp operon on interfamilial transfer. Nature 266:745–746

Nagahari K, Sakaguchi K (1978) RSF 1010 plasmid as a potentially useful vector in *Pseudomonas* species. J. Bacteriol 133:1527–1529

Nagahari K (1978) Deletion plasmids from transformants of *Pseudomonas aeruginosa trp* cells with the RSF 1010-*trp* hybrid plasmid and high levels of enzyme activity from the gene on the plasmid. J Bacteriol 136:312–317

Nagahari K (1980) Development of molecular cloning systems in *Pseudomonas*. In: Sakaguchi K, Okanishi M (eds) Molecular breeding and genetics of applied microorganisms. Kodansha-Academic Press, Tokyo New York San Francisco London, pp 124–127

Nagahari K (1981a) RSF 1010 as a vector for *Pseudomonas* species. Protein, nucleic acid and enzyme, gene engineering (in Japanese), vol 26, 4:487–490

Nagahari K (1981b) RSF 1010 plasmid as a candidate for a cloning vector in *Pseudomonas*. In: Panopoulos NJ (ed) Genetic engineering in the plant sciences. Praeger Pub, New York

Nakazawa T, Hayashi E, Yokota T, Ebina Y, Nakazawa A (1978) Isolation of TOL and RP4 recombinants by integrative suppression. J Bacteriol 134:270–277

Nakazawa T, Inouye S, Nakazawa A (1980) Physical and functional mapping of RP4-TOL plasmid recombinants: analysis of insertion and deletion mutants. J Bacteriol 144:222–231

Negoro S, Shinagawa H, Nakata A, Kinoshita S, Hatozaki T, Okada H (1980) Plasmid control of 6-aminohexanoic acid cyclic dimer degradation enzymes of *Flavobacterium* sp. KI72. J Bacteriol 143:238–245

Negoro S, Taniguchi T, Komoto T, Shinagawa H, Okada H (1981) Cloning and functional expression of nylon oligomer degradation genes in *Escherichia coli*. J Bacteriol (to be published)

Olson BH, Barkay T, Colwell RR (1979) Role of plasmids in mercury transformation by bacteria isolated from the aquatic environment. Appl Environ Microbiol 38:478–485

Palleroni NJ, Ballard RW, Ralston E, Doudoroff M (1972) DNA homologies among some *Pseudomonas* species. J Bacteriol 110:1–11

Proctor AR, Crawford IP (1975) Autogenous regulation of the inducible tryptophan synthetase of *Pseudomonas putida*. Proc Natl Acad Sci USA 72:1249–1253

Sakaguchi K (1980) Species barriers to the maintenance and expression of foreign DNA. In: Sakaguchi K, Okanishi M (eds) Molecular breeding and genetics of applied microorganisms. Kodansha-Academic Press, Tokyo New York San Francisco London, pp 1–7

Sano Y, Kageyama M (1977) Transformation of *Pseudomonas aeruginosa* by plasmid DNA. J Gen Appl Microbiol 23:183–186

Stanisich VA, Bennett PM, Richmond MH (1977) Characterization of a translocation unit coding resistance to mercuric ions that occurs on a nonconjugative plasmid in *Pseudomonas aeruginosa*. J Bacteriol 129:1227–1233

Tezuka T, Tonomura K (1976) Purification and properties of an enzyme catalyzing the splitting of carbon-mercury linkages from mercury-resistant *Pseudomonas* K-62 strain. I. Splitting enzyme 1. J Biochem 80:79–87

Tonomura K, Kanzaki F (1969) The reductive decomposition of organic mercurials by cell-free extract of a mercury-resistant pseudomonas. Biochim Biophys Acta 184:227–229

Yano K (1980) Degradative plasmids: aspect of microbial evolution. In: Sakaguchi K, Okanishi M (eds) Molecular breeding and genetics of applied microorganisms. Kodansha, Tokyo, Academic Press, New York San Francisco London, pp 47–60

Yano K, Nishi T (1980) pKJ1, a naturally occurring conjugative plasmid coding for toluene degradation and resistance to streptomycin and sulfonamides. J Bacteriol 143:552–560

Host:Vector Systems for Gene Cloning in Pseudomonas

Michael Bagdasarian* and Kenneth N. Timmis**

1 Introduction

Soil bacteria in general, and the pseudomonads in particular, exhibit an enormous wealth of metabolic activities (Table 1) that include their ability to (a) utilize as carbon and/or nitrogen sources exotic organic compounds, such as naphthalenes, terpenes, alkaloids and recalcitrant man-made aromatic compounds (*Clarke* and *Richmond* 1975; *Hütter* and *Leisinger* 1981), (b) carry out photosynthesis (*Clayton* and *Sistrom* 1978), (c) fix nitrogen (*Dalton* and *Mortenson* 1972; *Dixon* et al. 1981) and (d) produce disease in animals (*Lowbury* 1975; *Doggett* 1979) and plants (*Dickey* et al. 1978). Some species in addition possess unusual structural features [e.g. *Methylomonas* (*Davies* and *Whittenbury* 1970;) *Halobacterium* (*Bayley* and *Morton* 1978)], whereas others are able to tolerate or even require extreme growth conditions, such as high salt (e.g. *Halobacterium*, *Bayley* and *Morton* 1978), or extremes of pH (e.g. *Thiobacillus*, *Lundgren* and *Silver* 1980) or temperature (e.g. *Thermus*). Moreover, a number of microorganisms elaborate products that are or have been of considerable use to man (e.g. acetic acid production by *Acetobacter*). As a consequence, many metabolic activities of soil bacteria are of great scientific, medical, agricultural, environmental or economic importance and the genetic analysis and manipulation of such bacteria, particularly of pseudomonads, is now a major objective.

To date, little progress has been made towards the genetic analysis of soil bacteria that exhibit interesting biological properties because few are adequately characterized (some have not even been definitively identified) and even fewer appear to be amenable

* Max-Planck-Institute for Molecular Genetics, Berlin-Dahlem, West Germany
** Department of Medical Biochemistry, University of Geneva, Switzerland

Table 1. Some biological activities of soil and aquatic pseudomonads

Biological activity	Found in genus	Propagation of RSF1010[a]
Growth on exotic organic compounds	*Pseudomonas*	+
	Alcaligenes	+
	Flavobacterium	NK[b]
Photosynthesis	*Rhodopseudomonas*	+
Nitrogen-fixation-association	*Rhizobium*	+
Nitrogen-fixation-free living	*Azotobacter*	+
Nitrification $NH_3 \rightarrow NO_2^-$	*Nitrosomonas*	NK[b]
Sulphur-oxidizing	*Thiobacillus*	NK[b]
Hydrogen-oxidizing	*Pseudomonas*	+
Growth on methane/methanol	*Methylophilus*	+
Production of disease in animals	*Pseudomonas*	+
Production of disease in plants	*Pseudomonas*	+
Production of plant tumours	*Agrobacterium*	+

[a] Plasmid RSF1010 or one of its derivatives in bacteria of the indicated genus
[b] Not known

to genetic investigation (e.g. because convenient gene transfer systems have not yet been developed). Moreover, the wide diversity found within strains of individual genera of soil bacteria frequently prevents the use of genetic techniques developed for one well-characterized member of the genus in the analysis of another. During recent years, however, powerful new methods, such as gene cloning and transposition mutagenesis, have been developed for the analysis and manipulation of gene structure and function (*Kleckner* et al. 1977; *Sherratt* 1981; *Timmis* 1981b). Gene cloning methods are of particular value for poorly-characterized bacteria because they are universal, in the sense that their usage is not limited by the relatedness of the organism under investigation to any well-defined laboratory strains and does not depend upon the availability of detailed genetic information.

Until recently, the only extensively developed gene cloning systems consisted of the bacterium *Escherichia coli* K-12 and its plasmids and bacteriophage λ genomes (*Cohen* 1975; *Murray* 1976; *Collins* 1977; *Timmis* et al. 1978), although chapters in this volume testify to the rapid progress made with other systems in the last few years. While the *E. coli* K-12 host is generally very useful for the cloning, amplification and structural analysis of genes from a wide variety of organisms; it may be unsuitable for the cloning of genes for those interesting properties of soil bacteria discussed above, *when these must be functionally expressed after cloning,* because *E. coli* does not constitute an appropriate 'physiological background' for the expression of such properties (for example, *Pseudomonas* genes that encode enzymes for the catabolism of toluene/xylene are poorly expressed in this host, *Jacoby* et al. 1978; *Ribbons* et al. 1979; *Franklin* et al. 1981a). In particular, it exhibits a limited range of metabolic activities, is 'fermentative' rather than 'oxidative' and is structurally and therefore functionally distinct from the pseudomonads. On the other hand, some species of the genus *Pseudomonas*, particularly *P. aeruginosa* and *P. putida*, exhibit enormous metabolic versatility, are oxidative, share many properties with other pseudomonads and, importantly, have been well characterized genetically and

biochemically (*Clarke* and *Richmond* 1975). These bacteria are currently being utilized therefore in the development of host:vector systems for the cloning of genes of soil bacteria.

The two essential components of a gene cloning system are a well-characterized host strain into which DNA molecules can be efficiently introduced (transformed) and a cloning vector or vehicle, a plasmid or bacteriophage genome, into which DNA fragments may be inserted and which can stably propagate itself and inserted DNA fragments in host bacteria.

2 Host Strains

Bacterial strains that may be appropriate hosts for gene cloning experiments generally (a) can be made highly 'competent' for transformation with purified plasmid or bacteriophage DNA (enables efficient uptake of recombinant DNA molecules), (b) are defective in 'restriction', the ability to degrade by means of a polynucleotide sequence-specific endonuclease heterologous or foreign DNA that is not 'modified' by a corresponding sequence-specific polynucleotide methylase (prevents degradation of recombinant DNA molecules), (c) are defective in general recombination functions (*rec⁻*) (reduces loss of the cloned DNA from the extrachromosomal state, by recombination with the chromosome, and hence preserves its ability to be recovered from transformed cells) and (d) lack endogenous plasmids (which may confuse the analysis of recombinant DNA molecules isolated from host cells).

The restriction systems of *P. aeruginosa* and *P. putida* are highly effective barriers to the uptake of heterologous DNA (*Holloway* 1965; *Bagdasarian* et al. 1979). *Dunn* and *Holloway* (1971) have isolated after nitrosoguanidine mutagenesis a number of mutants of *P. aeruginosa* strain PAO that are defective in restriction or restriction and modification (*rmo*). One of these, *P. aeruginosa* PAO1162 *leu*-38, *rmo*-11 (*Dunn* and *Holloway* 1971), and KT2440 (*Franklin* et al. 1981a), a spontaneous *rmo⁻* derivative of *P. putida* mt-2 (Paw 85), are two strains that readily accept heterologous DNA, such as plasmid DNA isolated from *E. coli* (*Bagdasarian* et al. 1979; *Bagdasarian* et al. 1981).

A number of investigators have reported the transformation of cells of *P. putida* (*Chakrabarty* et al. 1975; *Nagahari* and *Sakaguchi* 1978) and *P. aeruginosa* (*Nagahari* and *Sakaguchi* 1978; *Mercer* and *Loutit*, 1979) with plasmid and bacteriophage DNA, using modifications of procedures described for the transformation and transfection of *E. coli* (*Mandel* and *Higa* 1970; *Cohen* et al. 1972). We have optimized for the transformation of strains PAO1162 and KT2440 with plasmid RSF1010 DNA (see below) the procedure of *Kushner* (1978), and routinely obtain $> 10^5$ transformants/µg DNA.[1] Although this value

---

[1] Our transformation procedure is as follows: a) Bacteria are grown in L-broth (*Miller* 1972) to exponential phase (about 2×10^8 cells/ml) and harvested by centrifugation at 0 °C; b) Cells are washed in an equal volume of cold buffer I (10 mM MOPS, pH 7.0–10 mM RbCl – 100 mM MgCl₂) by resuspension and centrifugation and are resuspended and held for 30 min at 0 °C in an equal volume of cold buffer II (100 mM MOPS, pH 6.5–10 mM RbCl – 100 mM CaCl₂); c) Cells are centrifuged and resuspended in $\frac{1}{10}$ volume of cold buffer II and 0.2 ml portions of the cell suspension incubated with purified plasmid DNA (0.2–1.0 µg) at 0 °C for 45 min, then at 42 °C for 1 min; d) Cell suspensions are diluted by addition of 3 ml of L-broth and incubated at 30 °C for 90 min with shaking, prior to dilution and plating on selective media

is an order of magnitude lower than that obtained with *E. coli* K-12, and needs to be substantially improved in the future, it nevertheless is high enough for many types of gene cloning experiments (e.g. the generation of miniplasmid derivatives of Rms149 in vitro; *Bagdasarian* et al. 1979).

Regarding the two other desirable features of host strains indicated above, progress is less far advanced. Recombination-deficient (*rec*A) strains of *P. aeruginosa* PAO certainly exist (*Chandler* and *Krishnapillai* 1974) but, so far, the *rec*A and *rmo* alleles have not been combined in a single host strain. This is not generally a major problem because recombination deficiency is required principally for 'self-cloning', i.e. the cloning of genes from one strain into bacteria of the same strain, where polynucleotide sequence homology between the chromosome and the cloned DNA segment is high, and restriction barriers are not generally a problem in such experiments.

Lastly, although the *P. putida* strain KT2440 does not seem to carry any endogenous plasmid, our derivatives of *P. aeruginosa* PAO1162 may carry the FP2 sex factor (*Dunn* and *Holloway* 1971). This plasmid is so large that it is not isolated by procedures routinely used to purify recombinant plasmids, and hence does not complicate their analysis.

3 Plasmid Cloning Vectors

General purpose cloning vectors are genetically and structurally well-characterized extrachromosomal elements, such as plasmids and bacteriophage genomes, that are readily purified in large quantities. Some examples of different types of extrachromosomal elements found in *Pseudomonas* strains are listed in Table 2. Although plasmids (predominantly relatives of *Col*E1; *Bolivar* and *Backman* 1979; *Kahn* et al. 1979; *Bernard* and *Helinski* 1980; *Timmis* 1981), single-stranded DNA phage genomes (mainly M13; *Barnes* 1980) and double-stranded DNA phage genomes (predominantly λ; *Brammar* 1979; *Williams* and *Blattner* 1980) have all been developed as useful vectors in *E. coli*, so far only plasmid vectors exist for gene cloning in *Pseudomonas*.

Versatile plasmid cloning vectors should (a) be small, (b) code for at least one property, and preferably two or three, that can be used to select bacteria which have taken up vector molecules during transformation, (c) contain unique cleavage sites for many of the commonly used restriction endonucleases, (d) possess one or more strong promoters that can actively transcribe cloned genes and (e) have properties that permit the detection or selection of hybrid molecules. Ordinarily, vectors contain two or three antibiotic resistance genes, each of which contains a unique cleavage site for one or more restriction endonucleases. Insertion of DNA fragments into such sites results in inactivation of the corresponding antibiotic resistance gene, thereby providing a useful phenotypic change for the detection of bacterial clones that carry hybrid molecules (*insertional inactivation; Timmis* et al. 1974; *Bolivar* et al. 1977; *Timmis* et al. 1978a; *Timmis* et al. 1978b), and may cause the cloned gene to be expressed from an active constitutive antibiotic resistance gene promoter.

Plasmid vectors most commonly used for gene cloning in *E. coli,* like pBR322 (*Bolivar* et al. 1977) and pACYC184 (*Chang* and *Cohen* 1978), are based on *Col*E1-type replicons. Such vectors, and others based on plasmid pSC101, *Col*D, F and R6-5, are not, however, useful for cloning in *Pseudomonas* because they cannot be introduced into and

Table 2. Some extrachromosomal elements of *Pseudomonas*[a]

Element	Host	Genome size (kb)	Incompatibility group	Properties[b]
I. *Temperate bacteriophages*				
B3	*P. aeruginosa*	30– 37		Transducing: linear DNA genome without cohesive ends; UV non-inducible; shows HCM
F116	*P. aeruginosa*	57		Transducing; linear DNA genome without cohesive ends; UV-inducible; no HCM
G101	*P. aeruginosa*	57– 62		Transducing; linear DNA genome without cohesive ends; UV non-inducible; shows HCM
D3	*P. aeruginosa*			Mediates conversion of bacterial surface antigens
II. *Filamentous bacteriophages*				
Pf1	*P. aeruginosa*	7.4		Single-stranded DNA; not sex specific
III. *Plasmids*				
RP1/RP4/RK2/R68	Broad host range	60	P-1	Cb Km Nm Tc Tra$^+$; copy number 2–3
CAM	*P. putida*	139–226	P-2	Ter camphor degradation Tra$^+$
OCT	*P. putida*	41– 43	P-2	Alkane degradation
RIP64	*P. aeruginosa*		P-3	Pb Cm Gm Su Tm Hg Tra$^+$
R1162 (R300B RSF1010)	Broad host range	9	P-4	Sm Su Tra$^-$ Mob$^+$; copy number 15–20
Rms163	*P. aeruginosa*		P-5	Cm Su Tc Bor Tra$^+$
Rms149	*P. aeruginosa*	54	P-6	Cb Gm Sm Su; replicates also in *E. coli*
Rms148	*P. aeruginosa*	143	P-7	Sm Tra$^+$
FP2	*P. aeruginosa*	89	P-8	Hg Pmr; mobilizes the chromosome of *P. aeruginosa*; copy number 1–2
SAL	*P. putida*	45– 65	P-9	Salicylate degradation; Tra$^-$
TOL	*P. putida*	117	P-9	Toluene/xylene degradation Tra$^+$; copy number 1–2; replicates also in *E. coli*
pSR1	*P. aeruginosa*	65	P-10	Gm Km Su Tm Hg Pmr Tra$^+$
pVS1	*P. aeruginosa*	30	unclassified	Hg Su

[a] Compiled and representative examples taken from *Holloway* and *Krishnapillai* (1975); *Jacoby* (1979) and *Shapiro* (1977)

[b] Abbreviations: Ap, ampicillin-; Cb, carbenicillin-; Cm, chloramphenicol-; Gm, gentamycin-; Km, kanamycin-; Nm, neomycin-; Sm, streptomycin-; Su, sulphonamide-; Tc, tetracycline-; Tm, tobramycin-; Bor, borate-; Hg, mercuric chloride-; Pmr, phenylmercuric acetate-; Ter, tellurite-resistance; UV, ultraviolet irradiation; HCM, host controlled modification; Tra, ability for conjugational self-transfer; Mob, ability to be mobilized by conjugative plasmids

Table 3. Ability of different plasmids to be introduced into *P. aeruginosa* and *P. putida*

Plasmid[a]	Ref.	Replicon	No. transformants/µg DNA		
			E. coli SK1592 *hsd*R4 *hsd*M$^+$	*P. aeruginosa* PA01162 *rmo*-11	*P. putida* KT2440 *rmo*
pBR322$_c$	(1)	pMB1	4.0×10^6	< 1	< 1
pACYC184$_c$	(2)	P15A	2.0×10^6	< 1	< 1
pML21$_c$	(3)	ColE1	1.0×10^6	< 1	< 1
pKT101$_c$	(4)	ColD	2.0×10^6	< 1	< 1
pSC101$_c$	(5, 6)	pSC101	4.0×10^4	< 1	< 1
pKT001$_c$	(7)	R6-5	1.0×10^5	< 1	< 1
pBK80$_c$	(8)	F	2.0×10^4	< 1	< 1
Rms149$_a$	(9)	Rms149	1.7×10^4	6.0×10^2	2.0×10^2
RSF1010$_p$	(10)	RSF1010	4.0×10^6	2.0×10^5	1.0×10^5
RSF1010$_c$	(10)	RSF1010	4.0×10^6	4.7×10^4	4.0×10^4
pKT230$_p$	(11)	RSF1010	5.7×10^5	1.0×10^5	3.0×10^4

[a] Subscripts refer to the origin of plasmid DNA: c, a and p correspond to *E. coli*, *P. aeruginosa* and *P. putida*, respectively. References: (1) *Bolivar* et al. (1977); (2) *Chang* and *Cohen* (1978); (3) *Lovett* and *Helinski* (1976); (4) *Timmis* et al. (1981); (5) *Cohen* and *Chang* (1973); (6) *Cohen* and *Chang* (1977); (7) *Timmis* et al. (1978a); (8) *Manis* and *Kline* (1978); (9) *Shapiro* (1977); (10) *Guerry* et al. (1974); (11) *Franklin* et al. (1981a).

stably propagated in either *P. aeruginosa* or *P. putida* (Table 3). These plasmids therefore exhibit narrow host range specificities.

Plasmids that are potentially useful for development as vectors for gene cloning in *Pseudomonas* may be either plasmids indigenous to this host (Table 2) or broad host range plasmids that exhibit wide host specificity, such as the IncW plasmid S-a (encodes resistance to chloramphenicol, gentamycin, kanamycin, streptomycin and sulphonamide) that was originally isolated in *Shigella*, and the RSF1010/R300B plasmids (encode resistance to streptomycin and sulphonamide) that were isolated in *Salmonella* and that are similar to or identical with the R1162 plasmid of *P. aeruginosa*. Inspection of Table 2 reveals that, with the exception of R1162, most *Pseudomonas* plasmids thus far characterized to any extent are exceedingly large and have low copy numbers and, as such, are not obvious candidates for cloning vectors. Nevertheless, attempts have been made by various groups to reduce in size all antibiotic resistance plasmids in Table 2 that are 60 kb in length or smaller, and S-a, and to develop the resulting miniplasmids as vectors. Because vectors derived from the R1162, RSF1010 and R300B plasmids are currently the most versatile, they will be described in some detail.

3.1 Broad Host Range Vectors Derived from IncQ Plasmids

The broad host range IncQ/P-4 group plasmids RSF1010, R300B and R1162 are 8–9 kb in length, specify resistance to streptomycin and sulphonamide (*Barth* and *Grinter* 1974; Fig. 1) and can be transferred among different bacterial strains by mobilization (i.e. are

Mob^+) if conjugal transfer functions are provided by a coexisting transfer-proficient (Tra^+) plasmid, such as RP1. A detailed physical and functional map of the RSF1010 plasmid, which includes the locations of restriction endonuclease cleavage sites (*Bagdasarian* et al. 1979; *Franklin* et al. 1981a), RNA polymerase binding sites (*Bagdasarian* et al. 1981), genes for streptomycin (Sm) and sulphonamide (Su) resistance (*Rubens* et al. 1976) and plasmid mobilization (*Mob; Bagdasarian* et al. 1981) and the origins of vegetative and transfer replication (*Ori* and *Nic*, respectively; *De Graaff* et al. 1978; *Nordheim* et al. 1980) is presented elsewhere (*Bagdasarian* et al. 1981). Its essential features are summarized in Fig. 3. Unfortunately, like most other broad host range plasmids, RSF1010 contains few restriction endonuclease cleavage sites and very few unique sites, namely *Eco*RI, *Sst*I, *Bst*EII and *Pst*I (two adjacent sites), for the endonucleases commonly used in gene cloning experiments. The RSF1010 molecule contains no sites for the *Hind*III, *Bam*HI, *Bgl*II, *Sal*I, *Cla*I, *Kpn*I, *Xba*I, *Xho*I and *Xma*I endonucleases. Moreover, sulphonamide resistance is not a convenient selection phenotype for plasmid-carrying bacteria, and DNA fragments inserted into the *Eco*RI or *Sst*I sites, or between the two *Pst*I sites, inactivate expression of streptomycin resistance, due to the fact that the Sm resistance gene is expressed from a promoter located upstream of the Su resistance gene which appears to be responsible for transcription of both genes (*Rubens* et al. 1976; *Bagdasarian* 1981).

To provide RSF1010 with additional restriction endonuclease cleavage sites and effective markers that can be used for the selection of plasmid-carrying transformant bacteria, a number of different DNA fragments that carry antibiotic resistance genes[2] have been incorporated into the plasmid (Table 4); *Bagdasarian* et al. 1979; *Bagdasarian* et al. 1981). The pKT210 derivative was constructed by replacement of the small *Pst*I-generated fragment of RSF1010 with a 3.6 kb *Pst*I-generated DNA fragment that carries the chloramphenicol (Cm) resistance gene of plasmid S-a and contains a single *Hind*III cleavage site (Fig. 1). The pKT210 vector may be used for cloning *Eco*RI-, *Sst*I- and *Hind*III-generated DNA fragments; transformant bacteria that contain hybrid plasmids with DNA insertions in the *Eco*RI or *Sst*I sites are selected by their Cm resistance and identified by their loss of Sm resistance, whereas those containing plasmids with insertions in the *Hind*III site are selected by their Cm or Sm resistance. The pKT248 vector was constructed by insertion of a 4.8 kb *Pst*I fragment carrying the Cm resistance gene of plasmid R621a1a into one of the *Pst*I cleavage sites of RSF1010. This vector contains a single *Sal*I cleavage site within the Cm gene; transformant bacteria that contain hybrid plasmids with DNA insertions in the *Sal*I site are selected by their Sm resistance and identified by their loss of Cm resistance, whereas those containing insertions in the

---

[2] It should be mentioned that some antibiotic resistance genes are more suitable than others for vector construction. Although the tetracycline (Tc) resistance genes of the pSC101 plasmid *(Cohen* and *Chang* 1973; *Cohen* and *Chang* 1977) and the Tn*10* transposon *(Jorgensen* et al. 1979) are attractive because they contain a number of important restriction endonuclease cleavage sites (*Bolivar* et al. 1977; *Jorgensen* et al. 1979), they can be unsuitable for the construction of high copy number vector plasmids, particularly those for *Pseudomonas*, because they may not be stably inherited in the absence of selection (*M. Bagdasarian*, unpublished experiments; see also *Wood* et al. 1981). Whether this instability results from gene loss from the plasmid or from plasmid loss from the bacterial culture is currently not known, but is presumed to be caused by Tc resistance protein-mediated cell membrane changes that are deleterious to the bacterial host. The use of Tc resistance genes in vectors for insertional inactivation can be particularly unsatisfactory (*M. Bagdasarian*, unpublished experiments)

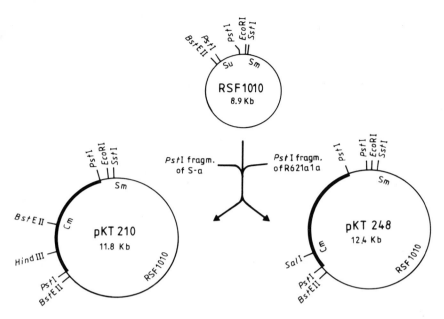

Fig. 1. Genealogy of cloning vectors derived from plasmid RSF1010 by insertion of chloramphenicol resistance determinants

*Eco*RI or *Sst*I sites are selected by their Cm resistance and identified by their loss of Sm resistance. The Cm gene of pKT248 is, as far as we are aware, the only insertional inactivation gene for *Sal*I-generated DNA fragments, other than the pSC101 Tc resistance gene.

Vectors that can be used to clone *Hind*III-, *Xho*I- and *Xma*I-generated DNA fragments, with insertional inactivation detection of recombinant plasmids, have been constructed by incorporation of the Tn601/Tn903 kanamycin (Km) resistance gene into the RSF1010 plasmid (Fig. 2). The pKT231 derivative was constructed by insertion of a 4.6 kb *Pst*I-generated fragment that contains the Km resistance gene from the R6-5 miniplasmid pKT105 (*K.N. Timmis*, unpublished). This vector may be used for cloning *Hind*III-, *Xma*I-, *Xho*I-, *Cla*I-, *Eco*RI- and *Sst*I-generated DNA fragments with insertional inactivation detection of hybrid plasmids containing inserts in all cloning sites except that of *Cla*I (Table 4). A physical and functional map of the pKT231 vector plasmid is shown in Fig. 3. A vector similar to pKT231 was obtained by cloning the pACYC177 plasmid (*Chang* and *Cohen* 1978), linearized by cleavage with the *Pst*I endonuclease, between the *Pst*I sites of RSF1010. This vector, which is designated pKT230, contains in addition to the cloning sites present in pKT231, a single *Bam*HI site (Fig. 2; Table 4).

Double replicons that are analogous to pKT230 have been constructed by linkage of a) R1162 and pBR322 (*Bolivar* et al. 1977) through their *Eco*RI cleavage sites and b) RSF1010 and pBR322 through their *Pst*I cleavage sites. The former vector, pFG7 (*Gautier* and *Bonewald* 1980), contains ampicillin (Ap) and tetracycline (Tc) resistance genes and *Cla*I, *Bam*HI, *Hind*III and *Sal*I cloning sites within the Tc resistance gene; whereas the latter, pMW79 (*Wood* et al. 1981), contains Tc and Sm resistance genes and *Bam*HI, *Sal*I

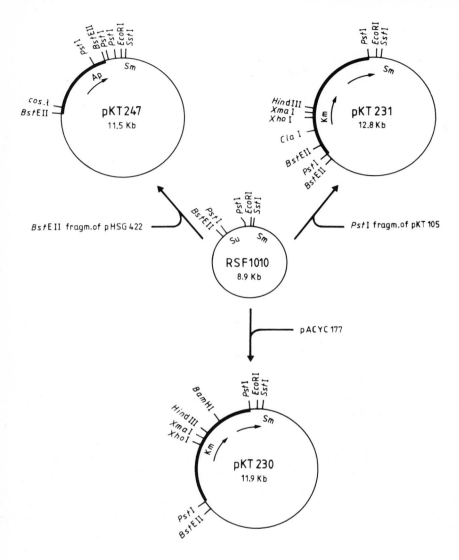

Fig. 2. Genealogy of cloning vectors derived from plasmid RSF1010 by insertion of kanamycin re-sistance determinants and the bacteriophage λ *cos* site. The *arrows inside the circles* indicate the origins and direction of transcription from relevant promoters

and *Hind*III cloning sites within the Tc resistance gene. The vector pMW79 was reported to be unstable in *Pseudomonas* (*Wood* et al. 1981), which is consistent with previous findings indicating that Tc resistance genes may be inappropriate for the construction of high copy number vector plasmids for *Pseudomonas*. Vector pTB70 is a somewhat larger derivative of the plasmid R300B and contains the Tn*5* transposon (*Windass* et al. 1980). It contains cloning sites for *Eco*RI-, *Bam*HI- and *Sal*I-generated DNA fragments.

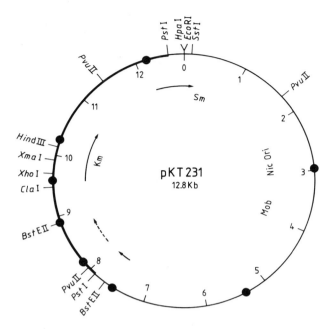

Fig. 3. Restriction endonuclease site cleavage and functional map of the vector pKT231. The *filled circles* represent RNA polymerase binding sites and the *arrows* the directions of transcription (*Bagdasarian* et al. 1981). The *thin portion of the circle* represents the large *Pst*I fragment of RSF1010 whereas the *thick line* indicates the cloned *Pst*I-Km resistance gene fragment of pKT105. Abbreviations: *Km*, resistance to kanamycin; *Sm*, resistance to streptomycin; *Mob*, ability to be mobilized by conjugative plasmids; *Nic*, the relaxation nick site (probably identical to the origin of transfer); *Ori*, origin of vegetative replication

The RSF1010/R1162/R300B plasmids and their derivatives exhibit an extremely broad host range specificity. They have been introduced into and shown to be stably maintained in a variety of gram-negative bacteria including *E. coli, Pseudomonas aeruginosa* and *P. putida (Nagahari* and *Sakaguchi* 1978), *Agrobacterium tumefaciens (M. van Montagu,* personal communication), *Rhizobium meliloti (M. David,* personal communication), *Acetobacter xylinum (S. Valla,* personal communication), *Azotobacter vinelandi (David* et al. 1981), *Alcaligenes eutrophus (F.C.H. Franklin,* unpublished data), *Rhodopseudomonas spheroides (W.T. Tucker,* personal communication) and *Methylophilus methylotrophus (Windass* et al. 1980). This property of the vectors enables them to be used for the cloning of DNA from and into almost any gram-negative bacterial strain that can be transformed with plasmid DNA. Moreover, all of the vectors thus far described are mobilized with very high efficiencies by conjugative plasmids[3], such as RP1, to many types of gram-negative bacteria. Gene cloning experiments with these vectors may there-

---

[3] The RSF1010-like plasmids are mobilized by a wide range of conjugative plasmids (*P. Barth,* personal communication). The use of a narrow host range conjugative plasmid, such as R64drd11 for mobilization from *E.coli,* circumvents the problem of the conjugative plasmid being cotransferred and maintained with the RSF1010 derivative in the recipient strain

fore be carried out either entirely within a particular strain of soil bacteria (cloning of homologous DNA) or in a more highly developed host system (cloning of heterologous DNA) such as *E. coli* or *P. putida,* followed by transfer of some or all the hybrid plasmids into the bacterial species of choice. Because of the high degree of efficiency and sophistication of the *E. coli* cloning system, the '2-stage' cloning procedure is frequently preferred, particularly where the direct selection of required genes is not possible.

3.2 Cosmids

In contrast to the joining of specific, purified or partially purified DNA segments to a vector molecule, the linkage of a large number of DNA fragments of different lengths (for instance, during the cloning of restriction endonuclease-generated fragments of an entire genome to obtain a gene bank) results in a preferential cloning of the shortest (0.5–3.0 kb). The number of transformant clones carrying hybrid plasmids required to ensure the presence of all sequences of an entire genome is consequently very large. Methods which bias the cloning in favour of long DNA fragments are therefore particularly suitable for the construction of gene banks. The in vitro packaging of gene cloning mixtures in phage particles that are packed by a 'head-full mechanism' provides an important biological selection for long recombinant molecules, 40–50 kb in size in the case of λ particles (*Hohn* and *Murray* 1977; *Sternberg* et al. 1977). Moreover, the method of introduction of packaged recombinant molecules into host bacteria, namely by infection, is much more efficient than transformation with naked DNA.

At least in the case of λ DNA packaging, the only DNA requirement is the presence of a packaging recognition sequence, *cos*, at appropriate intervals on long DNA molecules that are to be packaged. The λ*cos* site has been introduced into various *E.coli* vectors, including the pBR322 plasmid to produce the pHC79 'cosmid' vector, and these plasmids, when linked to long DNA fragments, can be efficiently packaged in vitro into λ heads (*Collins* and *Hohn* 1978; *Hohn* and *Collins* 1980; *Hohn* and *Hinnen* 1980), DNA fragments up to 40 kb in length can be readily cloned in this manner and the pHC79 plasmid has proven invaluable for the construction of gene banks.

Thus far, no method has been described for the in vitro packaging of DNA in *Pseudomonas* phage particles, and hence it is not currently possible to use cosmid packaging in the construction of gene banks directly in *Pseudomonas.* On the other hand, a number of RSF1010/R1162 broad host range vector derivatives that carry the λ*cos* site have been constructed and these may be used to generate gene banks in *E. coli* using the λ in vitro packaging system. The hybrid plasmids thereby generated may subsequently be transferred to *Pseudomonas* or another soil bacterium by mobilization (e.g. by RP1) or transformation. The pKT247 cosmid vector (*Bagdasarian* et al. 1981) was constructed from the RSF1010 plasmid by insertion of a 2.2 kb *Bst*EII fragment that contains the Tn*3* Ap resistance gene and the λ*cos* site from the pHSG422 cosmid vector (*Hashimoto-Gotoh* et al. 1981), and may be used for the cloning of *Eco*RI- and *Sst*I-generated DNA fragments. The pFG6 vector (*Gautier* and *Bonewald* 1980) is an analogous cosmid and was generated by linkage of the pHC79 cosmid to R1162. This vector can be used for cloning of *Cla*I-, *Bam*HI-, *Hind*III- and *Sal*I-generated DNA fragments. Data on the stability of the pFG6 Tc resistance plasmid in *Pseudomonas* have not thus far been reported.

Table 4. Properties of vector plasmids for cloning in *Pseudomonas*[a]

Vector	Replicon(s)	Size (kb)	Copy[b] No.	Cloning sites	Primary[c] selection	Inser-tional[c] inac-tivation	Remarks[d]	Reference
pKT210	RSF1010	11.8	15–20	*Eco*RI	Cm	Sm	Mob[+]	1
				*Sst*I	Cm	Sm		
				*Hin*dIII	Cm, Sm	–		
pKT248	RSF1010	12.4	15–20	*Eco*RI	Cm	Sm	Mob[+]	1, 2
				*Sst*I	Cm	Sm		
				*Sal*I	Sm	Cm		
pKT230	RSF1010 pACYC177	11.9	15–20	*Eco*RI	Km	Sm	Mob[+]	3
				*Sst*I	Km	Sm		
				*Hin*dIII	Sm	Km		
				*Xho*I	Sm	Km		
				*Xma*I	Sm	Km		
				*Bst*EII	Km, Sm	–		
				*Bam*HI	Km, Sm	–		
pKT256	Hydroxylamine-induced point mutant of pKT230[e]						Mob[–]	3
pKT231	RSF1010	12.8	15–20	*Eco*RI	Km	Sm	Mob[+]	3
				*Sst*I	Km	Sm		
				*Hin*dIII	Sm	Km		
				*Xho*I	Sm	Km		
				*Xma*I	Sm	Km		
				*Cla*I	Km, Sm	–		
pKT254	Hydroxylamine-induced point mutant of pKT231[e]						Mob[–]	3
pKT253	Hydroxylamine-induced point mutant of pKT231						Mob[+], Rep$_{ts}$	3
pKT247	RSF1010	11.5	15–20	*Eco*RI	Ap	Sm	Mob[+], cosmid	3
				*Sst*I	Ap	Sm		
pKT258	Hydroxylamine-induced point mutant of pKT247[e]						Mob[–], cosmid	3
pTB70	R300B	17.6	15–20	*Eco*RI	Km	Sm	Mob[+]	4
				*Bam*HI	Km, Sm	–		
				*Sal*I	Km, Sm	–		
pFG6	R1162 pHC79	15.3	15–20	*Cla*I	Ap	Tc	Mob[+], cosmid	5
				*Bam*HI	Ap	Tc		
				*Hin*dIII	Ap	Tc		
				*Sal*I	Ap	Tc		
pFG7	R1162 pBR322	13.3	15–20	*Cla*I	Ap	Tc	Mob[+]	5
				*Bam*HI	Ap	Tc		
				*Hin*dIII	Ap	Tc		
				*Sal*I	Ap	Tc		
pRP301	RP4	54.7	2– 4	*Bam*HI	Ap, Tc	–	Tra[+]	4
				*Eco*RI	Ap, Tc	–		
				*Hin*dIII	Ap, Tc	–		
				*Sal*I	Ap	Tc		
pRK290	RK2	20.0	2– 4	*Eco*RI	Tc	–	Mob[+]	6
				*Bgl*II	Tc	–		
pME448	pVS1	13.8		*Cla*I	Hg	–	Mob[–]	7
				*Hin*dIII	Hg	–		

Groups of hybrid plasmids that comprise a gene bank should ideally contain DNA fragments having random, or near-random termini. DNA fragments with random termini may be generated by DNA shearing, whereas fragments with near-random termini may be generated by partial cleavage with a 'frequent cutter' restriction endonuclease. The *Eco*RI* activity of the *Eco*RI endonuclease (*Polisky* et al. 1975) can be used to generate DNA fragments that are suitable for gene bank construction and that can be inserted in the *Eco*RI cleavage site of pKT247. Similarly, the *Sau*3A endonuclease can be used to generated DNA fragments suitable for insertion into the *Bam*HI site of pFG6.

3.3 Low Copy Number Vectors Derived from Large Conjugative Plasmids

The second group of broad host range plasmids that have been studied intensively are the IncP-1 conjugative plasmids RP1, RP4, RK2 and R68, which appear to be essentially identical to one another (*Burkardt* et al. 1979). A conjugative Ap and Tc resistance deletion derivative of an RP4:Tn7 hybrid plasmid that is 55 kb in length and contains cloning sites for *Bam*HI-, *Eco*RI-, *Hin*dIII- and *Sal*I-generated DNA fragments (the latter is located in the Tc resistance gene) has been proposed as a wide host range cloning vector (*Barth* 1979), but it was recently reported to be unstable in *Methylophilus methylotrophus* (*Windass* et al. 1980). A smaller Tra^- Mob^+ Tc resistance derivative of the RK2 plasmid, pRK290, is 20 kb in length and contains cloning sites for *Eco*RI- and *Bgl*II-generated DNA fragments (*Ditta* et al. 1980). This vector can be mobilized to a range of gram-negative bacteria by plasmids that specify the RK2-type plasmid transfer system. A miniderivative of the S-a broad host range plasmid, pGV1106, that contains unique cleavage sites for the *Bgl*II, *Bam*HI, *Eco*RI and *Kpn*I endonucleases, has also been generated (*De Wilde* et al. 1978; *M. van Montague*, personal communication). The small size of pGV1106, 8.8 kb, is an attractive feature of this vector. It should be emphasized that no quantitative data have until now been presented on the stability of the pGV1106 and pRK290 vectors in *Pseudomonas* and other gram-negative bacteria, and hence the extent of their usefulness for gene cloning in hosts other than *E. coli* remains to be determined.

Broad host range vector plasmids are generally attractive as universal vectors for gram-negative bacteria, and are particularly useful for experiments in which gene cloning is first carried out using a well-characterized host, such as *E. coli*, with the subsequent

---

[a] All vectors, with the exception of pME448, which can be used for gene cloning in *Pseudomonas* and *Aeromonas,* are broad host range plasmids that can be employed for gene cloning in a wide range of gram-negative bacteria

[b] Number of copies/genome equivalent

[c] Abbreviations as defined in Table 2

[d] Abbreviations: Mob+/Mob−, ability or inability, respectively, to be mobilized by a coexisting conjugative plasmid to other bacteria; Tra+, self-transmissible; Rep$_{ts}$, temperature-sensitive replication; cosmid, contains the bacteriophage λ *cos* site and may be used in conjunction with the in vitro λ DNA packaging procedure for the cloning of large DNA fragments in *E. coli* host cells

[e] All Mob− mutant plasmids were derived from a single hydroxylamine-induced point mutant of RSF1010

References: (1) *Bagdasarian* et al. (1979); (2) *Franklin* et al. (1981b); (3) *Bagdasarian* et al. (1981); (4) *Windass* et al. (1980); (5) *Gautier* and *Bonewald* (1980); (6) *Ditta* et al. (1980); (7) *J.M. Watson,* personal communication

transfer of specifically identified hybrids to the organism of choice. Such vectors have, however, no particular advantage for the cloning of DNA in a single, well-defined host, such as *P. aeruginosa*, and may be undesirable for the cloning of genes that specify potentially hazardous products. The only narrow host range vector plasmids thus far described for gene cloning in *Pseudomonas* are derivatives of plasmid pVSl (*Stanisich* et al. 1977) that are stably maintained in *P. aeruginosa, P. putida* and *Aeromonas formicans* but not *E. coli*. The smallest of these, pME447 and pME448, encode resistance to mercuric ions, are non-mobilizable and contain single cleavage sites for the *Cla*I, *Eco*RI, *Hind*III and *Sst*I endonucleases *(J.M. Watson*, personal communication).

4 Host:Vector Systems for Gene Cloning in Pseudomonas

Many species of gram-negative bacteria have been shown to exchange genetic information with one another by known physiological processes, in particular by R'-promoted conjugal DNA transfer. According to the National Institutes of Health (NIH) Recombinant DNA Research Guidelines, the cloning of DNA segments isolated from one bacterial species into another of this group is exempt from the Guidelines (*Federal Register* 1980). The group currently consists of

1. Genus *Escherichia*
2. Genus *Shigella*
3. Genus *Salmonella* (including *Arizona)*
4. Genus *Enterobacter*
5. Genus *Citrobacter* (including *Levinea)*
6. Genus *Klebsiella*
7. Genus *Erwinia*
8. *Pseudomonas aeruginosa, P. putida* and *P. fluorescens*
9. *Serratia marcescens*

Self-cloning experiments, in which the donor and recipient strains are one and the same, are also exempt from the Guidelines (Guidelines for Research Involving Recombinant DNA Molecules, 1980). The cloning in *Pseudomonas* bacteria of DNA from strains that have not yet been demonstrated to exchange genetic information with this host are, however, subject to the Guidelines and must be carried out in a certified host: vector (HV) system.

Host:vector systems are classified according to the level of biological containment they provide; HV1 systems have the lowest permissible level, namely that typical of EK1 systems, which consist of bacterium *E. coli* K-12 and its non-conjugative plasmid and λ vectors (Guidelines for Research Involving Recombinant DNA Molecules, 1980).

The *P. aeruginosa* PAO1162 host strain contains the conjugative plasmid FP2 (*Dunn and Holloway* (1971). This plasmid does not, however, mobilize RSF1010 or its derivatives and hence does not reduce the suitability of PAO1162 as a host for an HV1 system. No conjugative plasmid has thus far been detected in the *P. putida* mt-2 strain KT2440. Moreover, these strains are strict aerobes and the probability of their escape from the laboratory through laboratory personnel via the oral-faecal route is therefore low.

The conjugative plasmid pRP301 is able to transmit itself at high frequencies among bacteria (*Bath* 1979; *Windass* et al. 1980) and hence is unsuitable as the vector com-

ponent of an HV1 system. The remaining vectors listed in Table 4 are non-conjugative and hence are candidates as vector components of HV1 systems. The ability of the broad host range RSF1010-/R300B-/R1162-based vectors to be mobilized by RP1-type conjugative plasmids is, however, somewhat undesirable, particularly for cloning experiments involving DNA from pathogenic bacteria. This feature has been eliminated in the P. aeruginosa PAO1162 and RSF1010 vectors pKT254, pKT256 and pKT258, by introduction of a Mob^- (mobilization-defective) mutation that reduces the frequencies of RP4-promoted transfer of these plasmids by four to five orders of magnitude (*Bagdasarian* et al. 1981) and that substantially increases their biological containment. The RSF1010-derived vectors and the *P. putida* mt-2 KT2440 host strains have been submitted to the Office of Recombinant DNA Activities, NIH, for approval as *Pseudomonas* HV1 cloning systems. The pME448 vector is mobilization deficient and exhibits narrow host range specificity, two features that should significantly decrease its ability to spread to other bacteria and thereby increase its containment.

5 Gene Cloning with Broad Host Range Vectors

A number of the broad host range vectors described in Table 4 have been used to clone genes of *Pseudomonas* and other soil bacteria. One experiment, which took advantage of a direct selection procedure for desired hybrid plasmids, involved the cloning of DNA from the methanol-utilizing *Pseudomonas* strain AM1 in *E. coli* K-12 (RP4), using the plasmid R1162, and the mobilization of hybrids thereby isolated into a methanol dehydrogenase-negative mutant of AM1, with selection for methanol utilization (*Gautier* and *Bonewald* 1980). Two clones that could grow on methanol were obtained in this experiment.

An elegant example of effective 'enzyme recruitment' is the introduction of the glutamate dehydrogenase gene of *E. coli* into *Methylophilus methylotrophus* using the R300B-derived vector pTB70 (*Windass* et al. 1980). This gene, which has previously been cloned on an *E. coli* vector plasmid, enabled a glutamate synthase-defective mutant of *M. methylotrophus* to assimilate ammonia more efficiently via glutamate dehydrogenase than could the wild-type strain via glutamine synthetase and glutamate synthase, and thereby to convert more growth substrate (methanol) into cellular carbon during single cell protein manufacture.

More general cloning experiments that have been carried out with broad host range vectors have been the construction of gene banks. A binary vector system that consists of a mini-RK2 cloning vector, pRK290, and a plasmid which carries the RK2 transfer system, pRK2013, and which promotes the transfer of the former to various gram-negative bacteria, has been used to construct a gene bank in *E. coli* of the nitrogen-fixing symbiont *Rhizobium meliloti* (*Ditta* et al. 1980). One hybrid plasmid in the gene bank was shown to contain sequences that specifically hybridize with the cloned nitrogenase genes of *Klebsiella pneumoniae* and presumably therefore contains *R. meliloti* genes for nitrogen fixation.

One of the properties of *Pseudomonas* which has received considerable attention during recent years is its ability to catabolize a range of natural and synthetic organic compounds that include aliphatic and aromatic hydrocarbons and their sulphonated and halogenated derivatives (*Williams* 1978; *Knackmuss* 1981). In particular, the TOL plas-

mid-encoded pathway for toluene/xylene degradation has been subjected to intensive investigation (*Williams* 1978) and the organization of the catabolic genes on the plasmid was recently described (*Franklin* et al. 1981a). A number of these genes, including that of the first enzyme of the pathway *xylA* (xylene oxidase) and those of several enzymes of the *meta*-cleavage pathway, including *xylD* (toluate oxidase), *xylE* (catechol 2,3-oxyge-nase), *xylG* (hydroxymuconic semialdehyde dehydrogenase) and *xylF* (hydroxymuco-nic semialdehyde hydrolase), have now been cloned on the RSF1010-derived *Pseudomo-nas* vectors (*Franklin* et al. 1981a, b). Although the expression of these genes on the TOL plasmid is inducible and positively regulated by the products of the *xylR* and *xylS* genes, which interact with substrates of the catabolic pathway (xylene/toluene and their alcohols and toluate/benzoate, respectively), it is constitutive in the hybrid plasmids (*Franklin* et al. 1981a; Table 5).

Measurements of the catechol 2,3-oxygenase levels in bacteria carrying plasmids in which the *xylE* gene has been inserted into either the Sm of Km resistance genes of the pKT230 and pKT231 vectors have revealed the relative strengths of promoters of these two genes. Enzyme levels in acetate- or acetate plus m-toluate-grown bacteria carrying the *xylE* gene under control of the Sm resistance gene promoter are about 900 mUnits/mg protein, compared with 1500 mU/mg protein in acetate-grown TOL plas-mid-carrying cells induced with m-toluate and 3300 mU/mg protein in similar cells that were fully induced by growth on *m*-toluate as the sole carbon source (*Franklin* et al. 1981b). On the other hand, enzyme levels in bacteria carrying the *xylE* gene under control of the Km resistance gene promoter are about 7000 mU/mg protein, which indicates that the promoter of the Km resistance gene is about eightfold more active than that of the Sm resistance gene. This is of considerable importance in the cloning of catabolic genes for the purpose of elevating enzyme levels. It is interesting to note in this regard that the Km resistance gene of Tn*601/903* has been incorporated into a narrow host range. *E. coli* plas-mid, *ColD*, which undergoes extensive amplification of copy number when host bacteria enter stationary phase (*Timmis* et al. 1981). When the *xylE* gene is inserted into the Km resistance gene of this vector, enzyme levels that exceed 12 000 mU/mg protein can be obtained (*Franklin* et al. 1981a). According to published estimates of the specific activity of catechol 2,3-oxygenase, and preliminary experiments involving the radiolabelling of total protein of bacteria carrying the hybrid plasmid, this corresponds to 5%–10% of total cellular protein (*M. Bagdasarian*, unpublished data).

One hybrid plasmid that was recently constructed contains a DNA fragment which

Table 5. Expression of the TOL plasmid catechol 2, 3-oxygenase gene in the RSF1010-derived vec-tors in *P. putida*

Plasmid	Growth substrate	Enzyme activity[a]
pWWO (TOL)	Acetate	90
	Acetate + m-toluate	1500
pKT502 (pKT230/TOL *XhoI-I*)	Acetate	7300
	Acetate + m-toluate	6800
pKT505 (pKT231/TOL *SstI-D*)	Acetate	900
	Acetate + m-toluate	850

[a] mU activity/mg protein; data taken from *Franklin* et al. (1981a, b)

encodes all enzymes of the benzoate/m-toluate *meta* ring cleavage pathway (*Franklin* et al. 1981b). This fragment also contains *xylS*, the gene which encodes one positive regulator of the pathway; *P. putida* and *E. coli* bacteria that carry the hybrid plasmid only produce the catabolic enzymes after exposure to the specific inducer *m*-toluate, but not after exposure to the *xylR* product co-inducers *m*-xylene, toluene or methylbenzylalcohol (*Franklin* et al. 1981b). The *meta* cleavage pathway hybrid plasmid thus provides a unique opportunity to characterize in detail the *xylS* gene product and its mechanism of regulation of the catabolic pathway. Moreover, it will enable characterization of the *xylR* gene and its product, which regulates both the upper part (hydrocarbon → carboxylic acid) and lower part (*meta* cleavage) of the toluene/xylene catabolic pathway (*Williams* 1978). The *meta* cleavage pathway is an important and widespread route for the metabolism of aromatic compounds. Its cloned genes represent a crucial component demanded by gene manipulation strategies whose goals are to generate new hybrid pathways for the degradation of aromatic pollutants that accumulate in the environment, or for the production of key pathway intermediates that can be used in the synthesis of chemicals of industrial importance.

6 Summary and Outlook

Although host:vector systems currently available for gene cloning in *Pseudomonas* and other soil bacteria by no means have the sophistication of those developed in recent years for *E. coli*, they are nevertheless relatively versatile and of considerable utility, as evidenced by the cloning experiments described above. Well-characterized, restriction-negative host strains of *P. aeruginosa* and *P. putida* that can be transformed with plasmid DNA have been identified and several broad and narrow host specificity plasmid vectors that contain a range of cloning sites, many of which are located within antibiotic resistance genes, have been constructed.

The broad host specificity vectors are particularly useful in that they can be transferred to and stably propagated in gram-negative bacteria of many genera, and hence enable the powerful genetic engineering technology to be extended to a large number of poorly characterized but important bacterial species. Moreover, their ability to be used in *E. coli* permits the investigator to take full advantage of the wide array of genetic procedures available for this host prior to transfer of hybrid plasmids into a desired *Pseudomonas* strain. This feature of broad host range vectors is particularly important for the construction of gene banks, which is most readily accomplished with cosmids or phage vectors and an in vitro phage packaging system, combinations of which currently exist only in *E. coli*.

It is to be anticipated that efforts to improve the basic system available at present and to develop new systems will accomplish, inter alia, the following objectives:

1. The construction of cloning vectors from plasmids that undergo amplification of copy number when host cells are treated with protein synthesis inhibitors, to facilitate the isolation of large quantities of vector and hybrid plasmid DNA;

2. The development of vectors that enable the direct selection of hybrid molecules, to simplify the identification of transformant clones carrying hybrid plasmids;

3. The generation of conditional, high-level expression vectors, that enable cloned genes to be actively expressed upon exposure of host bacteria to an appropriate signal; and

4. The construction of phage genome vectors and the development of an in vitro phage genome packaging system, and the generation of appropriate cosmids that can be packaged by such a system, in order to permit the construction of gene banks directly in *Pseudomonas*.

The realization of these goals will be greatly facilitated by better characterization of the genetics and biochemistry of *Pseudomonas* and by increasing our level of understanding of the basic regulatory mechanisms that govern plasmid and bacteriophage biology and behaviour in this host, in particular the molecular basis of narrow and broad host range specificity.

Acknowledgments. We thank *F.C.H. Franklin* for stimulating discussions and for providing the data presented in Table 5, and *H. Markert* and *I. Schallehn* for secretarial assistance. Work carried out in the authors' laboratory was supported by grants from the Deutsche Forschungsgemeinschaft and the Bundesministerium für Forschung und Technologie.

References

Bagdasarian M, Bagdasarian MM, Coleman S, Timmis KN (1979) New vector plasmids for cloning in *Pseudomonas*. In: Timmis KN, Pühler A (eds) Plasmids of medical, environmental and commercial importance. Elsevier-North Holland, Amsterdam, pp 411–422

Bagdasarian M, Franklin FCH, Lurz R, Rückert B, Bagdasarian MM, Timmis KN (1981) Specific purpose cloning vectors. II. Broad host range, high copy number RSF1010-derived vectors, and a host:vector system for gene cloning in *Pseudomonas*. Gene 16 (in press)

Barnes WM (1980) DNA cloning with single-stranded phage vectors. In: Setlow JK, Hollaender A (eds) Genetic engineering, principles and methods, vol 2. Plenum Press, New York, pp 185–199

Barth PT (1979) RP4 and R300B as wide host-range plasmid cloning vehicles. In: Timmis KN, Pühler A (eds) Plasmids of medical, environmental and commercial importance. Elsevier/ North Holland, Amsterdam, pp 399–410

Barth PT, Grinter NJ (1974) Comparison of the deoxyribonucleic acid molecular weights and homologies of plasmids conferring linked resistance to streptomycin and sulphonamides. J Bacteriol 120:618–630

Bayley ST, Morton RA (1978) Recent development in the molecular biology of extremely halophylic bacteria. Crit Rev Microbiol 11:151–205

Bernard HU, Helinski DR (1980) Bacterial plasmid vehicles. In: Setlow JK, Hollaender A (eds) Genetic engineering, principles and methods, vol 2. Plenum Press, New York, pp 133–167

Bolivar F, Backman F (1979) Plasmids of *Escherichia coli* as cloning vectors. Methods in Enzymol 68:245–267

Bolivar R, Rodriguez RL, Greene PJ, Betlach MC, Heyneker HL, Boyer HB, Crosa JH, Falkow S (1977) Construction and characterization of new cloning vehicles. II. A multipurpose cloning system. Gene 2:95–113

Brammar WJ (1979) Safe and useful vector systems. In: Garland PB, Williamson R (eds) Biochemistry of genetic engineering. The Biochemical Society, London, pp 13–27

Burkardt HJ, Riess G, Pühler A (1979) Relationship of group P1 plasmids revealed by heteroduplex experiments: RP1, RP4, R68 and RK2 are identical. J Gen Microbiol 114:341–348

Chakrabarty AM, Mylroie JR, Vacca JG (1975) Transformation of *Pseudomonas putida* and *Escherichia coli* with plasmid-linked drug-resistance factor DNA. Proc Natl Acad Sci USA 72:3647–3651

Chandler PM, Krishnapillai V (1974) Isolation and properties of recombination-deficient mutants of *Pseudomonas aeruginosa*. Mutat Res 23:15–23

Chang ACY, Cohen SN (1978) Construction and characterization of amplifiable multicopy DNA cloning vesicles derived from the p15A cryptic plasmid. J Bacteriol 134:1141–1156

Clarke PH, Richmond MH (1975) Genetics and biochemistry of *Pseudomonas*. Wiley, London New York Sydney Toronto

Clayton RK, Sistrom WR (1978) The Photosynthetic bacteria. Plenum Press, New York London

Cohen SN (1975) The manipulation of genes. Scientific American 233:24–33

Cohen SN, Chang ACY (1973) Recircularization and autonomous replication of a sheared R-factor DNA segment in *Escherichia coli* transformants. Proc Natl Acad Sci USA 70:1293–1297

Cohen SN, Chang ACY (1977) Revised interpretation of the origin of the pSC101 plasmid. J Bacteriol 132:734–737

Cohen SN, Chang ACY, Hsu L (1972) Non-chromosomal antibiotic resistance in bacteria: genetic transformation of *Escherichia coli* by R-factor DNA. Proc Natl Acad Sci USA 69:2110–2114

Collins J (1977) Gene cloning with small plasmids. Curr Top Microbiol Immunol 78:121–170

Collins J, Hohn B (1978) Cosmid: a type of plasmid gene-cloning vector that is packageable *in vitro* in bacteriophage λ heads. Proc Natl Acad Sci USA 75:4242–4246

Dalton H, Mortenson L (1972) Dinitrogen (N_2) fixation (with biochemical emphasis). Bacteriol Rev 36:231–262

David M, Tronchet M, Dénarié JL (1981) Transformation of *Azotobacter vinelandi* with DNA of plasmids RP4 (IncP-1 group) and RSF1010 (IncQ group). J Bacteriol 146:1154–1157

Davies SL, Whittenbury R (1970) Fine structure of methane and other hydrocarbon-utilizing bacteria. J Gen Microbiol 61:227–232

De Graaff J, Crosa JH, Heffron F, Falkow S (1978) Replication of the non-conjugative plasmid RSF1010 in *Escherichia coli* K-12. J Bacteriol 134:1117–1122

De Wilde M, Depicker A, De Vos G, De Beukelaer M, Van Haute E, Van Montagu M, Schell J (1978) Molecular cloning as a tool to the analysis of the Ti-plasmids of *Agrobacterium tumefaciens*. Ann Microbiol (Paris) 129B:531–532

Dickey RS, Goto M, Klement Z, Hayward AC, Maas Geesteranus HP, Ridé M (eds) (1978) Proceedings of the IVth International Conference on Plant Pathogenic Bacteria, Station de Pathologie Végétale et Phytobacteriologie I.N.R.A. Angers, France

Ditta G, Stanfield S, Corbin D, Helinski DR (1980) Broad host range DNA cloning system for Gram-negative bacteria: construction of a gene bank of *Rhizobium meliloti*. Proc Natl Acad Sci USA 77:7347–7351

Dixon R, Kennedy C, Merrick M (1981) Genetic control of nitrogen fixation. In: Glover SW, Hopwood DA (eds) Genetics as a tool in microbiology. Cambridge Univ Press, pp 161–185

Doggett RG (ed) (1979) *Pseudomonas aeruginosa:* chemical manifestations of infection and current therapy. Academic Press, New York San Francisco London

Dunn NW, Holloway BW (1971) Pleiotropy of p-fluorophenyl-alanine-resistant and antibiotic hypersensitive mutants of *Pseudomonas aeruginosa*. Genet Res 18:185–197

Federal Register (1980) 45:77405

Franklin FCH, Bagdasarian M, Timmis KN (1981a) Manipulation of degradative genes of soil bacteria. In: Hütter R, Leisinger T (eds) Microbial Degradation of Xenobiotics and Recalcitrant Compounds. Academic Press, London (in press)

Franklin FCH, Bagdasarian M, Bagdasarian MM, Timmis KN (1981b) A molecular and functional analysis of the TOL plasmid pWWO from *Pseudomonas putida* and cloning of genes for the entire regulated aromatic ring *meta* cleavage pathway. Proc Natl Acad Sci USA (in press)

Gautier F, Bonewald R (1980) The use of plasmid R1162 and derivatives for gene cloning in the methanol-utilizing *Pseudomonas* AM1. MGG 178:375–380

Guerry P, Van Embden J, Falkow S (1974) Molecular nature of two non-conjugative plasmids carrying drug resistance genes. J Bacteriol 117:619–630

Guidelines for research involving recombinant DNA molecules (1980) Federal Register 45.

Hashimoto-Gotoh T, Nordheim A, Timmis KN (1981) Specific purpose plasmid cloning vectors. I. Low copy number temperature-sensitive pSC101-derived containment vector. Gene 16 (in press)

Hohn B, Collins J (1980) A small cosmid for efficient cloning of large DNA fragments. Gene 11: 291–298

Hohn B, Hinnen A (1980) Cloning with cosmids in *E. coli* and yeast. In: Setlow JK, Hollaender A (eds) Genetic engineering, principles and methods, vol 2. Plenum Press, New York, pp 169–183

Hohn B, Murray K (1977) Packaging recombinant DNA molecules into bacteriophage particles in vitro. Proc Natl Acad Sci USA 74:3259–3263

Holloway BW (1965) Variations in restriction and modification of bacteriophage following the increase in growth temperature of *Pseudomonas aeruginosa*. Virology 25:634–642

Holloway BW, Krishnapillai V (1975) Bacteriophages and bacteriocins. In: Clarke PH, Richmond MH (eds) Genetics and biochemistry of *Pseudomonas*. Wiley, London New York Sydney Toronto, pp 99–132

Hütter R, Leisinger T (eds) (1981) Microbial degradation of xenobiotics and recalcitrant compounds. Academic Press, London

Jacoby GA (1979) Plasmids of *Pseudomonas aeruginosa*. In: Doggett RG (ed) *Pseudomonas aeruginosa:* Clinical manifestations of infection and current therapy. Academic Press, New York San Francisco London, pp 272–310

Jacoby GA, Rogers JE, Jacob AE, Hedges RW (1978) Transposition of *Pseudomonas* toluene-degrading genes and expression in *Escherichia coli*. Nature 274:179–180

Jorgensen RA, Berg DE, Allet B, Reznikoff WS (1979) Restriction enzyme cleavage map of Tn10, a transposon which encodes tetracycline resistance. J Bacteriol 137:681–685

Kahn M, Kolter R, Thomas C, Figurski D, Meyer R, Remaut E, Helinski DR (1979) Plasmid cloning vehicles derived from plasmids ColE1, F, R6K and RK2. Methods Enzymol 68:268–342

Kleckner N, Roth J, Botstein D (1977) Genetic engineering in vivo using translocatable drug-resistance elements. New methods in bacterial genetics. J Mol Biol 116:125–159

Knackmuss HJ (1981) Degradation of halogenated and sulfonated hydrocarbons. In: Hütter R, Leisinger T (eds) Microbial degradation of xenobiotics and recalcitrant compounds. Academic Press, London (in press)

Kushner SR (1978) An improved method for transformation of *Escherichia coli* with ColE1 derived plasmids. In: Boyer HW, Nicosia S (eds) Genetic engineering. Elsevier/North Holland, Amsterdam, pp 17–23

Lovett MA, Helinski DR (1976) Method for the isolation of the replication region of a bacterial replicon: construction of a mini-F' Km plasmid. J Bacteriol 127:982–987

Lowbury JL (1975) Biological importance of *Pseudomonas aeruginosa:* medical aspects. In: Clarke PH, Richmond MH (eds) Genetics and biochemistry of *Pseudomonas*. Wiley, London New York Sydney Toronto, pp 37–65

Lundgren DG, Silver M (1980) Ore leaching by bacteria. Annu Rev Microbiol 34:263–283

Mandel M, Higa A (1970) Calcium-dependent bacteriophage DNA infection. J Mol Biol 53:159–162

Manis JJ, Kline BC (1978) F plasmid incompatibility and copy number genes: their map locations and interactions. Plasmid 1:492–507

Mercer A, Loutit JS (1979) Transformation and transfection of *Pseudomonas aeruginosa:* effects of metal ions. J Bacteriol 140:37–42

Miller JH (1972) Experiments in molecular genetics. Cold Spring Harbor Laboratory, Cold Spring Harbor

Murray K (1976) Biochemical manipulation of genes. Endeavour 126:129–133

Nagahari K, Sakaguchi K (1978) RSF1010 plasmid as a potentially useful vector in *Pseudomonas* species. J Bacteriol 133:1527–1529

Nordheim A, Hashimoto-Gotoh T, Timmis KN (1980) Location of two relaxation nick sites in R6K and single sites in pSC101 and RSF1010 close to origins of vegetative replication: implication for conjugal transfer of plasmid DNA. J Bacteriol 144:923–932

Polisky B, Greene P, Garfin DE, McCarthy BJ, Goodman HM, Boyer HW (1975) Specificity of substrate recognition by the *Eco*RI restriction endonuclease. Proc Natl Acad Sci USA 72:3310–3314

Ribbons DW, Wigmore GJ, Chakrabarty AM (1979) Expression of TOL genes in *Pseudomonas putida* and *Escherichia coli*. Soc Gen Microbiol Quart 6:24–25

Rubens C, Heffron F, Falkow S (1976) Transposition of a plasmid deoxyribonucleic acid sequence that mediates ampicillin resistance: independence from host *rec* functions and orientation of insertion. J Bacteriol 128:425–434

Shapiro JA (1977) Bacterial plasmids. 1. Tables. In: Bukhari AJ, Adhyia SL (eds) DNA insertion elements, plasmids and episomes. Cold Spring Harbor Laboratory, Cold Spring Harbor, pp 601–670

Sherratt D (1981) *In vivo* Genetic manipulation in bacteria. In: Glover SW, Hopwood DA (eds) Genetics as a tool in microbiology. Cambridge Univ Press, pp 35–48

Stanisich VA, Bennett PM, Richmond MH (1977) Characterization of a translocation unit encoding the resistance to mercuric ions that occurs on a nonconjugative plasmid in *Pseudomonas aeruginosa*. J Bacteriol 129:1227–1233

Sternberg N, Tiemeier D, Enquist L (1977) In vitro packaging of a λ *Dam* vector containing endo R. *Eco*RI DNA fragments of *Escherichia coli* and phage P1. Gene 1:255–280

Timmis KN (1981) Gene manipulation in vitro. In: Glover SW, Hopwood DA (eds) Genetics as a tool in microbiology. Cambridge Univ Press, pp 49–109

Timmis KN, Cabello F, Cohen SN (1974) Utilization of two distinct modes of replication by a hybrid plasmid constructed in vitro from separate replicons. Proc Natl Acad Sci USA 71:4556–4560

Timmis KN, Cabello F, Cohen SN (1978a) Cloning and characterization of *Eco*RI and *Hind*III restriction endonuclease-generated fragments of antibiotic resistance plasmids R6-5 and R6. MGG 162:121–137

Timmis KN, Cohen SN, Cabello FC (1978b) DNA cloning and the analysis of plasmid structure and function. Prog Mol Subcell Biol 6:1–58

Timmis KN, Bagdasarian M, Brady G, Franklin FCH (1981) Special purpose cloning vectors. III. A multicopy, self-amplifying ColD-derived vector containing multiple cloning sites (in preparation)

Williams BG, Blattner FR (1980) Bacteriophage lambda vectors for DNA cloning. In: Setlow JK, Hollaender A (eds) Genetic engineering, principles and methods, vol 2. Plenum Press, New York, pp 201–281

Williams PA (1978) Microbial genetics relating to hydrocarbon degradation. In: Watkinson RJ (ed) Developments in biodegradation of hydrocarbons. Applied Science Publishers, London, pp 135–164

Windass JD, Worsey MJ, Pioli EM, Pioli D, Barth PT, Atherton KT, Dart EC, Byrom D, Powell K, Senior PJ (1980) Improved conversion of methanol to single-cell protein by *Methylophilus methylotrophus*. Nature 287:396–401

Wood DO, Hollinger MF, Tindol MB (1981) Versatile cloning vector for *Pseudomonas aeruginosa*. J Bacteriol 14:1448–1451

Gene Cloning in Streptomyces

K.F. CHATER**, D.A. HOPWOOD**, T. KIESER** AND C.J. THOMPSON**

1 Introduction

Our laboratory is interested in aspects of morphological differentiation and antibiotic production displayed by *Streptomyces* and is using recombinant DNA technology to study these processes as they occur in the streptomycete background; because of their ver-

---

** John Innes Institute, Colney Lane, Norwich NR4 7UH, England

satile biosynthetic capabilities, the streptomycetes are also well suited for the efficient production or bioconversion of economically important molecules.

The typical life cycle of a streptomycete on a solid substrate begins with the germination of a spore to give rise to a branched septate substrate mycelium. Side branches of the substrate hyphae later give rise to an aerial mycelium consisting of hyphae which are straight, helically coiled, or branched in characteristic ways depending on the strain. In submerged cultures sporulation does not usually occur. After the completion of vegetative growth, cultures remain biosynthetically active and can produce large amounts of secondary metabolites or carry out bioconversions of specific precursors.

The biosynthetic versatility of secondary metabolism in streptomycetes is illustrated by the fact that, of the 4973 natural antibiotics described by 1978, 2769 were of streptomycete origin (*Bérdy* 1980). They include a great number of chemically different structures – aminoglycosides, macrolides, tetracyclines, polyethers, ansamycins, β-lactams, oligopeptides, etc. A single strain often (perhaps typically) produces several diverse chemical classes of antibiotics, as well as numerous minor variants of a particular antibiotic family; often similar or identical antibiotics are produced by taxonomically different species.

Insertion of cloned genes into the structural genes of extracellular enzymes has been used to extract commercially important products from *E. coli*. Correlated with the streptomycete mycelial growth habit, which has doubtless evolved as an adaptation to the colonisation and dissolution of solid food sources (*Chater* and *Merrick* 1979), is the well-developed capacity of streptomycetes to produce extracellular hydrolytic enzymes of many kinds, including proteases, amylases, cellulases, lipases, nucleases and others. Streptomycetes contain a wider variety of extracellular enzymes than *E. coli* and may therefore be more adaptable to the excretion and isolation of a wider range of products. Industrialists have accumulated a large body of information regarding the growth and handling of streptomycetes over the 3 decades in which streptomycete fermentations have been used to produce antibiotics. The use of cloning in streptomycetes for production purposes might thus be technically easier for the industrialist using existing facilities.

No streptomycete is known to cause disease in or be a commensal of man or animals, while only one (*S. scabies*) is a plant pathogen, and regulations on the use of streptomycetes for cloning purposes are no longer stringent. For example, the British Genetic Manipulation Advisory Group (GMAG) has taken the view that the whole genus should be regarded as a single entity for the purposes of the categorisation of recombinant DNA work; all DNA transfers within and between nonpathogenic streptomycetes are regarded as "self-cloning", and may be freely performed on a laboratory scale in Great Britain, provided that "good microbiological practice" is observed. Many transfers of foreign DNA into *Streptomyces* hosts can also be done under these conditions. In the United States the N.I.H. Guidelines are also gradually reducing the levels of containment specified for experiments with *Streptomyces*.

2 Genetic Systems

In addition to allowing gene transfers by recombinant DNA techniques, which are the subject of this chapter, the streptomycetes are quite versatile genetically. Many strains have natural systems of gene exchange through conjugation, a process which requires

cell contact and results in the transfer of large segments of chromosome from a donor to a recipient. Moreover, it probably always depends on the activity of sex plasmids. The mechanisms of sex factor activity in *Streptomyces* are almost unstudied, but the conjugation system itself is genetically much simpler than that mediated by the well-known F plasmid of *E. coli* (see later). Most streptomycetes probably possess sex plasmids. At least two are active in *Streptomyces coelicolor* A3(2), genetically the most intensively studied strain. SCP1 is large (c. 150 kb; *Westpheling* 1980) and occurs autonomously in the wild-type strain, but can integrate into the chromosome or form SCP1-prime derivatives; and SCP2 is smaller (c. 30 kb) and has so far been found only autonomously (*Bibb* et al. 1977). *S. lividans* 66 has a sex plasmid, SLP2, so far not defined physically, and can act as a host for SCP1 and SCP2, and several plasmids from other related strains. Two further plasmids, SLP1 and SLP4, have emerged in *S. lividans* after interspecific crosses with *S. coelicolor* A3(2); both act as fertility factors but only SLP1 has been studied physically (*Bibb* et al. to be published).

The recombination promoted by various of these plasmids, either singly or in combination, can be used to generate a circular chromosomal linkage map. The *S. coelicolor* A3(2) map has been developed over many years and carries about 150 markers (*Chater* and *Hopwood* to be published). A similar, but so far only sparsely marked map, has recently been constructed for *S. lividans* 66 (*D.A. Hopwood* and *H.M. Wright*, unpublished results).

Generalised transduction is not available for these strains, although transduction has been demonstrated in *Streptomyces venezuelae* (*Stuttard* 1979). A natural system of competence for transformation is also lacking. However, artificially promoted recombination through the use of protoplasts is very efficient. Protoplast fusion between marked derivatives of *S. coelicolor* A3(2), or of *S. lividans* 66 (or any of several other species), yields generalised intra-strain chromosomal recombinants at frequencies of 10% or more (*Hopwood* et al. 1977, *Hopwood* and *Wright* 1978, *Hopwood* 1981b); it is therefore a powerful tool for the construction of complex genotypes involving non-selectable markers. Liposomes containing chromosomal DNA of *S. coelicolor* A3(2) (or *S. clavuligerus*) can be fused with protoplasts of the same strain to yield transformation frequencies of 10% or so per marker (*Makins* and *Holt* to be published). This technique may, in the future, be useful for fine structure mapping, or for in vitro mutagenesis of particular chromosomal regions, in addition to several (as yet unexplored) potential uses as an adjunct to gene cloning.

All recombinational events described above presumably involve homologous recombination; there is still no compelling evidence that the many unstable genetic phenomena and DNA rearrangements found in *Streptomyces* involve the action of transposable elements (for a discussion of existing claims, see *Chater* and *Hopwood* to be published).

Restriction modification systems, several of them associated with specific type II restriction endonucleases, are widespread in the genus *Streptomyces* (*Lomovskaya* et al. 1980) and could limit any interspecific gene transfer. Circumstantial evidence that *S. coelicolor* A3(2) is a restricting strain rests on the markedly lower frequency of conjugal transfer of SCP1 from *S. lividans* to *S. coelicolor* than in the opposite direction (*Hopwood* and *Wright* 1973) and on the failure to transfect *S. coelicolor* protoplasts with a pBR322-ϕC31 chimaeric molecule propagated in *E. coli*, even though the same DNA preparation quite efficiently transfected *S. lividans* (*J.E. Suarez*, unpublished results).

3 Cultural Procedures

Although methods have been devised for the reproducible fragmentation of the vegetative mycelium of certain streptomycetes, rendering the use of spores unnecessary even in genetic studies (eg *Baltz*, 1978), we rely on sporulation for much of our work. The efficiency of sporulation depends on both the strain and the medium. Spore suspensions are made by harvesting the spores in water, filtering, centrifuging and re-suspending, typically in 20% glycerol (this allows suspensions to be stored almost indefinitely at –20 °C with little loss of viability). Spores plated on solid media develop into colonies which are visible after about 2 days and sporulate after a few days more. After sporulation, colonies can be readily and very precisely replica plated by the normal procedure using velvet; growth of replicas is usually assessed 2 days later. Liquid cultures are started from an inoculum of spores and grown with vigorous aeration on a shaker. The degree to which dispersed growth occurs – important for the extraction of chromosomal or plasmid DNA as well as for biochemical work – depends on the medium and the strain. The mycelium of some strains constantly fragments, while in others the individual germlings actually adhere together to form a small number of large mycelial clumps. Baffling of flasks – most easily achieved by inserting a coiled stainless steel spring into the bottom of the flask – greatly aids dispersion. Although the cells in these cultures are presumably physiologically heterogeneous, approximately logarithmic growth is usually obtained for several mass doublings, usually from about 10–45 h after inoculation under optimal growth conditions.

From these remarks it is apparent that genetic work with *Streptomyces* takes longer than with a fast-growing unicellular bacterium like *E. coli*. However, experience allows reasonably rapid progress to be made.

4 DNA Preparation Techniques

4.1 Total DNA

Our current procedure for obtaining total DNA preparations is in Table 1. In some preparations, ccc plasmid DNA co-purifies with chromosomal DNA and may be seen as sharp bands ahead of or behind the more diffuse main chromosomal DNA band during agarose gel electrophoresis. DNA fragments isolated by this technique must largely be more than 40 kb in length, since a band corresponding to entire prophage ϕC31 can be visualised after appropriate restriction enzyme digestion of DNA of a lysogen. In general, *Streptomyces* DNA has relatively few recognition sites for enzymes such as *Eco*RI, *Hin*dIII and *Hpa*I which cut hexanucleotide sequences containing four A + T base pairs, and relatively frequent sites for enzymes such as *Sal*GI, *Sal*PI (= *Pst*I), *Bam*HI and *Sst*II which recognise sites rich in G + C. This presumably reflects the high (about 73%) overall G + C content of *Streptomyces* DNA. Several observations of particularly intense bands on agarose gels with certain DNA preparations digested with particular restriction enzymes have suggested that amplified or reiterated sequences of unknown function may be present in some variant strains (*M. Robinson, E. Lewis* and *E. Napier*, personal communication; *H. Schrempf*, personal communication). The *Streptomyces* genome size has

Table 1. Isolation of *Streptomyces* total DNA

1. Harvest mycelium from 40–48 h culture (e.g. YEME, 34% sucrose, 0.005 M MgCl$_2$) and wash in 10% glycerol.
2. Suspend 2 g (wet weight) mycelium in 25% sucrose, 0.05 M TRIS, pH 8 to 6 ml final volume.
3. Add 0.6 ml lysozyme (10 mg/ml) and incubate at 30 °C, triturating by pipette at 15-min intervals, until a drop is completely cleared by adding a drop of 10% SDS.
4. Add 1.2 ml 0.5 M EDTA and 0.7 ml pronase (predigested), 10 mg/ml, and incubate 5 min at 30 °C.
5. Add 3.6 ml SDS (3.3%), tilt immediately, then incubate 2 h at 37 °C.
6. Shake with 6 ml phenol (containing 0.1% 8-hydroxyquinoline and saturated with TE containing 0.1 M NaCl) for at least 10 min.
7. Add 6 ml chloroform and shake for 5 min, centrifuge to separate phases, remove aqueous phase (slowly, to reduce shearing) and re-extract phenol/chloroform phase with half its volume of TE containing 0.1 M NaCl. Centrifuge and pool the aqueous phase with the previous aqueous phase.
8. Re-extract the pooled aqueous phases with phenol and chloroform (twice if necessary, to reduce interphase material) and extract twice with chloroform to remove phenol.
9. Add RNase (preheated to eliminate DNase) to 40 µg/ml, and incubate 1 h at 37 °C.
10. Add 0.2 volumes 5 M NaCl, mix gently, and add PEG 6000 (30%) to 10% final concentration; leave at 0–4 °C 6 h or overnight.
11. Centrifuge gently and discard supernatant.
12. Dissolve in 5 ml TE containing 0.1 M NaCl, shaking gently at 30 °C for up to 24 h.
13. Add sodium acetate (to 0.3 M) and 2.2 volumes ethanol. Leave 1 h at –70 °C or overnight at –20 °C.
14. Sediment DNA by centrifugation, and wash the pellet with ethanol until OD$_{270}$ of ethanol is zero.
15. Dry the DNA pellet *in vacuo* and dissolve it in sterile TE.

Composition of media and buffers. YEME: Difco yeast extract, 0.3%; Difco peptone, 0.5%; Oxoid malt extract, 0.3%; glucose, 1%. TE: 10 mM TRIS HCl, 1 mM EDTA, pH 8.0

been estimated as 10 000 kb (about 2.5 times the size of the *E. coli* genome) *(Benigni* et al. 1975). Based on this value, Table 2 indicates the numbers of clones needed to obtain high probabilities of *Streptomyces* gene representation in gene libraries.

For some cloning purposes it may be useful to size fractionate digests of total DNA. We have found that the procedure of *Girvitz* et al. (1980), in which DNA is electrophoresed from agarose gels into dialysis-membrane-backed filter paper, is very convenient for this purpose, giving fragments which can effectively be restricted or ligated.

Table 2. Number of clones in a complete genomic library for a streptomycete of genome size 10^4 kb (from *Hopwood* 1981a)

Length of cloned fragments (kb)	Percent probability of complete library			
	80	90	95	99.99
5	3218	4604	5988	18414
10	1609	2302	2994	9206
15	1072	1534	1998	6136
20	804	1150	1496	4601
25	643	920	1148	3679
30	536	766	997	3065

4.2 Plasmid DNA

Since the first physical characterisation of a *Streptomyces* plasmid (*Schrempf* et al. 1975), *Streptomyces* plasmids have been successfully isolated in many laboratories. The methods that we routinely employ all use lysozyme treatment to render the mycelium susceptible to lysis by SDS. Either SDS is added to a culture which is just about to lyse, or stable protoplasts are formed in a sucrose solution containing Ca^{2+} and Mg^{2+} necessary for their stabilisation: lysis is then induced by the addition of a mixture of SDS and EDTA or CDTA. The separation of plasmid DNA from chromosomal DNA is achieved either by salt-SDS precipitation followed by CsCl-ethidium bromide centrifugation or agarose gel electrophoresis, or by denaturing the DNA with NaOH followed by precipitation with salt-SDS or phenol. Alkaline lysis is rapid and gives excellent results with the smaller plasmids which are generally used as cloning vectors. With experience it is often possible to remove virtually all chromosomal DNA from the plasmid preparation. Salt-SDS precipitation after alkaline lysis sometimes gives preparations that produce a fluorescent smear on agarose gels, obscuring plasmids bands. These smeary preparations, after removal of residual SDS by repeated ethanol precipitation, can still be used for transformations. Removal of chromosomal DNA with phenol and chloroform generally gives more satisfactory preparations and samples are ready for agarose gel electrophoresis c. 30 min after lysis. The removal of phenol for restriction digestion, ligation or transformation is, however, rather laborious.

   The protocols in Tables 3 and 5 give two examples of neutral lysis procedures. For the isolation of large plasmids such as pSV1 (a plasmid from *S. violaceoruber* which resembles

---

Table 3. Large scale preparation of plasmid DNA from *Streptomyces* (*Bibb* et al. 1977)

---

1. Harvest mycelium from a 500 ml 40–48 h 30 °C culture (YEME, 34% sucrose, 0.005 M MgCl$_2$) and wash in 10% glycerol.
2. Suspend mycelium in 50 ml TE containing 34% sucrose at 30 °C and transfer to a 1-litre beaker; add 10 ml 0.25 M EDTA (pH 8.0) and 5 ml lysozyme solution (50 mg/ml in 0.01 M Tris HCl, pH 8.0), and incubate 10–30 min at 30 °C (until protoplasts begin to form: time depends on strain and growth phase).
3. Add 50 ml ice-cold TE containing 34% sucrose; place in an ice bath, and add 30 ml ice-cold 0.25 M EDTA (pH 8.0) and 5 ml ice-cold 0.01 M TRIS HCl (pH 8.0).
4. Distribute into precooled 50-ml centrifuge tubes (25 ml per tube) and add 3.6 ml 10% SDS to each tube. Invert several times.
5. Add 7.2 ml ice-cold 5 M NaCl to each tube and invert slowly several times; leave 2 h to overnight at 4 °C.
6. Centrifuge (e.g. 16 000 r.p.m., Sorval SS34 rotor, for 30 min) at 4 °C and pool supernatants.
7. Add cold 30% PEG to 10% final concentration and leave 2 h to overnight at 4 °C.
8. Centrifuge gently (e.g. 4000 r.p.m., Sorvall GSA rotor, 4 min) at 4 °C and dissolve pellet in 6 ml of 0.03 M TRIS HCl, 0.005 M EDTA, 0.05 M NaCl (pH 8.0) at 4 °C.
9. Add CsCl (1.05 g per ml DNA solution) and dissolve.
10. Add ethidium bromide solution (10 mg/ml) to 500 μg/ml and adjust refractive index to 1.3925 at 20 °C.
11. Centrifuge (e.g. Beckman 40 rotor, 60 h 20°, 36 000 r.p.m.) and collect CCC DNA band from gradient (using near u.v. light to visualise DNA).
12. Remove ethidium bromide by extraction with propanol saturated with solution used in step 8, saturated with CsCl and dialyse against TE.

For composition of media and buffers, see Table 1

Table 4. Rapid plasmid preparation by alkaline lysis of protoplasts followed by phenol/chloroform extraction (developed for *S. coelicolor* and *S. lividans* from methods of *Birnboim* and *Doly* 1979, and *McMaster* et al. 1980)

1. Inoculate 50 ml TSB (Tryptone Soya Broth, Oxoid CM 129) with c. 10^7 spores and heat to 50 °C for 10 min before incubating for 16–20 h at 30–33 °C.
2. Harvest mycelium and wash once with 10.3% sucrose.
3. Resuspend mycelium in 2 ml lysozyme solution and incubate at 30 °C for 15–60 min. Mix by pipetting.
4. Add 2 ml of hot lytic mix and pipette up and down vigorously. Incubate at room temperature for c. 15 min, mixing at intervals by pipetting.
5. Add 4 ml phenol/chloroform and agitate on a vortex mixer for 30 s.
6. Centrifuge to separate phases (e.g. 10 min at 10 000 r.p.m. in Sorvall SS34 rotor).
7. The upper (aqueous) phase can be loaded directly on an agarose gel without addition of sucrose or dye.
8. Remove the upper phase to a new tube and extract three times with ether.
9. Concentrate DNA and remove residual SDS by repeated ethanol precipitation.

Reagents: Lysozyme solution (can be stored indefinitely until lysozyme is added): 10.3% sucrose, 100 ml; 2.5% K_2SO_4, 1 ml; 0.5% KH_2PO_4, 1 ml; 2.5 M $MgCl_2$, 0.1 ml; 0.25 M $CaCl_2$, 1 ml; 1 M TRIS HCl (pH 8), 2.5 ml; 50% glucose, 4 ml; lysozyme, 100 mg. Lytic mix (warm to 50 °C to dissolve SDS before use): 5 M NaOH, 10 ml; 10% SDS, 20 ml; 0.25 M CDTA (cyclohexane diamine tetracetate), 25 ml; water, 45 ml. Phenol/chloroform (use lower organic phase): phenol (AnalaR), 500 g; hydroxyquinoline, 0.5 g; 0.5 M NaCl, 200 ml; chloroform, 500 ml

SCP1 both in size and in encoding production of and resistance to methylenomycin: *Aguilar* and *Hopwood* 1981), the procedure in Table 5 is followed, only a small part of the gel being irradiated to locate the plasmid band. Plasmid DNA from the unirradiated section of the gel is then electrophoresed into filter paper backed with dialysis membrane (*Girvitz* et al. 1980), from which it is subsequently eluted. The resulting DNA has usually acquired nicks, and so will not enter agarose gels unless broken into linear fragments, for example by treatment with restriction enzymes (for which it is a perfectly good substrate: *A. Aguilar,* personal communication). The protocol in Table 4 is for alkaline lysis of protoplasts. This method is based on the procedures of *Birnboim* and *Doly* (1979) and *McMaster* et al. (1980).

Table 5. Preparation of plasmid DNA for purification by gel electrophoresis (after *Westpheling* 1980)

1. Centrifuge mycelium from 25 ml culture (e.g. YEME, 34% sucrose, 0.005 M $MgCl_2$).
2. Suspend in 10 ml TE containing 15% sucrose (pH 8.0).
3. Centrifuge mycelium, suspend in 4 ml 0.05 M TRIS HCl, pH 8.0 containing 15% sucrose (pH 8.0), and add 1 ml lysozyme solution (5 mg/ml in 0.25 M TRIS HCl, pH 8.0).
4. Incubate at 30 °C until appearance becomes "fluffy" but before full lysis occurs.
5. Add 2 ml 0.25 M EDTA (pH 8.0) and 3 ml 20% SDS (special pure grade), and allow the suspension to become clear. (0.2 ml diethylpyrocarbonate sometimes added before step 5.)
6. Transfer to a 50 ml centrifuge tube, add 3 ml 5 M NaCl and leave at 4 °C for 2 h to overnight.
7. Centrifuge at 16 000 r.p.m. for 20 min at 4 °C.
8. Add 4 ml sterile 50% PEG 6000 to supernatant and leave for 2 h to overnight.
9. Centrifuge gently and dissolve pellet in 1 ml sterile TE.

For composition of media and buffers, see Table 1

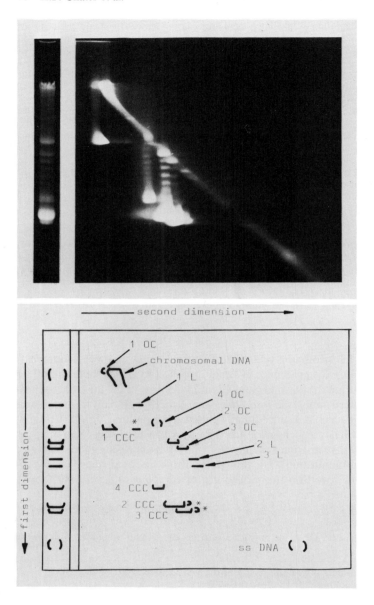

Fig. 1. Two-dimensional agarose-gel electrophoresis (see *Hintermann* et al. 1981) of plasmid DNA isolated from the strain in which plasmids pIJ101–104 were found. After running in the first dimension (0.7% agarose, 90 mM Tris-borate, 10 hrs at 1.5 V/cm) the gel was stained with 0.5 μg/ml ethidium bromide for 2 hrs and photographed under 300 nm UV illumination. In the presence of ethidium bromide, DNA is nicked by the UV, causing supercoiled (*CCC*) molecules to relax to the open circular (*OC*) form but including insignificant changes in OC or linear DNA. The gel track was cut out, cast into a new agarose gel and run in the second dimension (0.8% agarose, 90 mM Tris-borate, 8 hrs at 2.5 V/cm). The increase in the agarose concentration and in the voltage makes the linear DNA molecules run slightly faster relative to the circular form. This allowed the separation of

Table 6. Large-scale preparation of *Streptomyces* phage DNA

1. Prepare large, just confluently lysed, top layers on Difco Nutrient Agar containing glucose (0.5%) and appropriate divalent cations (0.2–0.5 m^2 of surface area).
2. Scrape top layers into 2.5 volumes Difco Nutrient Broth and leave at 20 °C for 2 h.
3. Centrifuge to remove agar.
4. Sediment phage particles (e.g. Beckman 30 rotor, 25 000 r.p.m. 4 °C, 75 min) and suspend gently in SM buffer to final volume of about 6 ml, using a rotary shaker for at least 2 h.
5. Add CsCl (0.85 g per millilitre suspension).
6. Centrifuge (e.g. Beckman 40 rotor, 36 000 r.p. m., 20 °C, 18 h).
7. Remove blueish phage band from gradient (e.g. by syringe).
8. Dialyse 4 h against SM buffer (one change).
9. Add one-tenth volume 10 × SSC, heat to 60 °C for 10 min, and cool to room temperature.
10. Add 4 M NaCl to 0.25 M final concentration.
11. Add equal volume of phenol (containing 0.5% 8-hydroxyquinoline and saturated with TE buffer containing 0.25 M NaCl), invert 50 times and centrifuge to separate phases.
12. Re-extract upper layer twice with phenol as in 11.
13. Re-extract phenol layers with TE containing 0.25 M NaCl.
14. Pool the aqueous layers of 12 and 13 and dialyse against TE.

Media and buffers. SM buffer: 0.02 M TRIS HCl pH 7.5, 0.001 M MgSO$_4$, 0.1 M NaCl, 1% gelatin. SSC: 0.15 M NaCl, 0.015 M radium citrate. For TE, see Table 1

Small multicopy plasmids isolated from *Streptomyces* cultures often give multiple bands on agarose gels. Two-dimensional electrophoresis (*Hintermann* et al. 1981) and restriction analysis suggest that the cells contain plasmid DNA with different numbers of supercoil turns (Fig. 1).

4.3 Phage DNA

Large and small scale procedures for phage DNA isolation are in Tables 6 and 7. Although we have generally used just confluently lysed top layers as the source of high titre phage preparations, shaken liquid cultures can often equally well be used (*Dowding* 1973). In general, phage DNA preparation is somewhat more laborious than plasmid preparation, especially for screening purposes, when ten plasmid DNA isolations can be done for one phage DNA isolation. On the other hand, the yields of phage DNA are relatively high compared with those of low copy number plasmids.

the OC form of plasmid pIJ101 (*1 OC*) from the chromosomal DNA. On the second-dimension gel several DNA species can be identified. Four supercoiled species (*1–4 CCC*), recognisable by their slower migration speed compared to the first dimension, represent plasmids pIJ101–4 respectively. The corresponding OC-forms have run the same distance as the UV-nicked supercoiled molecules. The bands between each CCC form and its OC form, not included in the diagrammatic representation of the gel, are molecules with different numbers of supercoil turns. The other bands represent linear forms of the plasmids (*L* marks the linear species that were in the original preparation, and * marks the linear forms generated by nicking from the CCC forms). The *fastest running band* possibly represents single-stranded DNA (ssDNA) (from Kieser et al., to be published)

Table 7. Small-scale preparation of *Streptomyces* phage DNA (*Suarez* and *Chater* 1981)

1. Scrape into 2.5 volumes of Difco Nutrient Broth six just confluently lysed top layers (8-cm Petri dishes: see Table 6 for medium) and leave for 2 h.
2. Centrifuge to remove agar.
3. Centrifuge to sediment phage (see Table 6).
4. Suspend gently (see Table 6) in 4 ml SM buffer containing RNase (40 μg/ml preheated to remove DNase).
5. Add 0.8 ml 0.25 M EDTA (pH 7.4), 0.5 M Tris HCl (pH 9.6), 2.5% SDS, and heat (70 °C, 30 min).
6. Add 1 ml 8 M potassium acetate and leave in ice 15 min.
7. Centrifuge (17 000 r.p.m., 30 min, 2 °C) and decant (if any SDS still present, centrifuge again).
8. Dialyse supernatant against 100 volumes TE (overnight, one change of buffer).
9. Add sodium acetate to 0.28 M final concentration and 2 volumes cold ethanol and place at –20 °C overnight.
10. Centrifuge (15 min, 8000 r.p.m., 2°–4 °C), discard the supernatant and dry the pellet.
11. Suspend in 0.4 ml TE. Add ammonium acetate to 0.28 M final concentration and transfer to microcentrifuge tube. Add 1.0 ml ethanol, and place at –70 °C for 20 min.
12. Centrifuge, discard the supernatant, dry the pellet and dissolve it in about 50 μl TE.

For composition of TE, see Table 1, and of SM, see Table 6

4.4 Ligation Conditions

We use the T4 ligase reaction mixture as described by *Hepburn* and *Hindley* (1979). DNA concentrations routinely used in shotgun experiments are as follows: for SLP1-derived vectors, vector 5 μg/ml, donor DNA 35 μg/ml; and for øC31-derived vectors, vector 200 μg/ml, donor DNA about 50 μg/ml. In some cases, the mixture has been slowly cooled from room temperature to 4 °C, the reaction being terminated after 1–5 days; and in others, the reaction mixture has been kept at 12 °C for at least 24 h.

5 The Reintroduction of DNA into Cells

Both transformation with plasmid DNA and transfection depend on the use of polyethylene glycol (PEG) and protoplasts (*Bibb* et al. 1978; *Suarez* and *Chater* 1980a).

5.1 Preparation of Protoplasts

A current protocol for *S. lividans* protoplast preparation is in Table 8. The growth stage is important for obtaining protoplasts giving high transformation or transfection frequencies: thus, although very late "logarithmic" mycelium (30–36 h) of *S. lividans* gave the highest transformation frequencies per DNA molecule for SLP1 derivatives (*Thompson* et al. to be published a), protoplasts from *S. parvulus* or *S. albus* G could be transfected only when prepared from young (18 h) cultures (protoplasts from older cultures of *S. parvulus*, but not of *S. albus* G, could be transformed efficiently with plasmid DNA) (*Suarez* and *Chater*, 1980a, 1981; *H.M. Wright* and *D.A. Hopwood* unpublished results).

Protoplasts can be used for several days after preparation if stored at 4 °C (but with slight loss of transformation efficiency), and for many months if stored at –70 °C (using

Table 8. Preparation of *Streptomyces lividans* protoplasts

---

1. Centrifuge 30–36 h culture (25 ml YEME, 34% sucrose, 0.005 M MgCl$_2$, 0.5% glycine, in 250 ml baffled flask).
2. Suspend pellet in 10.3% sucrose and centrifuge again.
3. Repeat 2.
4. Suspend mycelium in 4 ml lysozyme solution (1 mg/ml in P medium or, for slightly more efficient plasmid transformation, in P medium with CaCl$_2$ and MgCl$_2$ concentrations reduced to 0.0025 M and incubate at 30 °C for 15–60 min.
5. Mix by pipetting three times in a 5-ml pipette and incubate for further 15 min.
6. Add 5 ml P medium and mix by pipetting as in 5.
7. Filter through cotton wool and sediment protoplasts gently (e.g. 7 min at 800 x G) in a bench centrifuge.
8. Suspend protoplasts in 4 ml P medium and centrifuge again.
9. Repeat step 8 and suspend protoplasts in the drop of P medium left after pouring off the supernatant (for transformation); for storage add 2 ml P medium to the suspended pellet.

Media and buffers. P medium: dissolve 10.3 g sucrose, 0.025 g K$_2$SO$_4$, 0.203 g MgCl$_2$.6H$_2$O and 0.2 ml trace element solution in 80 ml distilled water and autoclave. Then add, in order, 1 ml KH$_2$PO$_4$ (0.5%), 10 ml CaCl$_2$,2H$_2$O (3.68%), and 10 ml TES buffer (0.25 M, pH 7.2). Trace element solution (per litre): ZnCl$_2$, 40 mg; FeCl$_3$.6H$_2$O, 200 mg; CuCl$_2$.2H$_2$.O, 10 mg; MnCl$_2$.4H$_2$O, 10 mg; Na$_2$B$_4$O$_7$ .10H$_2$O, 10 mg; (NH$_4$)$_6$ Mo$_7$O$_{24}$.4H$_2$O, 10 mg. For YEME, see Table 1

slow freezing and rapid thawing); even repeated freezing at –20 °C and thawing seems to be satisfactory. Frozen *S. lividans* protoplasts have about tenfold reduced ability to be transfected, but are virtually unaffected in their efficiency as recipients of plasmid DNA.

5.2 Polyethylene Glycol Assisted DNA Uptake

The protocol for transformation/transfection is in Table 9. The 20% PEG optimum has been noted for both transformation (*Bibb* et al. 1978) and transfection (*Suarez* and *Chater*

Table 9. Transformation and transfection of *S. lividans* protoplasts (*Thompson* et al., to be published

---

1. Centrifuge protoplast suspension and suspend protoplast pellet (4×10^9 protoplasts; viability normally 1%) in the drop of P medium left after pouring off the supernatant.
2. Add DNA in less than 20 µl TE (in many cases an ethanol precipitation is used both to reduce volume and to ensure sterility).
3. Immediately add 0.5 ml PEG 1000 solution (2.5 g PEG dissolved in 7.5 ml of P buffer or, for more efficient transformation, in 7.5 ml of 2.5% sucrose, 0.0014 M K$_2$SO$_4$, 0.1 M CaCl$_2$, 0.05 M TRIS-maleic acid pH 8.0, plus trace elements) and pipette once to mix the components.
4. After 60 s, add 5 ml P medium and sediment the protoplasts by gentle centrifugation.
5. Suspend the pellet in 1 ml P medium.
6. Plate out 0.1 ml on R2YE plates (for transformation dry plates to 85% of their fresh weight (e.g. in a laminar flow cabinet); for transfection, dry plates superficially only and overlay with molten R2YE diluted with an equal volume of P medium and containing *S. lividans* spores at 10^7/ml).
7. Incubate at 30 °C. Score plaques after 18 h, and pocks after 2–3 days.

Media and buffers. R2YE: dissolve 10.3 g sucrose, 0.025 g K$_2$SO$_4$. 1.012 g MgCl$_2$.6H$_2$O, 1 g glucose, 0.01 g Difco casamino acids, and 2.2 g Difco agar in 80 ml distilled water and autoclave. Add sequentially 0.2 ml trace element solution (Table 8), 1 ml 0.5% KH$_2$PO$_4$, 8.02 ml 3.68% CaCl$_2$.2H$_2$O; 1.5 ml 20% L-proline; 10 ml 0.25 M TES buffer, pH 7.2; 0.5 ml 1 N NaOH, 5 ml 10%, yeast extract. See Table 1 for the composition of TE and Table 8 for that of P medium

1980a). We employ PEG 1000 and have not attempted to use PEG of different molecular weights, but *Krügel* et al. (1980) reported the use of PEG 6000 in *S. lividans* transfections, again with an optimum concentration of 20%. Although up to 3 min may be allowed for the transformation to take place (before the PEG is diluted and washed away), the reaction is probably completed in less than 1 minute (*Suarez* and *Chater* 1980a).

Quantities of DNA to be used will depend on the nature of the experiment. However, typically a shotgun cloning experiment using 1 µg SLP1-derived vector yields about 10^4–10^5 transformants (about 100–1000-fold lower than with 1 µg uncut vector). In model shotgun cloning experiments with ϕC31 vectors, similar reductions in transfection frequency compared with uncut vector have been observed. Preliminary results suggest that at least a tenfold stimulation of plaque numbers can be obtained by the addition of liposomes to transfection mixtures (*R. Rodicio* and *K.F. Chater,* unpublished results). Transformation and transfection treatments do not usually substantially reduce the viable counts of protoplasts.

5.3 The Detection of Transformants/Transfectants

After exposure to DNA and PEG, protoplasts are deposited on agar regeneration medium, which for *S. lividans* 66 and *S. coelicolor* A3(2) is R2 medium (*Okanishi* et al. 1974) supplemented with 0.5% yeast extract (R2YE). Highest transformant yields are obtained using R2YE plates dried to 85% of their initial weight. *Baltz* and *Matsushima* (1980) have also reported that use of a temperature which is suboptimal for growth is important for optimal regeneration in certain strains. The transformants are detected either by the formation of "pocks" by plasmid-containing regenerants in a plasmid-free background (see Sect. 6.1) or by their acquisition of drug resistance (e.g. thiostrepton in the cases of pIJ41 or pIJ350). Drug resistance may be detected by replication of the regenerated protoplasts after sporulation or (possibly more efficiently) by overlaying plates with soft agar containing the appropriate drug at a suitable time after inoculation to allow for phenotypic expression (8–24 h for neomycin, viomycin or thiostrepton resistance). Transfection events are detected by incorporating a sample of the transfection mixture into overlays of soft R2 medium (half-strength agar) seeded with *S. lividans* spores, which are poured on R2YE base plates.

Further discussion of the detection of clones will be given after the descriptions of vectors currently in use.

6 Plasmid Vectors

6.1 General Features of Streptomyces Plasmids

A property of many conjugative *Streptomyces* plasmids is their capacity to give rise to a visible reaction in plate cultures when a strain bearing a plasmid grows in contact with a strain lacking the corresponding plasmid, and transfers its plasmid to the recipient strain: each spore of the plasmid-bearing strain, when surrounded by a population of plasmid-free spores, gives rise to a circular zone of retarded growth called a "pock" (Fig. 2). A corresponding reaction appears as a narrow zone of inhibition when a plasmid-bearing

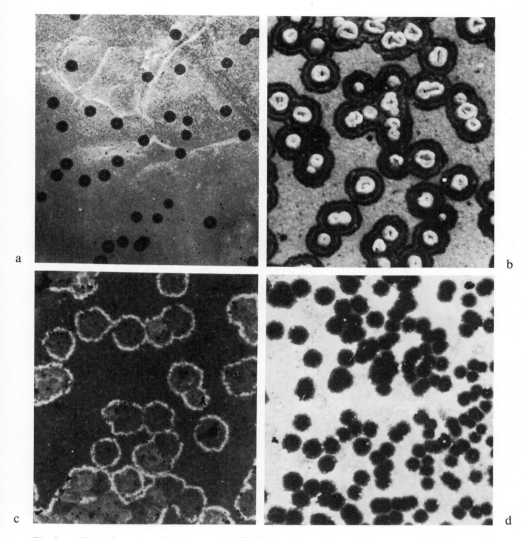

Fig. 2a–c. "Pocks" produced by the growth of individual plasmid-containing spores in backgrounds of plasmid-free organisms. *a* SCP2* in *Streptomyces coelicolor* A3(2). *b* SCP2* in *S. coelicolor* A3(2) with SCP1 present in both cultures. *c* pIJ101 in *S. lividans* 66. *d* pIJ101 in *S. parvulus* ATCC 12434 (*a* and *c* from *Bibb* and *Hopwood* 1981)

culture is replica plated to, or streaked on, a lawn of the plasmid-free strain. This inhibition reaction has been called "lethal zygosis" (Ltz) *(Bibb* et al. 1977) by analogy with the phenomenon of that name in *E. coli (Skurray* and *Reeves* 1973), but this analogy may turn out to be superficial: it is not known if entry of a plasmid into plasmid-free *Streptomyces* cells ever results in their death. The Ltz reaction, and particularly the production of pocks, is very useful in the detection of transformants and in the isolation of plasmid clones *(Bibb* et al. 1978).

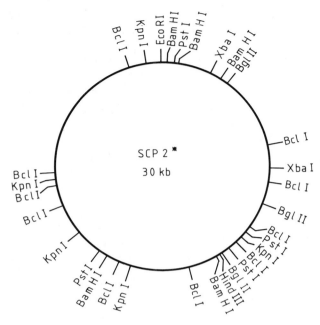

Fig. 3. Restriction map of SCP2* DNA (see *Chater* and *Hopwood*, to be published, for sources of data)

There are numerous examples of physically characterised *Streptomyces* plasmids. However, only three groups of plasmids have so far been developed as cloning vectors to our knowledge – SCP2*, the SLP1 family, and pIJ101 – and these will be described in turn. They are all conjugative and spores or protoplasts carrying them give rise to pocks in plasmid-free cultures. However, derivatives of pIJ101 lacking particular segments of its DNA, either natural variants such as pIJ102, pIJ103 and pIJ104 which coexist with pIJ101 in its wild-type host, or derivatives generated in vitro, are non-conjugative and Ltz⁻ (*Kieser* et al. to be published).

6.2 SCP2 and SCP2*

SCP2 (or pSH1), the first *Streptomyces* plasmid to be studied physically (*Schrempf* et al. 1975), was later shown to be a sex plasmid (*Bibb* et al. 1977). The ability to promote chromosomal recombination, and to show the Ltz reaction, is greatly enhanced in a variant form of the plasmid called SCP2*, which has not been physically distinguished from SCP2 by restriction analysis (*Bibb* and *Hopwood* 1981). The copy number of both plasmids is somewhere between 1 and 5. A restriction map of the plasmid is in Fig. 3. Functions required for plasmid replication, transfer, lethal zygosis and chromosomal recombination are located in the left hand half of the plasmid, since derivatives carrying only the largest *Pst*I fragment, or the largest *Bam*HI fragment, show all these properties (*Bibb* et al. 1980a). These plasmid derivatives are readily lost from the host cultures,

whereas derivatives consisting of both the largest and second largest *Pst*I fragments (in either orientation) are stable, suggesting the presence of a plasmid partition function on the smaller *Pst*I fragment (*Bibb* et al. 1980a).

SCP2 has been used to clone the methylenomycin resistance conferred by the SCP1 plasmid of *S. coelicolor* (*Bibb* et al. 1980a), and several biosynthetic genes of the same species (*Thompson* et al. to be published a), but has not yet been refined as a cloning vector. It shows promise as a low copy number vector for *S. coelicolor* A3(2) and can also replicate in *S. lividans* 66 and *S. parvulus* ATCC 12434.

6.3 The SLP1 Family of Plasmids

Close to the *strA* locus on the *S. coelicolor* A3(2) chromosome is a DNA sequence which can form, together with variable segments of the chromosome on either side of this sequence, a series of plasmids capable of replicating autonomously in *S. lividans* 66 (*Bibb* et al. to be published). They are revealed as pocks when the *S. lividans* progeny of matings between the two strains are plated in a background of *S. lividans*; from a proportion of the Ltz$^+$ strains, plasmid DNA can be isolated, its size varying in different isolates. The plasmids have been named SLP1.1, SLP1.2, etc. Further members of the family have been generated in vitro by cleaving and ligating total *S. coelicolor* DNA and transforming the ligation mixture into *S. lividans* (*Bibb* et al. to be published). The SLP1 plasmids can readily be transferred by mating or transformation within *S. lividans*. *S. coelicolor* can be transformed only at very low frequency by marked SLP1 derivatives, probably because of incompatibility exerted by the resident integrated SLP1 sequences. *S. parvulus* has not been successfully infected with SLP1, and this may also be due to the presence of an endogenous plasmid incompatible with SLP1 (*T. Kieser* and *D.A. Hopwood*, unpublished results). Outside of this group of strains the host range of the SLP1 plasmids is not well defined; they can certainly be introduced successfully into *S. reticuli* (*H. Schrempf*, personal communication).

The copy number of the autonomous SLP1 plasmids in *S. lividans* is about 4–5 (*Bibb* et al. 1980b). SLP1.2 (Fig. 4), the largest naturally occurring member of the series identified to date, has been used in cloning because the segment of dispensable DNA which distinguishes SLP1.2 from SLP1.6 (the smallest member) contains some suitable sites for insertion of foreign DNA, notably the unique *Bam*HI site and the only three *Pst*I sites (originally thought to be just two) in the entire plasmid. *Bibb* et al. (1980b) cloned a *Pst*I fragment of the SCP1 plasmid conferring methylenomycin resistance into SLP1.2 to replace the region bordered by the outermost *Pst*I sites. *Thompson* et al. (1980), using several combinations of restriction enzymes, cloned genes for resistance to neomycin mediated by either an acetyltransferase or a phosphotransferase from total *S. fradiae* DNA and thiostreption resistance mediated by a ribosomal RNA pentose methylase from *S. azureus*. Later, viomycin resitance, mediated by a phosphotransferase from *S. vinaceus* and erythromycin resistance, mediated by a ribosomal RNA adenine methylase of *S. erythreus,* were added to the list (*Graham* and *Weisblum* 1979; *Skinner* and *Cundliffe* 1980; *Thompson* et al. to be published b). The cloning sites in SLP1.2 were either the *Bam*HI site or a combination of the *Bam*HI site and the *Pst*I site furthest from it. In some experiments, donor DNA was cut with *Mbo*I (recognition site ↓GATC) instead of *Bam*HI (G↓GATCC) to help to randomise the population of donor DNA fragments in the lig-

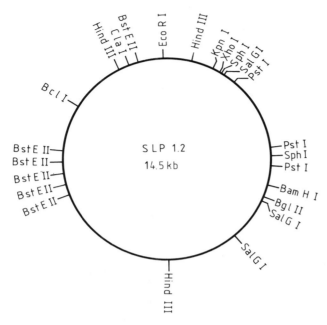

Fig. 4. Restriction map of SLP1.2 DNA. The map is based on that of *Bibb* et al., to be published, with the addition of further data of *Thompson* et al., to be published a; differences in the estimated size of the plasmid reflect the use of different molecular weight size standards

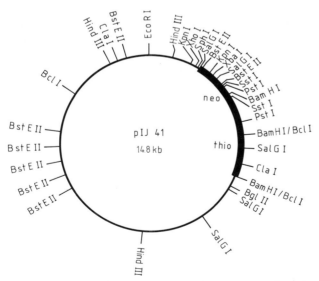

Fig. 5. Restriction map of pIJ41 DNA (*Thompson* et al., to be published c). The plasmid was derived by in vitro replacement of a segment of SLP1.2 DNA (see Fig. 6) with DNA (*heavy line*) from *S. fradiae* and *S. azureus* coding for neomycin resistance and thiostrepton resistance respectively

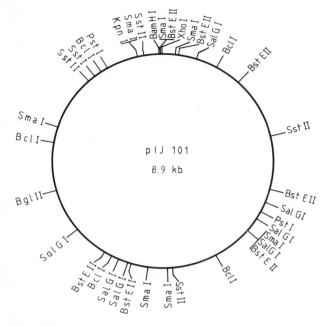

Fig. 6. Restriction map of the DNA of the pIJ101 series of plasmids (data of *Kieser* et al., to be published). The map is that of pIJ101. pIJ102, pIJ103 and pIJ104 lack various segments of pIJ101 DNA from the lower segment of the map

ation mixture. The experiments made available a set of DNA segments bearing resistance genes for neomycin, thiostrepton or viomycin (the erythromycin clones yielded little plasmid DNA for unknown reasons) and bordered by various restriction enzyme target sites. These have been used to mark other plasmids and phages and to develop a second generation of cloning vectors based on SLP1.2.

pIJ41 (Fig. 5) is one such vector (*Thompson* et al. to be published c). It carries neomy-

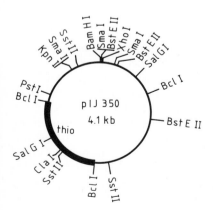

Fig. 7. Restriction map of pIJ350 (data of *Kieser* et al., to be published). The plasmid was derived from pIJ102 by in vitro replacement of a *Bcl*I fragment by DNA from *S. azureus* coding for thiostrepton resistance. The plasmid has a high copy number, broad host range and is non-conjugative

cin and thiostrepton resistance genes on a replicon derived from SLP1.2. The unique *Bam*HI site within the neomycin phosphotransferase gene is suitable for clone recognition by insertional inactivation. By selecting thiostrepton resistant transformants and screening for neomycin-sensitive clones, pIJ41 has already proved satisfactory as a primary shotgun cloning vector. pIJ41 can be used for cloning with many other enzymes. *Xho*I, Sph*I* and *Xba*I have unique sites within DNA dispensable for plasmid replication and transfer and should therefore be available. The *Eco*RI and *Bcl*I sites, although in a region not known to be dispensable, have been shown empirically to be cloning sites (*C.J. Thompson* and *E. Katz*, unpublished results). Sst*I*, Pst*I* and *Kpn*I should each eliminate a dispensable fragment; in the first two cases neomycin sensitivity could result. Several cloning experiments using the *Bgl*II site have failed.

6.4 The pIJ101 Series of Plasmids

In a survey of 24 strains in the *S. violaceoruber* species group, several were found to harbour easily extractable plasmid DNA. One strain was of particular interest because it contained four CCC plasmid species, at least two in high copy number; the plasmids have been named pIJ101 (8.9 kb), pIJ102 (4.2 kb), pIJ103 (3.9 kb) and pIJ104 (4.7 kb) (*Hopwood* et al. to be published). Restriction enzyme mapping showed the three smaller members of the series to be subsets of particular sequences present in the largest plasmid (Fig. 6); perhaps they are natural deletion variants of it which coexist with it in the wild-type host. While pIJ101 is conjugative and Ltz$^+$, the other three plasmids are non-conjugative and Ltz$^-$ (*Kieser* et al. to be published). The host range of pIJ101 is wide in the genus *Streptomyces,* members of several species having been successfully transformed with it, including strains of *S. acrimycini, S. albus, S. azureus, S. fradiae, S. glaucescens, S. griseus, S. parvulus, S. pristinaespiralis* and *S. rimosus*. Derivatives of this series of plasmids have obvious potential as cloning vectors with broad host range, high copy number, and conjugative or non-conjugative properties as desired. pIJ350 (Fig. 7) is an early example of such a vector; only the single Pst*I* site has been proved to be available for cloning, but the *Kpn*I and *Cla*I sites are likely to be. On the other hand the *Bam*HI and *Xho*I sites appear to be in essential DNA.

6.5 Bifunctional Plasmid Vectors

The construction of plasmids able to replicate in *Streptomyces* and one or more alternative hosts, for studies of heterologous gene expression and other uses, presents no particular conceptual or general practical problems. An example of a bifunctional *Streptomyces – E. coli* plasmid is pIJ28, which is a fusion of *Bam*HI-cleaved pBR322 with *Bcl*I-cleaved pIJ2 (pIJ2 is SLP1.2 with a neomycin resistance determinant inserted at its *Bam*HI site, and contains a single *Bcl*I site). This vector was initially constructed by introducing the ligated DNA into *E. coli*, hybrid molecules being selected by transforming *S. lividans* with total plasmid DNA isolated from the *E. coli* transformants. One of these (pIJ28) transformed *E. coli* quite efficiently, and DNA re-isolated from these *E. coli* transformants was used to infect *S. lividans*, giving a transformation frequency about 10 times lower than that obtained when the pIJ28 came from *S. lividans* (*C.J. Thompson* and

J.M. Ward, unpublished results) (a similar result was obtained with the ⌀C31-pBR322 bifunctional replicon described below: *Suarez* and *Chater*, 1980b). It appears that pIJ28 (and other comparable bifunctional replicons described by *Schottel* et al. 1981) is stable within its alternative hosts. Aberrations may not be uncommon among the *S. lividans* transformants when DNA from *E. coli* is used.

We detected no evidence that the neomycin phosphotransferase gene of pIJ28 was expressed in *E. coli*. However, at least one other resistance gene (that encoding viomycin phosphotransferase from *S. vinaceus*) is expressed in *E. coli* when inserted into the *Bam*HI site of pBR322 (*J.M. Ward* and *C.J. Thompson*, unpublished results). Expression of some *E. coli* genes has also been detected in streptomycetes, including the chloramphenicol resistance determinant of pACYC184 and the kanamycin resistance determinant of pACYC177 (*Schottel* et al. 1981) and the *tet* gene of pBR322 (using the ⌀C31:pBR322 hybrid described below).

We find bifunctional replicons conceptually appealing because they should allow the exploitation of complementary benefits of the different host/vector systems. However, we do not envisage the extensive study of *Streptomyces* antibiotic biosynthesis and differentiation in *E. coli* or other heterologous systems, because of the complexity of these processes as they occur in streptomycetes.

7 Bacteriophage Vectors

Two hetero-immune *Streptomyces* temperate phages, ⌀C31 and R4, have been considered by us as potential cloning vectors. They both form plaques on a fairly wide range of *Streptomyces* spp. (12 out 20 strains tested for ⌀C31; 18 out of 36 for R4) and might therefore find general applications. Nearly all our work has been on ⌀C31.

7.1 Relevant Features of ⌀C31 Biology and Genetics

A detailed consideration of these aspects of ⌀C31 and other phages has recently been published (*Lomovskaya* et al. 1980). ⌀C31 is usually propagated on *S. lividans* 66, up to 10^{11} p.f.u. being obtainable from a single confluently lysed plate. It does not usually form plaques on *S. coelicolor* A3(2), but ⌀C31-sensitive mutants can easily be isolated (about 1% of the survirvors of u.v. irradiation to 99% killing) on which plaques are formed with high efficiency. This was previously thought to reflect curing of a defective ⌀C31 prophage, but we have recently found by Southern blotting that ⌀C31-related DNA sequences are absent from both resistant and sensitive strains (*C.J. Bruton* and *K.F. Chater*, unpublished results). Both kinds of strain can be lysogenised, and a chromosomal *att* site has been mapped (*Lomovskaya* et al. 1980). The prophage state of ⌀C31 is controlled by a repressor specified by the *c* gene, and clear plaque (*c*) mutants are easily recognisable. Mutants with a heat inactivated repressor (*c*ts) have also been obtained.

Extensive genetic analysis of temperature-sensitive (ts) ⌀C31 mutants has so far identified 19 complementation groups, 14 of which have been mapped by conventional phage crosses on a linear non-permuted linkage map which also includes host range and plaque morphology mutations (*Chinenova* and *Lomovskaya* 1975; *Lomovskaya* et al. 1980). Most of these ts mutations were expressed late in the phage growth cycle, and

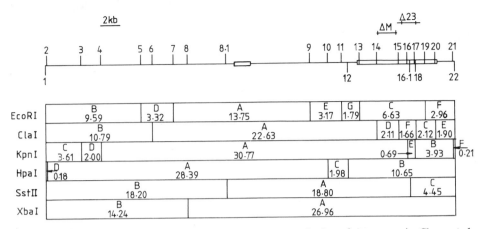

Fig. 8. Restriction map of øC31 DNA. The mapping and numbering of sites are as in *Chater* et al. (1981) with the addition of *Sst*II sites (8.1 and 16.1). The numbers in the *lower part* of the figure are fragment lengths in kilobases. There are no sites for *Bam*HI, *Pst*I, *Xho*I or *Pvu*II; seven for *Sph*I (*J. Harris,* personal communication); and 13 or more for *Hind*III, *Bcl*I, *Sal*GI, *Sma*I, *Fsp*A1 (= *Bst*E11) and *Hgi*AI. The overall genome length is 41.2 kb. Functions encoded in the *boxed segments* are normally inessential for plaque formation. ΔM and Δ23 are deletions which were important in vector development

mapped to the left of the *c* gene. A host range mutation was located within this array of late genes. Presumably many structural components of the mature virion are encoded in this region of the DNA. The only mutation mapping to the right of the *c* gene, L24, was also the earliest expressed.

7.2 øC31 DNA

Electron microscopy (*Sladkova* et al. 1977) showed øC31 DNA to comprise about 39.4 kb of non-permuted ds DNA, with cohesive ends (this value allows for the fact that the original estimate of 37.7 kb was done with a phage later shown to contain a deletion – ΔM – of about 1.7 kb: *Lomovskaya* et al. 1980). Restriction enzyme analysis (*Chater* et al. 1981) has given the slightly higher estimate of 41.2 kb. A restriction map of the wild-type phage is given in Fig. 8.

The DNA of øC31 prophage, studied by hybridisation of radioactive øC31 DNA to Southern blots of restriction enzyme digested DNA of lysogens, is in a different permutation from that in the mature phage: the cohesive end fragments are covalently joined, and the position (*attP*) at which the phage genome interacts with the chromosomal (*attC*) site is between sites 18 and 19 on the restriction map (*C.J. Bruton* and *K.F. Chater,* unpublished results).

7.3 The Isolation of Deletion Mutants

Like many other phages whose DNA packaging involves enzymatic cutting to give cohesive ends, øC31 is sensitive to chelating agents. Mutants with increased resistance

Fig. 9. Restriction map of DNA of a øC31-pBR322 chimaera. pBR322 (indicated by a *thicker line*) was inserted into øC31 ΔMΔ23 (see Fig. 8) (*Suarez* and *Chater*, 1980b). The overall genome length is 42.4 kb. ΔW12 and ΔW17 are deletions discussed in the text (see also Fig. 10)

to, for example, sodium pyrophosphate, invariably contain deletions. Some of the mutants have clear plaques, and in these the deletions are always located near the centre of the linear phage DNA (*Lomovskaya* et al. 1980). Thus the position of the *c* gene close to one end of the genetic map defined by ts mutants is not reflected in the physical map. Deletion mutants with turbid plaques have invariably lost DNA segments near the right-hand end of the DNA. In some of them (*lyg* mutants) the ability to lysogenise stably has been lost, and these have deletions in the *attP* region (*Lomovskaya* et al. 1980).

7.4 A øC31-pBR322 Chimaeric Molecule

Suarez and *Chater* (1980b) described the insertion of pBR322 into site 13 (an *Eco*RI site: Figs. 8 and 9) of a doubly deleted (ΔMΔ23) derivative of øC31 *c*ts1, giving a chimaeric molecule acting as a temperate phage in *Streptomyces* and a multicopy plasmid in *E. coli*. (The detection of the recombinant involved the successful application of the plaque hybridisation technique commonly used in coliphage lambda work: *Benton* and *Davis* 1977). This molecule contains single *Bam*HI and *Pst*I target sites, in its pBR322 segment, which can potentially be used for inserting DNA fragments. Several features of the øC31-pBR322 chimaera have or may have relevance to its use as a progenitor of cloning vectors.

  1. It showed no evidence of DNA rearrangements or deletions when transformed between *E. coli* and *S. lividans*, in either direction.

  2. It is capable of lysogenising *Streptomyces*, though at a rather low efficiency because of the *c*ts 1 mutation (recombinant c^+ derivatives have now been constructed).

  3. It is subject to restriction by *S. albus* P, which contains a restriction enzyme – *Sal*PI – that recognises the *Pst*I target site (*Chater* 1977; *Carter* et al. 1980).

  4. It gives a low e.o.p. and small plaques (the Rm^S phenotype, see below) on *S. albus* G (restrictionless mutants of *S. albus* G have to be used for this test because øC31 DNA contains more than 20 target sites for *Sal*GI, a restriction enzyme normally found in *S. albus* G: *Chater* and *Wilde* 1980).

  5. The *tet* gene of pBR322 causes increased tetracycline resistance in lysogens of *S. albus* strains G and P, from about 0.02 μg/ml to 0.5 μg/ml. Many other streptomycetes are too resistant to tetracycline to allow measurement of the expression of this pBR322 gene (*K.F. Chater, J.E. Suarez* and *C.J. Bruton*, unpublished results).

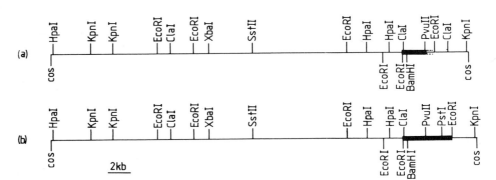

Fig. 10. a, b. Restriction maps of DNA of øC31::pBR322 deletion mutants (*Chater* et al., to be published, b). Thicker lines indicate pBR322 DNA. *a* øC31::pBR322 ΔW12 (37.6 kb); *b* øC31::pBR322 ΔW17 (38.2 kb)

7.5 Deletion Mutants of øC31 *cts* 1ΔMΔ23::pBR322

Deletion mutants of the chimaeric molecule were easily obtained by selecting for pyrophosphate resistance (*Chater* et al. to be published b). All those examined had lost segments between target sites 13 and 20 and the combined results showed that the whole of this 8 kb segment is genetically dispensable for plaque formation. This is consistent with the paucity of essential genes in this region discovered by genetic studies (see above).

A second collection of five deletion mutants was assembled by selecting for large plaque (Rm[R]) phenotype on *S. albus* G. All five (Rm[R]2–6) contained deletions in the region of the *Bam*HI site of pBR322. Of seven more Rm[R] mutants isolated from a derivative of the øC31::pBR322 chimaera already containing a 3.5 kb deletion (ΔW3), six had no apparent deletion and the seventh had a small deletion (smaller than any of the Rm[R]2–6 deletions), again removing the *Bam*HI site.

From these results we were able to make a significant observation: regardless of the size of the starting genome, and despite the fact that a continuous segment of 8 kb (20%) is genetically dispensable, the smallest genome obtained in any set of deletion mutants was about 91.5% of the wild-type genome size. This indicates that at least 37.5 kb of DNA is needed *physically* for the formation of a mature infectious virion. We do not know what is the *maximum* amount of DNA that can be stably inserted into a mature øC31 virion in normal experimental conditions, though the initial øC31::pBR322 chimaera – which is stable – contains c. 3% more than the wild-type øC31.

7.6 Expression of Cloned Genes in øC31 Vectors

A *Bam*HI fragment of 1.9 kb encoding the viomycin phosphotransferase of *S. vinaceus* was subcloned from a primary plasmid clone (*Thompson* et al. to be published a) into the *Bam*HI site of øC31 *c*[+]ΔMΔ23::pBR322ΔW12 (Fig. 10a). (This vector contains a 4.8 kb deletion (ΔW12) removing part of pBR322 and part of the *Eco*RI-C[ΔMΔ23] fragment,

without eliminating functions needed for integration into the chromosome.) Replication of plaques of this phage to spores of a viomycin-sensitive, øC31-sensitive indicator strain leads to viomycin-resistant lysogenic growth corresponding to the plaques. Thus the phenotypic identification of genes cloned into this vector appears to present no problem, given a suitable indicator strain. However, lysogens cannot be formed directly by phages which are deleted for the *attP* region, one of which (ΔW17, Fig. 10b) may be particularly useful as a vector. This problem has been solved by using øC31 lysogens as indicator strains. The resident prophage provides a region of DNA sequence homology into which the superinfecting phage can recombine. Such recombination events occur at a rather low frequency: only one or two recombinant colonies, on average, are obtained when a plaque is replica plated to a lawn of spores of a lysogen. However, it has proved easy to increase this number to 10–100, giving a reliable detection system, by mixing the lysogenic indicator spores with an equal number of otherwise isogenic non-lysogenic spores, which provide a means of amplifying the replicated phages and therefore of increasing the number of superinfected lysogens.

7.7 The Efficiency of a øC31 Vector in "Shotgun Cloning"

In order to establish that DNA can indeed be efficiently introduced into øC31 for the purposes of shotgun cloning, total DNA of *S. coelicolor* A3(2), partially digested sequentially with *Mbo*I and *Pst*I to give a population of fragments largely in the size range 2–8 kb, was ligated with DNA of øC31 *cts*ΔMΔ23::pBR322ΔW17 cut with *Bam*HI and *Pst*I. Of the plaques obtained after transfection with the ligated DNA, more than half contained insertions of donor DNA, in the range 2–6 kb. With removal of the smaller *Bam*HI – *Pst*I segment of the vector prior to its ligation with donor DNA, essentially every plaque contained inserted DNA. High frequencies of insertion were also obtained with *Mbo*I-cleaved DNA into the *Bam*HI site and with *Pst*I-cleaved DNA into the *Pst*I site. Recent experiments aimed at increasing the low level number of plaques obtained in these first attempts at shotgun cloning with this vector have indicated that the addition of liposomes to transfection mixtures containing shotgun ligation mixtures is extremely effective (*R. Rodicio* and *K.F. Chater* unpublished results), and we are now able to obtain enough plaques for essentially complete genome representation using only a few micrograms of vector DNA.

7.8 Summary of Current Cloning Strategy Using øC31-Related Vectors

Our present favoured vector is the ΔW17 deletion of øC31 *cts*1ΔMΔ23::pBR322 (Fig. 10b). The vector DNA can be isolated either from phage preparations or from *E. coli*, where it replicates as a multicopy plasmid (pIJ502). It is prepared by cutting with *Bam*HI and *Pst*I, after which the 3.23-kb pBR322 segment between the *Bam*HI and *Pst*I sites can easily be removed to leave a 35-kb segment with *Bam*HI and *Pst*I ends which can initiate plaques only when a fragment of donor DNA of about 2–8 kb is inserted into it. This necessitates the somewhat tedious preparation of donor DNA partially digested with *Pst*I and an enzyme such as *Mbo*I giving 5′-GATC cohesive ends. The plaques so obtained (on *S. lividans* 66) may be replicated (using velvet) to spores of an appropriate øC31 sen-

sitive indicator strain in a 1:1 lysogen:non-lysogen mixture. This allows the *attP⁻* phage to integrate by homologous recombination into the lysogen and thus the cloned DNA can be detected phenotypically provided that it is capable of phage-independent expression.

7.9 Future Directions

In order to circumvent the use of two restriction enzymes in shotgun cloning, we are currently developing *Bam*HI, *Pst*I and *Xho*I replacement vectors, which contain replaceable segments separating the respective restriction enzyme sites (as in λ replacement vectors: *Williams* and *Blattner* 1979). We also hope to exploit the ability of ØC31 to integrate very stably into the chromosome as a route for stable strain construction.

8 The Evaluation of DNA Cloning in Streptomyces

The streptomycetes are organisms of rich interest for both basic and applied research. In basic research, the most obvious primary objectives are the cloning of genes for antibiotic synthesis and differentiation, but many other areas of ancillary or independent interest will also benefit from the cloning approach, such as: the structure of regulatory regions of the *Streptomyces* genome; the elucidation of primary metabolic regulatory systems such as catabolite repression; the molecular genetics of restriction-modification systems; the unravelling of the genetic basis of various unstable characteristics; and the study of excreted enzymes. We may divide these topics into: those in which the introduction of the cloned gene(s) could in principle be detected in many *Streptomyces* hosts (such as restriction-modification genes, genes for certain exoenzymes and drug-resistance genes) and those in which clone recognition will probably depend on the "complementation" of mutations (such as differentiation genes, antibiotic synthesis genes, regulatory genes and unstable genes). Another categorisation (which fortuitously produces exactly the same division of topics) concerns the question of whether multicopy or low copy number vectors would be more appropriate (often either would be suitable).

Where any host would serve, and copy number is not important, the present vector of choice is pIJ41 (derived from the SLP1.2 plasmid), the host is *S. lividans* 66, and the primary strategy is to clone 5′ GATC-ended fragments into the *Bam*HI site to permit the use of insertional inactivation. In the near future, we envisage the use of vectors derived from ØC31 and pIJ101 for the same purpose, again using *S. lividans* 66 as the host.

Where the use of mutants is demanded, these will (in our work) usually be in *S. coelicolor* A3(2). Here the use of SLP1-related replicons is still not feasible, and one has the choice of three as yet only partially developed vectors, based on: the multicopy plasmid pIJ101; the low copy number plasmid SCP2; and ØC31 derivatives.

For cloning that requires the use of other hosts, evidently the question of the host range of the existing four vector systems will be crucial and will need empirical definition.

Many of these topics may also be relevant to industrial problems, but an approach peculiar to industry is the use of random cloning for strain improvement and for new products (*Hopwood* and *Chater* 1980; *Hopwood* 1981a). In principle, a few thousand colonies from a "self-cloning" experiment using either a production strain or an

undeveloped wild strain should yield clones in which increased copy numbers of genes for rate limiting steps results in a substantial yield increase. The availability of cloned "relevant" genes would facilitate their further manipulation (e.g. by in vitro mutagenesis) and introduction into other independently improved lines. If the cloning experiment is interspecific, then one may hope that genes for enzymes able to introduce variations into the resident antibiotic biosynthetic pathways will be cloned, resulting in the formation of new products (clearly the efficacy of this approach relies greatly on the efficiency of the antibiotic screening). An interesting question for the near future will be whether products traditionally made by other systems, or completely new "foreign" products, might also profitably be made by this benign group of organisms.

References

Aguilar A, Hopwood DA (1981) Isolation and characterization of a plasmid from *Streptomyces violaceus-ruber* and its relation to methylenomycin production. Soc Gen Microbiol Quarterly 8:135–136

Baltz RH (1978) Genetic recombination in *Streptomyces fradiae* by protoplast fusion and cell regeneration. J Gen Microbiol 107:93–102

Baltz RH, Matsushima P (1980) Applications of protoplast fusion, site directed mutagenesis and gene amplification to antibiotic yield improvement in *Streptomyces*. Proceedings Joint US/USSR Seminar on the Genetics of Actinomycetes. U.S. Department of Commerce, Nat Technical Inf Service, pp 124–148

Benigni R, Antonov RP, Carere A (1975) Estimate of the genome size by renaturation studies in *Streptomyces*. Appl Microbiol 30:324–326

Benton WD, Davis RW (1977) Screening λgt recombinant clones by hybridization to single plaques *in situ*. Science, N.Y. 196:180–182

Bérdy J (1980) Recent advances in and prospects of antibiotic research. Process Biochem, Oct/Nov, pp 28–35

Bibb MJ, Freeman RF, Hopwood DA (1977) Physical and genetical characterisation of a second sex factor, SCP2, for *Streptomyces coelicolor*. MGG 154:155–166

Bibb MJ, Ward JM, Hopwood DA (1978) Transformation of plasmid DNA into *Streptomyces* at high frequency. Nature 274:398–400

Bibb MJ, Schottel JL, Cohen SN (1980a) A DNA cloning system for interspecies gene transfer in antibiotic-producing *Streptomyces*. Nature 284:526–531

Bibb MJ, Ward JM, Hopwood DA (1980b) The development of a cloning system for *Streptomyces*. Dev Ind Microbiol 21:55–64

Bibb MJ, Hopwood DA (1981) Genetic studies of the fertility plasmid SCP2 and its SCP2* variants in *Streptomyces coelicolor* A3(2). J Gen Microbiol 126:427–442

Bibb MJ, Ward JM, Kieser T, Cohen SN, Hopwood DA (1981) Excision of chromosomal DNA sequences from *Streptomyces coelicolor* forms a novel family of plasmids detectable in *Streptomyces lividans*. MGG (to be published)

Birnboim HC, Doly J (1979) A rapid alkaline extraction procedure for screening recombinant plasmid DNA. Nucleic Acids Res 7:1513–1523

Carter JA, Chater KF, Bruton CJ, Brown NL (1980) A comparison of DNA cleavage by the restriction enzymes *Sal*PI and *Pst*I. Nucleic Acids Res 8:4943–4954

Chater KF (1977) A site-specific endodeoxyribonuclease from *Streptomyces albus* CMI 52766 sharing site-specificity with *Providencia stuartii* endonuclease *Pst*I. Nucleic Acids Res 4:1989–1998

Chater KF, Hopwood DA (1981) *Streptomyces* genetics. In: Goodfellow M, Mordarski M, Williams ST (eds) Biology of the actinomycetes. Academic Press, London (to be published)

Chater KF, Merrick MJ (1979) Streptomycetes. In: Parish JH (ed) Developmental biology of prokaryotes Blackwell Scientific Publications Oxford pp 93–114

Chater KF, Wilde LC (1980) *Streptomyces albus* G mutants defective in the *Sal*GI restriction-modification system. J Gen Microbiol 116:323–334

Chater KF, Bruton CJ, Suarez JE, Springer W (1981a) *Streptomyces* phages and their applications in DNA cloning. In: Schlessinger D (ed) Microbiology. American Society for Microbiology, Washington DC (to be published a)

Chater KF, Bruton CJ, Suarez JE (1981) Restriction mapping of the DNA of the *Streptomyces* temperate phage ϕC31 and its derivatives. Gene 14:183–194

Chater KF, Bruton CJ, Springer W, Suarez JE (1981b) Dispensable sequences and packaging constraints of DNA from the *Streptomyces* temperate phage ϕ C31. Gene (to be published b)

Chinenova TA, Lomovskaya ND (1975) (Temperature-sensitive mutants of actinophage ϕC31 of *Streptomyces coelicolor* A3(2). Genetika 11:132–141

Dowding JE (1973) Characterization of a bacteriophage virulent for *Streptomyces coelicolor* A3(2). J Gen Microbiol 76:163–176

Girvitz SC, Bacchetti S, Rainbow AJ, Graham FL (1980) A rapid and efficient procedure for the purification of DNA from agarose gels. Anal Biochem 106:492–496

Graham MY, Weisblum B (1979) Ribosomal ribonucleic acid of macrolide-producing streptomycetes contains methylated adenine. J Bacteriol 137:1464–1467

Hepburn AG, Hindley J (1979) Small-scale techniques for the analysis of recombinant plasmids. J Biochem Biophys Methods 1:299–308

Hintermann G, Fischer HM, Crameri R, Hütter R (1981) Simple procedure for distinguishing CCC, OC and L forms of plasmid DNA by agarose gel electrophoresis. Plasmid 5:371–373

Hopwood DA (1981a) Future possibilities for the discovery of new antibiotics by genetic engineering. In: Salton MRJ, Shockman GD (eds) β-lactam antibiotics. Academic Press, New York London pp 585–598

Hopwood DA (1981b) Genetic studies with bacterial protoplasts. Ann Rev Microbiol 35

Hopwood DA, Chater KF (1980) Fresh approaches to antibiotic production. Phil Trans Royal Soc B 290:313–328

Hopwood DA, Wright HM (1973) A plasmid of *Streptomyces coelicolor* carrying a chromosomal locus and its inter-specific transfer. J Gen Microbiol 79:331–342

Hopwood DA, Wright HM (1978) Bacterial protoplast fusion: recombination in fused protoplasts of *Streptomyces coelicolor*. MGG 162:307–317

Hopwood DA, Wright HM, Bibb MJ, Cohen SN (1977) Genetic recombination through protoplast fusion in *Streptomyces*. Nature 268:171–174

Hopwood DA, Thompson CJ, Kieser T, Ward JM, Wright HM (1981) Progress in the development of plasmid cloning vectors for *Streptomyces*. In: Schlessinger D (ed) Microbiology. American Society for Microbiology, Washington DC (to be published)

Kieser T, Hopwood DA, Wright HM, Thompson CJ (1981) A family of multi-copy, broad host-range *Streptomyces* plasmids and their development as DNA cloning vectors (to be published)

Krügel H, Fiedler G, Noack D (1980) Transfection of protoplasts from *Streptomyces lividans* 66 with actinophage SH10 DNA. MGG 177:297–300

Lomovskaya ND, Chater KF, Mkrtumian NM (1980) Genetics and molecular biology of *Streptomyces* bacteriophages. Bacteriol Rev 44:206–229

Makins JF, Holt G (1981) Liposome-protoplast interactions. J Chem Tech and Biotech (to be published)

McMaster GK, Samulski RJ, Stein JL, Stein GS (1980) Rapid purification of covalently closed circular DNAs of bacterial plasmids and animal tumor viruses. Anal Biochem 109:47–54

Okanishi M, Suzuki K, Umezawa H (1974) Formation and reversion of streptomycete protoplasts: cultural conditions and morphological study. J Gen Microbiol 80:389–400

Schottel JL, Bibb MJ, Cohen SN (1981) Cloning and expression in *Streptomyces lividans* of antibiotic resistance genes derived from *Escherichia coli*. J Bacteriol 146:360–368

Schrempf H, Bujard H, Hopwood DA, Goebel W (1975) Isolation of covalently closed circular deoxyribonucleic acid from *Streptomyces coelicolor*. J Bacteriol 121:416–421

Skinner RH, Cundliffe E (1980) Resistance to the antibiotics viomycin and capreomycin in *Streptomyces* which produce them. J Gen Microbiol 120:95–104

Skurray RA, Reeves P (1973) Characterization of lethal zygosis associated with conjugation in *Escherichia coli* K-12. J Bacteriol 113:58–70

Sladkova IA, Lomovskaya ND, Chinenova TA (1977) The structure and size of the genome of actinophage ϕC31 of *Streptomyces coelicolor* A3(2). Genetika 13:342–244

Stuttard CS (1979) Transduction of auxotrophic markers in a chloramphenicol-producing strain of *Streptomyces*. J Gen Microbiol 110:479–482

Suarez JE, Chater KF (1980a) Polyethylene glycol-assisted transfection of *Streptomyces* protoplasts. J Bacteriol 142:8–14

Suarez JE, Chater KF (1980b) DNA cloning in *Streptomyces*: a bifunctional replicon comprising pBR322 inserted into a *Streptomyces* phage. Nature 286:527–529

Suarez JE, Chater KF (1981) Development of a DNA cloning system in *Streptomyces* using phage vectors. Cienc Biol (Portugal) 6:99–110

Thompson CJ, Ward JM, Hopwood DA (1980) DNA cloning in *Streptomyces*: resistance genes from antibiotic-producing species. Nature 286:525–527

Thompson CJ, Ward JM, Hopwood DA (to be published a) Cloning of antibiotic-resistance and nutritional genes in *Streptomyces*.

Thompson CJ, Skinner RH, Thompson J, Ward JM, Hopwood DA, Cundliffe E (to be published b) Biochemical characterisation of resistance determinants cloned from antibiotic-producing streptomycetes.

Thompson CJ, Kieser T, Ward JM, Hopwood DA (to be published c) Physical analysis of antibiotic-resistance genes from *Streptomyces* and their use in vector construction.

Westpheling J (1980) Physical studies of *Streptomyces* plasmids. Ph.D. Thesis, University of East Anglia, Norwich

Williams BG, Blattner FR (1980) Bacteriophage lambda vectors for DNA cloning. In: Setlow JK and Hollaender A (eds) Genetic Engineering: principles and methods. II. Plenum Press, New York, pp 201–281

Gene Cloning in Neurospora crassa

Daniel Vapnek* and Mary Case*

1 Introduction

Although a great wealth of genetic and biochemical information, as well as a transformation system, exist for *N. crassa,* cloning in this organism is still in its infancy. The reasons for this are twofold: a lack of suitable cloning vectors and a lack of cloned genes which can be selected for in *Neurospora.* This review will focus on what is known about the *N. crassa* transformation system and describe some of the approaches that have been used in attempts to develop an efficient cloning system.

2 Transformation in N. crassa

Transformation in *Neurospora* was first described by *Mishra* 1973. However, a reproducible system for transformation was not available until recently. In 1979 *Case* and co-workers demonstrated transformation in *Neurospora* using an *E. coli* chimeric plasmid (*Case* et al. 1979). This plasmid carried the catabolic dehydroquinase gene (*qa-2*) of *N. crassa* cloned in the *Pst*I site of pBR322 (*Alton* et al. 1978). The transformation procedure used was a modification of the PEG-CaCl$_2$ method developed by *Hinnen* et al. (1978) for yeast and gave a low but reproducible level of transformation of a *qa-2⁻* strain to *qa-2⁺*. Recent modifications of the procedure have resulted in significantly higher levels of transformation (10^4 transformants/µg of DNA) (*Schweizer* et al. 1981a).

Genetic and Southern gel analyses of the *qa-2⁺* transformants demonstrated three different types: 1. replacements where the *qa-2⁻* allele had been replaced by the *qa-2⁺* allele; 2. linked insertions where the *qa-2⁺* allele was adjacent to the *qa-2⁻* allele; and 3.

* Department of Molecular and Population Genetics, University of Georgia, Athens, Georgia, 30602, U.S.A.

unlinked insertions were the *qa-2*[+] allele was found in a region of the genome unlinked to the *qa-2* locus. The vector, pBR322, was found to be integrated in about 25% of the linked and unlinked insertions (*Case* et al. 1979).

3 The Quinic Acid Gene Cluster

The catabolic dehydroquinase gene used as a selectable marker in the transformation experiments described in Sect. 2 is part of the quinic acid gene cluster (*qa* cluster). This cluster is involved in the catabolism of quinic acid and encodes three closely linked structural genes (*qa-2, qa-3,* and *qa-4*) and a positively acting regulatory gene (*qa-1*) (*Giles* et al. 1978). The *qa-2* gene was isolated by cloning restriction fragments of *N. crassa* chromosomal DNA into pBR322 followed by transformation into a *aroD⁻* strain of *E. coli* (*Vapnek* et al. 1977). This strain lacks the biosynthetic enzyme which carries out the conversion of dehydroquinic to dehydroshikimic acid in *E. coli*. Transformants able to grow on minimal medium were found to carry the *N. crassa qa-2* gene. Further experiments demonstrated that the catabolic dehydroquinase gene was efficiently transcribed and translated in *E. coli* (*Alton* et al. 1978).

4 Expression of Other Genes

Recently, the other genes of the *qa* cluster have been cloned (*Schweizer* et al. 1981b). Neither *qa-3*[+] (quinate dehydrogenase) nor *qa-4*[+] (5-dehydroshikimate dehydratase) have been found to be expressed in *E. coli*. In addition, negative results were obtained when a gene bank of *N. crassa* chromosomal DNA constructed in a cosmid vector was tested for complementation of a number of different *E. coli* auxotrophs (*Kushner*, personal communication).

A priori it might be expected that yeast, a lower eukaryote like *Neurospora*, would be a better host than *E. coli* for the expression of *N. crassa* DNA. However, this has not proved to be the case. Attempts to obtain expression of the *qa-2* gene in yeast have been unsuccessful (*Huiet*, personal communication), as have preliminary attempts to complement yeast auxotrophs with *N. crassa* DNA.

What are the barriers to the expression of *N. crassa* DNA in other hosts? Although a number of reasons could be postulated, several of the more obvious ones are the presence of intervening sequences in most *N. crassa* genes, translational start or stop sequences which are not efficiently recognized in other organisms, or transcriptional barriers.

Since the structural gene for the *N. crassa* catabolic dehydrogenase is efficiently expressed in *E. coli,* it presumably has no intervening sequences. The nucleotide sequence of this gene and its flanking region is currently being determined (*Alton* and *Vapnek,* personal communication) and should give some insight into the nature of the transcriptional and translational start and stop sites in *Neurospora*.

As mentioned previously, the other two genes of the *qa* cluster are not expressed in *E. coli,* but molecular analysis of these genes is also in progress. Hopefully, this analysis will reveal the nature of the barriers to the expression of these genes in *E. coli* and suggest strategies that could be used to overcome these barriers.

The importance of obtaining expression of additional *Neurospora* genes in other hosts is that such expression would greatly facilitate the development of shuttle vectors. These vectors, which carry genes that can be expressed in two different hosts, have played an important role in the development of the yeast cloning systems (*Struhl* et al. 1979). An approach to the isolation of genes suitable for shuttle vectors is to test *E. coli* (or yeast) genes for expression in *Neurospora*. Two antibiotic resistance genes from *E. coli* are being tested — the chloramphenicol acetyl transferase gene which imparts resistance to chloramphenicol and the aminoglycoside phosphotransferase gene which confers resistance to the 2-deoxystreptamine antibiotic G418. Both of these genes are functionally expressed in yeast and can be used for the direct selection of the resistance phenotype (*Cohen* et al. 1980, *Jiminez* and *Davies* 1980). Experiments to test for the expression of these genes in *N. crassa* have been hampered by the high rate of spontaneous reversion to both chloramphenicol and G418 resistance. To overcome this problem, plasmids carrying both of these markers, in addition to the *N. crassa qa-2* gene, have been constructed (*Vapnek*, unpublished data). These plasmids are currently being tested for their ability to impart resistance to both of these antibiotics in *Neurospora* (*Case* and *Vapnek*, unpublished data).

5 Development of Vector Systems

The second major problem in developing a generalized cloning system in *Neurospora* has been the lack of autonomously replicating plasmids in this organism. As referred to in Sect. 2 plasmid pBR322 is not self-replicating in *Neurospora*.

Three different types of replicons, which can be used as cloning vectors, exist in yeast. These are autonomously replicating sequences (*ars*), the yeast 2-μ circle, and centromeric fragments (*Struhl* et al. 1979; *Stinchcomb* et al. 1980; *Beggs* 1978; *Clarke* and *Carbon* 1980). The *Neurospora* chromosome contains sequences which can function as autonomously replicating sequences in yeast. This was demonstrated by *Stinchcomb* et al. (1980). They ligated restriction fragments of *N. crassa* DNA into plasmid YIp5, a pBR322 derivative that carries the *ura3* allele of yeast but is not capable of autonomous replication in yeast. The mixture was transformed into *S. cerevisiae*, and a plasmid was isolated which was able to undergo autonomous replication in yeast due to the presence of the inserted *Neurospora* DNA (YIp5-*N. crassa ars*). To determine whether this plasmid could replicate autonomously in *Neurospora*, the *qa-2* gene was cloned into the YIp5-*N. crassa ars* plasmid (*Selker*, personal communication). This plasmid was transformed into a *qa-2⁻* strain. Southern gel analysis of the stable *qa-2⁺* transformants indicated that the plasmid had integrated into the *N. crassa* chromosome (*Huiet* and *Case*, personal communication).

Experiments to test both the yeast 2-μ circle and the autonomously replicating centromeric fragments for their ability to function as replicons in *Neurospora* are currently in progress.

Do autonomously replicating plasmids exist in *Neurospora*? Although to our knowledge a systematic search for such plasmids has not been carried out, plasmids have not been detected in normal laboratory strains. Interestingly, a plasmid (3.6 kb) has been isolated from the mitochondrion of a wild-type *N. crassa* strain (*Collins* et al. 1981). A small plasmid (2.4 kb) has also been found in the mitochondrial fraction from the ascomycete *Podospora anserina* (*Stahl* et al. 1980). Although the function of these plasmids is un-

known, it will be of interest to determine whether they can be used as cloning vectors in *Neurospora*.

6 Summary

Gene cloning in *Neurospora* is comparable to the situation that existed in yeast some years ago. However, a number of laboratories are currently pursuing cloning studies with this organism and it will hopefully only be a matter of time before appropriate vectors and selectable genes are available for molecular cloning experiments with *Neurospora*.

Acknowledgments. We thank Layne Huiet for helpful discussions during preparation of this manuscript and Dr. Norman Giles for careful reading of the manuscript.

References

Alton NK, Hautala JA, Giles NH, Kushner SR, Vapnek D (1978) Transcription and translation in *E. coli* of hybrid plasmids containing the catabolic dehydroquinase gene for *Neurospora crassa*, Gene 4:241–259.

Beggs JD (1978) Transformation of yeast by a replicating hybrid plasmid. Nature 275:108–108

Case ME, Schweizer M, Kushner SR, Giles NH (1979) Efficient transformation of *Neurospora crassa* utilizing hybrid plasmid DNA. Proc Natl Acad Sci USA 76:5259–5263

Clark L, Carbon J (1980) Isolation of a yeast centromere and construction of functional small circular chromosomes. Nature 287:504–509

Cohen JD, Eccleshall TR, Needleman RB, Federoff H, Buchferer BA, Marmur J (1980) Functional expression in yeast of the *Escherichia coli* plasmid gene coding for chloramphenicol acetyltransferase. Proc Natl Acad Sci USA 77:1078–1082

Collins RA, Stohl LL, Cole MD, Lambowitz AN (1981) Characterization of a novel plasmid DNA found in mitochondria of *Neurospora crassa*. Cell 24:443–452

Giles NH, Alton NK, Case ME, Hautala JA, Jacobson JW, Kushner SR, Patel VB, Reinert WR, Stroman P, Vapnek D (1978) The organization of the *qa* gene cluster in *Neurospora crassa* and its expression in *Escherichia coli*. Stadler Symposium, vol 10, University of Missouri, Columbia, MO, pp 49–63

Hinnen A, Hicks JB, Fink GR (1978) Transformation of yeast. Proc Natl Acad Sci USA 75:1929–1933

Jimenez A, Davies J (1980) Expression of a transposable antibiotic resistance element in *Saccharomyces*. Nature 287:869–871

Mishra NC, Tatum EL (1973) Non-Mendelian inheritance of DNA induces inositol independence in *Neurospora crassa*. Proc Natl Acad Sci USA 70:3873–3879

Schweizer M, Case ME, Dykstra CC, Giles NH, Kushner SR (1981a) Identification and characterization of recombinant plasmids carrying the complete *qa* gene cluster from *Neurospora crassa* including the *qa-I+* regulatory gene. Proc Natl Acad Sci USA 78:5086–5090

Schweizer M, Case ME, Dykstra CC, Giles NH, Kushner SR (1981b) Cloning of the Quinic Acid (*qa*) Gene Cluster from *Neurospora crassa:* Identification of recombinant plasmids containing both *qa-2+* and *qa-3+*. Gene 14:11–21

Stahl W, Kuck U, Tudzynski PT, Esser K (1980) Characterization and cloning of plasmid-like DNA of the ascomycete *Podospora anserina*. MGG 178:639–646

Stinchcomb DT, Thomas M, Kelley J, Selker E, Davis RW (1980) DNA segments capable of autonomous replication in yeast. Proc Natl Acad Sci USA 77:4559–4563

Struhl K, Stinchcomb DT, Scherer S, Davis RW (1979) High frequency transformation of yeast: autonomous replication of hybrid DNA molecules. Proc Natl Acad Sci USA 76:1035–1039

Vapnek D, Hautala JA, Jacobson JW, Giles NH, Kushner SR (1977) Expression in *Escherichia coli* K-12 of the structural gene for catabolic dehydroquinase of *Neurospora crassa*. Proc Natl Acad Sci USA 74:3508–3512

Vectors for Cloning in Yeast

ALBERT HINNEN* UND BERND MEYHACK*

1 Introduction

Understanding how genes function is an essential part of molecular biology, since the expression of genetic information is the basis of all cellular activities. Unraveling of the mysteries of these fundamental biological processes requires the isolation and characterization of a large number of genes from many different organisms. Molecular cloning techniques allow the isolation of almost any gene sequence provided adequate detection systems can be found (e.g., hybridization probes, antibodies against gene products, biological assay systems). Molecular techniques combined with genetic approaches have led to a detailed understanding of the function of several prokaryotic genes. This is largely due to the fact that isolated genes can be put back into the original host cells by transformation, where defined genetic alterations can be studied. Transformation systems are crucial for our understanding of gene expression because they provide the link between gene structure determined in vitro and gene function assayed in vivo. Until recently, only a few prokaryotic organisms were amenable to this analysis; however, in vitro recombination techniques have rapidly changed this situation. With the help of purified genes, isolated by genetic engineering methods, several eukaryotic transfor-

* Friedrich Miescher-Institut, P.O. Box 273, CH-4002 Basel, Switzerland

mation systems have been established (*Hinnen* et al. 1978; *Wigler* et al. 1979; *Case* et al. 1979; *Beach* and *Nurse* 1981). This review describes the transformation system of the yeast *Saccharomyces cerevisiae*, a eukaryotic fungus. The wealth of genetic and biochemical data which exist for this organism coupled with recombinant DNA technology and the yeast transformation system make *S. cerevisiae* the organism of choice of studying yeast genes in a homologous environment. Furthermore, yeast cells might serve as an expression system for genes of other eukaryotic origin. (For a recent review on the molecular genetics of yeast the reader is referred to *Petes* 1980.)

2 The Transformation System

2.1 Methodology

The yeast transformation protocol is based on the coprecipitation of calcium-treated yeast spheroplasts and transforming DNA by polyethylene glycol. The spheroplasts are plated in a layer of high percentage agar where regeneration of the cells and selection for the transformants occurs.

Rather than provide detailed experimental techniques (*Beggs* 1978; *Hinnen* et al. 1978; *Gerbaud* et al. 1979; *Hsiao* and *Carbon* 1979; *Ilgen* et al. 1979) we prefer to give a few hints which we consider to be of additional help. It is important to note that transformation frequencies vary with strains and can range from no transformants at all to 10^4–10^5 transformants/µg DNA. We have also observed that frequencies can depend on the plasmid type and selection system. The yeast cell wall represents an absolute barrier for the uptake of DNA and it must be removed by enzymatic digestion with glucanases, most being commercially available as crude enzyme mixtures. *Hinnen* et al. (1978) used a crude extract of snail gut enzymes in their original experiments (Glusulase, Endo laboratory, United States). The same type of enzyme mixture (Helicase, Industrie Biologique Française, Clichy, France) was employed by *Beggs* (1978), whereas *Struhl* et al. (1979) obtained good results with lyticase (*Scott* and *Schekman* 1980), a glucanase/protease mixture isolated from *Oerskovia xanthineolytica*. An enzyme preparation isolated from *Arthrobacter luteus* consisting of glucanases, proteases, phosphatases, etc. (Zymolyase-5 000 or Zymolyase-60 000) seems to be related to the lyticase and can be used as well. Often a pretreatment of the cell wall with thiol reagents such as β-mercaptoethanol and dithiotreitol, is necessary to make the glycosidic bonds available for cleavage. Glusulase and Helicase have the advantage that overdigestion is virtually impossible. Zyomylase treatment on the other hand has to be timed more accurately because a long exposure can seriously effect the yeast cell metabolism and thus lead to a poor regeneration of the spheroplasts. Since yeast spheroplasts do not divide, high regeneration frequencies are a prerequisite for optimal transformation. Checking the cell wall digestion by monitoring the osmotic stability of the cells is recommended. This can be done by diluting aliquots in water and measuring the decrease in optical density in a spectrophotometer or simply by analysis under the microscope. The degree of digestion has to be adjusted to give optimal results. Ideally, one tries to produce many spheroplasts able to take up DNA but at the same time allow a high frequency of regeneration to colony-forming yeast cells. Protoplasts devoid of any cell wall remains give low transformation frequencies due to poor regeneration. In addition, good spheroplasts are essential but not sufficient for good

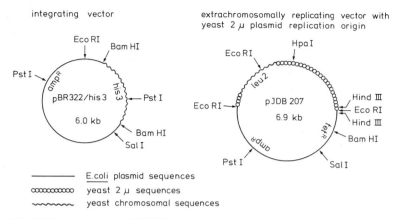

Fig. 1. Integrating vector (pBR322/*his3*) and extrachromosomally replicating 2-µ vector (pJDB207; *Beggs,* personal communication)

transformation. The basis for the ability of a strain to be transformed is not known; however, genetic data from crosses between strains which give high and low frequencies of transformation suggest that several genetic loci are involved.

2.2 Vectors

The plasmids currently in use for yeast transformation are almost exclusively hybrids between *Escherichia coli* plasmid DNA and yeast DNA. Typically, an *E. coli* plasmid replication origin and a selective genetic marker for *E. coli* are combined by in vitro manipulation with a yeast gene which can be selected in yeast. Depending on the mode of replication in yeast two main classes of yeast vectors can be distinguished: integrating vectors and extrachromosomally replicating vectors (Fig. 1).

2.2.1 Integrating Vectors

In the first transformation experiments described by *Hinnen* et al. (1978) yeast cells with a nonrevertible mutation in the *leu2* gene were transformed with the recombinant plasmid pYe*leu*10 (*Ratzkin* and *Carbon* 1977). This hybrid consists of yeast DNA cloned into the bacterial *Co*lE1 plasmid and was known to contain sequences of the wild-type *leu2* gene, for it complemented *E. coli leuB* mutations and hybridized to sequences of yeast chromosome III (*Hicks* and *Fink* 1977). By Southern hybridization analysis (*Southern* 1975) and genetic crosses several types of *Leu*⁺ transformants could be distinguished. In about 70% of all cases the whole pYe*leu*10 plasmid was integrated into the *leu2* region of chromosome III (Fig. 2). The data suggested that a homologous recombination event had taken place leading to a duplication of sequences around *leu2*, 20% of the transformants showed integration into other chromosomal locations and 10% could be interpreted as substitution events. A similar analysis was done with a plasmid containing the *his3* gene (pYe*his*1) (*Ratzkin* and *Carbon* 1977). With this plasmid integration occurred exclusively in the *his3* region of yeast chromosome XV (*Hicks* et al. 1979a). The ability of

Integration of plasmid pYeleu 10 into the leu 2 region of chromosome III

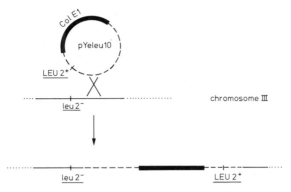

Fig. 2. Integration into a homologous chromosomal region creates a tandem duplication of the cloned yeast sequence

the *leu2*-containing plasmid to integrate into chromosomal locations other than the *leu2* region could later be explained by the presence of repeated sequences on the pYe*leu*10 plasmid (*Kingsman* et al. 1981). All available evidence suggests that these plasmids integrate by a homologous recombination event. In fact, it is even possible to direct integration into a nonyeast sequence. *Hinnen* and *Fink* (unpublished results) have transformed a *his3, leu2* mutant with the *his3*-containing plasmid pYe*his*1, thereby introducing *Col*El sequences at *his3* on chromosome XV. In a subsequent transformation experiment with pYe*leu*10 this plasmid was then found to integrate into the same location by way of the *Col*El homology. This specificity of integration can be used to map cloned genes (*Nasmyth* and *Reed* 1980). In an analogous experiment we have recently mapped yeast sequences which complement a *phoE phoC* double mutation. By recloning the genes into an integrating plasmid with *his3* as the selective marker we could show that the isolated sequences derive from the *phoE phoC* region of chromosome II (see Sect. 3.2).

The integration of plasmids into homologous sequences creates duplications of the structure shown in Fig. 2. These tandem duplications are unstable and can give rise to segregational products at a frequency of 0.1%–1% per cell division. The segregation frequency seems to be correlated with the amount of homology between resident and incoming DNA, extensive homology resulting in higher transformation as well as higher segregation frequencies. An easily detectable segregation product is the appearance of leucine auxotrophic cells arising by a reversion of the integration process. On the molecular level we can assume that a homologous recombination between tandem duplicated sequences leads to a "pop out" of one sequence, including the *E. coli* vector. If this recombination event can occur anywhere between the two repeated regions the end result can be a replacement of endogenous sequences with cloned DNA introduced by transformation. Theoretically all integrating vectors, because of their specific integration behavior, have the potential to replace DNA segments; however, in practice this event can be rather rare and is often not easy to detect. Nevertheless, the principle has been successfully applied by *Scherer* and *Davis* (1979) to construct an internal deletion of the *his3* gene.

Replacement vector (schematic)

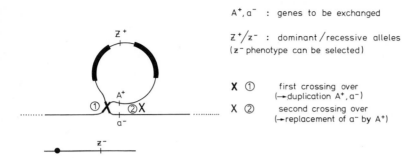

A^+, a^- : genes to be exchanged

z^+/z^- : dominant / recessive alleles
(z^- phenotype can be selected)

X ①   first crossing over
         (→duplication A^+, a^-)

X ②   second crossing over
         (→replacement of a^- by A^+)

Fig. 3. Replacement of the chromosomal a^- allele by the A^+ allele introduced by the replacement vector

The specific integration and excision mechanism seems to be unique for eukaryotic organisms and clearly has important implications for the functional analysis of yeast genes. It provides the possibility of mutating a given promoter or structural gene sequence in a defined manner and then testing the gene activity by replacing the resident gene with the in vitro mutated gene without having the problem of interference from plasmid sequences of other less-defined flanking regions. Also, long-range effects on gene expression, possibly brought about by distortion of nucleosome phasing, often thought to effect the expression of genes introduced in higher eukaryotic cells (*Corden* et al. 1980; *Pellicer* et al. 1980), are thus completely eliminated. The only disadvantage of this method is the relatively low frequency of segregation. It is, however, possible to use genetic methods to select for pop outs. Since the loss of one of the two repeated sequences is always accompanied with a loss of the *E. coli* vector sequences between them, we can exert selective pressure against sequences between the repeats and thereby concomitantly lose one of the two repeated areas. This is schematically shown in Fig. 3. Every gene from which recessive mutations can be selected is a potential genetic marker for a "replacement vector". If the dominant wild-type allele of the "replacement marker", Z^+, is combined with the yeast DNA sequence A^+ to be placed into the genome, the initial transformation event will result in an integration of the whole plasmid adjacent to a^- (under appropriate conditions, such as low extent of homology between Z^+ and z^-, integration can be directed primarily into the a^- region). Having the selectable recessive mutation z^- present in the yeast genome will now allow selection for a secondary segregation event. Of course, the exact location of the second crossing over cannot be predetermined, which means that the outcome can be the reconstruction of the original situation or the transfer of the mutation into the genome. Some auxotrophic mutations can be obtained selectively (*lys2*, uracil auxotrophs) and are therefore potential genetic markers for replacement vectors. *Lacroute* (personal communication) has successfully worked with a vector containing the *ura3$^+$* gene in a *ura3$^-$* genetic background. We have tried to make use of the cloned *sup4* gene (ocher suppressing tRNA gene, *Goodman* et al. 1977), combining it with a suppressible canavanine resistance (*canl-100*) mutation in the host genome. However, this selection system produced a high background of gene

conversions which obscured the results. Gene conversions are encountered frequently in yeast under any type of heterozygous condition. They affect the integration-excision cycle and can lead to an undirectional transfer of genetic information. They can, e.g., change the heterozygous +/– configuration in the duplicated area to +/+ or –/– without any loss of the intervening *E. coli* sequence or give rise to transformants without incorporation of additional vector sequences (*Klein* and *Petes* 1981).

2.2.2 Extrachromosomally Replicating Vectors

The first yeast transformation experiments with integrating plasmids gave relatively low transformation efficiencies (1–10 transformants/µg DNA). However, several laboratories soon realized that certain types of recombinant plasmids showed transformation frequencies about two or three orders of magnitudes higher than pYe*leu*10 (*Beggs* 1978, *Hicks* et al. 1979a; *Struhl* et al. 1979). Plasmids with a high frequency of transformation can be grouped into two categories, which are discussed separately below.

2.2.2.1 Vectors with Yeast 2 µ Plasmid Sequences

Plasmid pYe*leu*10 is always found covalently linked to chromosomal DNA, suggesting that the integration step is crucial for the maintenance of the transformed phenotype. In addition, the genetic data presented by *Hinnen* et al. (1978) clearly demonstrate the classical Mendelian 2:2 segregation of chromosomally inherited eukaryotic genes. At the same time, however, it seemed likely that integration is not needed for genetic expression. Evidence for complementation without recombination comes from the appearance of abortive transformants, i.e., small colonies on transformation plates with pYe*leu*10 and other integrating plasmids, which are unable to grow any larger and do not grow when transferred to new plates (Fig. 4, *Hicks* et al. 1979a) The amount of abortive transformants outnumbers the stably transformed cells by a factor of about 100. This behavior suggested the need for a replication apparatus and at the same time pointed to the integration event as the rate-limiting step in the transformation process. Since most yeast strains harbor a small plasmid of the length of about 2 µ (*Hollenberg* et al. 1970; *Royer* and *Hollenberg* 1977; *Kielland* 1980), which is maintained independently of the yeast chromosomes, it became obvious to attach the presumptive 2-µ plasmid replication origin to a selective yeast marker. This approach was used independently by *Beggs* (1978) and *Hicks* et al. (1979a), and it marked the beginning of the construction of a wide variety of yeast cloning vectors. At first it seemed rather striking that virtually any part of the 2-µ plasmid gave more or less the same high level of transformation frequency and the question whether the yeast plasmid contained more than one replication origin was discussed. By the use of 2-µ plasmid-free recipient cells it became clear that only a unique replication origin was present (*Broach* and *Hicks* 1980; *McNeil* et al. 1980; *Hollenberg* this volume). Since most laboratory yeast strains carry 50–100 2 µ plasmid copies, hybrids without 2-µ replication origins can efficiently integrate into endogenous 2-µ circles via homologous recombination to yield high transformation frequencies. Southern analysis (*Beggs* 1978) clearly showed supercoiled copies of the transforming plasmid and genetic analysis (*Hicks* et al. 1979a) revealed a $4^+:0^-$ segregation pattern indicative of plasmid-mediated inheritance. Both results showed that the recombinant plasmids behave like self-replicating units which are not normally found integrated into the chromosomes.

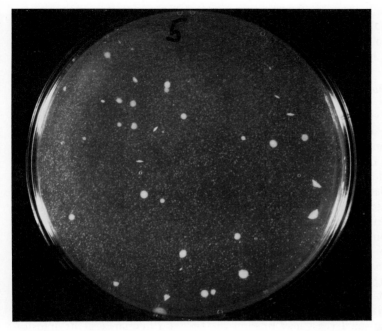

Fig. 4. Agar petri dish showing yeast colonies transformed with 5 μg of integrating pYe*his*4 plasmid (*Hinnen* et al. 1979). Large colonies (*stable transformants*) appear in a background of *small abortive transformants*

The exception to this rule has been presented by *Holmberg* et al. (1980) who found chromosomally integrated into 2-μ sequences. The transformants, however, were extremely unstable and gave rise to gross chromosomal aberrations. The available evidence suggests that integrated 2-μ DNA creates an element of instability which might be attributed to the inverted repeat structure present in these molecules.

All yeast vectors constructed on the basis of 2-μ DNA are associated with two types of instability: Firstly, they tend to give rise to sequence rearrangements (structural instability), especially in hybrids with large inserts. Recombination between incoming and resident 2-μ sequences are frequently observed (*Dobsen* et al. 1980a, b), these events being a prerequisite for the maintenance of 2 μ vectors without replication origin. Secondly, the transformed phenotype can be lost at relatively high frequencies (segregational instability). The loss is in most cases due to the loss of the whole hybrid plasmid and can vary between frequencies of about 1%–5% segregation per cell division for plasmid pJDB207 (Fig. 1) (*Beggs*, personal communication) and 40% for plasmid pYep6 (*Struhl* et al. 1979).

It is likely that segregational instability is associated with copy number, since plasmid pJDB207 has a copy number of about 50, whereas with plasmid pYep6 only 5–10 copies/cell are found. This means that pYep6 should always be kept in yeast under selective conditions, whereas pJDB207 can be maintained for many generations in a complex medium. It should be emphasized, however, that instability is increased with additionally inserted DNA. Segregational instability can be substantially reduced by using 2-μ-free

mutants as host cells (*Blanc* et al. 1979). A plausible explanation for this behavior is that endogenous 2-μ DNA competes for replication with the incoming transforming DNA. *Hollenberg* (this volume) has devised a method to obtain 2-μ-free strains by "outreplicating" endogenous 2-μ copies with hybrid 2-μ plasmids introduced by transformation.

2.2.2.2 Vectors with Chromosomal Replication Origins

A second type of vector with a high frequency of transformation has been described by *Struhl* et al. (1979) and *Kingsman* et al. (1979). They found that a hybrid consisting of the yeast *trpl* gene cloned into a bacterial plasmid gave transformation frequencies of 10^3–10^4 transformants/μg DNA. The Trp^+ phenotype in these transformants was extremely unstable (95%–99%) loss of Trp^+ cells after ten generations) and the loss of the Trp^+ character was always associated with the loss of the entire *E. coli*/yeast hybrid. Very infrequently the hybrid integrated into a yeast chromosome leading to stable transformation. Normally, however, no integrated copies of these hybrids could be detected, suggesting the ability for autonomous replication. It was possible to correlate the "high frequency of transformation" behavior with a specific yeast DNA sequence linked to *trpl* (*Stinchcomb* et al. 1979; *Tschumper* and *Carbon* 1980). These molecules behave like self-replicating units and are thought to contain their own replication origin. This interpretation was substantiated by the fact that other plasmids could be found with similar properties. *Hsiao* and *Carbon* (1979) showed that a hybrid *Col*El plasmid containing DNA from the yeast *arg4* region transformed yeast *arg4* mutants to Arg^+ at a frequency of 10^3 transformants/μg. In addition, it is possible to insert randomly cut yeast DNA into an integrating yeast vector and to select for functional replication origins simply by yeast transformation (*Beach* et al. 1980; *Stinchcomb* et al. 1980). The sequences which provide the replication function are obviously frequent enough to be picked up without a significant background of integrating molecules. This approach has even allowed the isolation of DNA sequences from various other eukaryotes (*Neurospora crassa, Dictyostelium discoideum, Caenorhabditis elegans, Drosophila melanogaster, Zea mays*) which have the property to act as replication origins in yeast (*Stinchcomb* et al. 1980). The question of whether these sequences behave as true replication origins in their natural environment is still unanswered. In specific instances, known replication origin sequences from eukaryotic organisms have failed to function in yeast (cauliflower mosaic virus; B. *Hohn*, personal communication).

An artificial circular minichromosome has recently been constructed by *Clarke* and *Carbon* (1980a). In the process of shotgun cloning of the centromere-linked *cdc*10 locus with a yeast vector containing a chromosomal replication origin they were able to isolate the centromere of chromosome III. The plasmid obtained behaves mitotically and meiotically like a minichromosome leading to a stably inherited additional linkage group. Such a plasmid has interesting properties for gene cloning in yeast, since it allows the stable introduction of a single gene dosage on a vector which can be easily manipulated.

2.3 Selection System

It is clear from the preceding section that any yeast gene which can be selected can act as a selective marker in a yeast transformation experiment. Since the first yeast genes

isolated were genes coding for steps in amino acid biosynthesis, these genes have been used extensively in the construction of yeast vectors (*Struhl* et al. 1976; *Ratzkin* and *Carbon* 1977; *Bach* et al. 1979; *Struhl* et al. 1979). By combining these vectors with the corresponding auxotrophic recipient strains, several host vector systems have been developed.

The selective pressure which can be applied in these systems depends entirely upon the reversion frequencies of the auxotrophic mutations. Unfortunately, these mutations usually have reversion frequencies in the order of 10^{-6}–10^{-7} resulting in the appearance of a sizeable number of revertants, which can be a serious obstacle to the identification of transformed cells, especially when integrating vectors are needed. To overcome this problem stable mutations have been constructed for most of the commonly used auxotrophic markers. Using genetic tricks double mutations have been obtained for *leu2* and *his3* (*Hinnen* and *Fink*, unpublished) and a triple mutation is available for *ura3* (*Lacroute*, personal communication). In addition, in vitro constructed deletions for *his3* (*Scherer* and *Davis* 1979) and *ura2* (*Potier*, personal communication) have been inserted into the yeast genome by replacement of the wild-type gene (see under Sect. 2.2.1). Selection based on these stable mutations is free of the problem of reversion but obviously has the disadvantage of requiring a genetic cross for the auxotrophy to be introduced into a specific genetic background. Drug resistance would alleviate this requirement but yeast drug resistance markers which are dominant are not available at present. The lack of suitable yeast resistance genes has prompted some groups to look into the possibility of using *E. coli* genes. Since several bacterial resistance genes have been found to be expressed in yeast (*Chevallier* and *Aigle* 1979; *Hollenberg* 1979; *Cohen* et al. 1980; *Jimenez* and *Davies* 1980), this seems to be a promising alternative. A potentially very useful system has been worked out by *Jimenez* and *Davies* (1980), who used the kanamycin-related aminoglycoside G418 in combination with the *E. coli* transposon Tn*601*. Yeast growth is completely inhibited by 500 µg/ml G418, but transformation with Tn*601* inserted into a 2-µ vector allows the production of enough phosphotransferase activity to destroy the drug (for more details see *Hollenberg*, this volume).

2.4 Double Transformation

In a double transformation experiment two DNA molecules carrying different genetic markers interact with a single cell. When the transformation frequencies are expressed relative to the number of competent cells, the double transformation frequency should be the product of the two single transformation frequencies. However, when yeast spheroplasts are transformed simultaneously with two different self-replicating vectors, the double transformation frequency is usually in the range of 10%–50% of the frequency of a single transformation experiment. High cotransformation frequencies are indicative of a low level of competence in a given cell population (for yeast, about 0.01%–0.1% of the cells).

The high degree of cotransformation can be exploited in many ways. If a gene on a yeast vector with low transformation efficiency (e.g., an integrating vector) needs to be introduced into a yeast strain at a high efficiency, it can be cotransformed together with the transformation vector at high frequency (e.g., the 2-µ vector). If the two different vectors have DNA sequences in common (these can be bacterial sequences present on the

vectors), recombination can take place leading to a single extrachromosomally inherited hybrid. This is at the same time a very efficient way to transfer genetic information from a single copy vector to a high copy vector. It is interesting to note that only one of the two vectors needs a selectable genetic marker. This procedure therefore provides a quick way to introduce genes cloned on a "pure" *E. coli* plasmid into yeast with the help of suitable "yeast helper vectors".

We have observed that cotransformation with DNA which is not selected for can greatly increase transformation efficiency. Adding a 20-fold excess of a *leu2*-containing integrating vector reproducibly increases the transformation frequency obtained with the 2-μ vector pYep6 by a factor of 10. We do not know at the moment whether there is a specificity with respect to the nature of the helper DNA.

2.5 Transformation with Linear DNA

When hybrid yeast vectors are linearized by cutting with restriction enzymes, a different effect is observed for integrating vectors and 2-μ vectors. Surprisingly, integrating vectors such as pYe*leu*10 (*Ratzkin* and *Carbon* 1977; *Hinnen* et al. 1978) give 5- to 20-fold higher transformation frequencies when cut with a variety of restriction enzymes (*Hicks* et al. 1979). Although conventional models predict a double crossover event to integrate a linear piece of DNA compared to a single crossover event for a circular molecule, similar observations have been made with prokaryotic transformation systems (*Duncan* et al. 1977). The situation is reversed for self-replicating 2-μ vectors. After linearizing the frequency of transformation is reduced, however, only to about 30%–50%. This shows an extremely high capacity of the yeast cell for religation. We have even observed that nonhomologous ends (*Bam*HI/*Sal*I) can be sealed with similar efficiencies as homologous ends (*Bam*HI/*Bam*HI or *Sal*I/*Sal*I).

3 Gene Isolation and Genetic Expression of Cloned Genes in Yeast

3.1 Choosing the Right Vector

All the vectors described above can serve as vehicles for inserting genetic material into yeast and thus for isolating yeast genes by complementation of yeast mutations (*Broach* et al. 1979; *Hicks* et al. 1979b; *Hinnen* et al. 1979; *Clarke* and *Carbon* 1980; *Guerry-Kopecko* und *Wickner* 1980; *Nasmyth* and *Reed* 1980; *Nasmyth* and *Tatchel* 1980; *Williamson* et al. 1980; *Fried* and *Warner* 1981). The type of yeast vector one chooses is clearly dictated by the specific cloning problem. If a high level of expression of an already isolated gene is desired, a 2-μ derived vector with high copy number is advisable. As discussed under Sect. 2.2.2, the stability of these hybrids can be quite good (especially in a 2-μ free host), although larger molecules can rearrange to some extent and a selection for smaller (e.g., faster replicating) derivatives can occur. At the moment, structural and segregational stability is best with integrating vectors. This has to be weighed, however, against a low copy number (1 per genome) and a low transformation frequency. For the isolation of genes from gene libraries a vector with a high frequency of transformation should be used, because a high yield of transformants is preferable to stability in the first step of a cloning ex-

periment. This is particularly true if selective pressure resides entirely on the desired gene. With high copy number vectors potential deleterious effects from high gene dosage also have to be considered and might in some cases favor the use of a low copy number 2-µ vector. *Gallwitz* (personal communication) has found that a functional yeast actin gene cannot be introduced into yeast on a high copy number 2-µ vector. Of course, as with all isolation systems based on direct complementation, the structure of a positive clone has to be compared with the structure obtained under nonselective conditions.

3.2 Isolation of Two Acid Phosphatase Structural Genes by Complementation

In this section we present as an example the protocol we used for the cloning of two genes of the yeast acid phosphatase system. *phoE* and *phoC* code for a repressible and a constitutive yeast acid phosphatase, respectively (*Toh-e* et al. 1975; *Bostian* et al. 1980). The repressible gene (*phoE*) is turned on only under inorganic phosphate starvation, whereas the genetically linked *phoC* gene is expressed constitutively. Our primary interest in this system stems from the unusually high extent of derepression of *phoE* (but not *phoC*) under appropriate physiological conditions, which makes this gene system an ideal candidate for regulatory studies. Since *phoE* and *phoC* are genetically linked, we hoped to be able to isolate both genes in the same cloning experiment. Formally, we had to construct a yeast gene library with wild-type yeast DNA cloned into a 2-µ vector and transform a *phoE⁻, phoC⁻* yeast host to *PHOE⁺, PHOC⁺*. Unfortunately, direct transformation from *pho⁻* to *PHO⁺* is not possible, since *phoE* and *phoC* are not essential for growth and an indirect approach had to be chosen. We decided to insert randomly cloned yeast DNA into a *phoE⁻, phoC⁻* background using an amino acid gene as a selection marker and then to test the transformants for the complementation of the phosphatase mutations using an in vivo staining assay. Since the number of yeast cells which have to be screened is inversely proportional to the size of the yeast DNA insert, we made use of the cosmid cloning technology developed by *Collins* and *Hohn* (1978) and adapted for yeast by *Hohn* and *Hinnen* (1980). Briefly, this technique allows cloning of very large DNA fragments (30–40 kb) with the help of the bacteriophage λ packaging system. When cosmids (i.e., plasmids containing the cohesive end site of λ) and DNA fragments are ligated under appropriate conditions, long concatemers consisting of vector and inserts are produced. These can be packaged in vitro into λ phage heads, provided the cohesive end regions present on the cosmids are positioned in the correct orientation about 40–50 kb apart. This last requirement leads to the selective cloning of hybrids of approximately λ size. By comparing the yeast genome size of about 14 000 kb (*Lauer* et al. 1977) with the cosmid insert size of 30 kb or more, we can conclude that under the assumption of random cutting of the inserted DNA 500 clones correspond to 1 genome equivalent. This means a 63% probability of having any particular gene represented in these 500 hybrids or a 95% probability of finding any specific yeast gene among 1 500 clones (*Clarke* and *Carbon* 1976). This is a very manageable figure compared with 5 000–10 000 clones for a conventional plasmid yeast gene bank and it allows even quite elaborate screening for specific hybrids. In the case of the *phoE* gene the yeast transformants had simply to be induced for phosphatase activity, by transferring the colonies to a phosphate-free medium by replica plating prior to the in vivo staining assay. (If only 5 ml of agar is used 80%–100% of yeast transformants growing inside the regeneration agar can be transferred to a new petri dish by replica plating.)

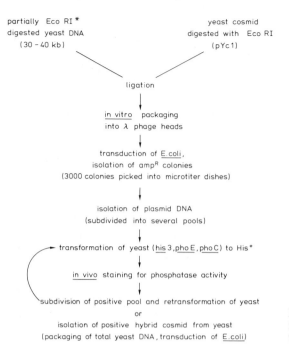

partially Eco RI*
digested yeast DNA
(30 – 40 kb)

yeast cosmid
digested with Eco RI
(pYc1)

ligation

in vitro packaging
into λ phage heads

transduction of E.coli,
isolation of amp^R colonies
(3000 colonies picked into microtiter dishes)

isolation of plasmid DNA
(subdivided into several pools)

transformation of yeast (his 3, pho E, pho C) to His⁺

in vivo staining for phosphatase activity

subdivision of positive pool and retransformation of yeast
or
isolation of positive hybrid cosmid from yeast
(packaging of total yeast DNA, transduction of E.coli)

Fig. 5. Cloning scheme for the isolation of *phoE* and *phoC* by yeast transformation

An outline of our cloning procedure is given in Fig. 5. We constructed a yeast gene bank by partially digesting high molecular weight yeast DNA (200 kb) with restriction enzyme *Eco*RI under *Eco*RI* conditions (*Mayer* 1978) and ligating the size fractionated DNA with the yeast cosmid pYcl (*Hohn* and *Hinnen* 1980). After transduction of bacterial strain HB101 we picked 3000 colonies into microtiter dishes and stored them at –70 °C in glycerol. From this gene library the bacterial colonies were grown by replica plating onto fresh agar plates, cells were scraped off the plates and plasmid DNA isolated by conventional techniques (*Guerry* et al. 1973). Pools of 500 plasmids thus isolated batchwise were used to transform yeast strain AH216 (*his3*-11, 3-15, *leu2*-3, 2-112, *phoE, phoC*) to His⁺. Afterwards the colonies were checked for phosphatase activity and, in a first round of screening, three positive clones were obtained. (A more detailed account of this work will be given elsewhere by *Meyhack* and *Hinnen*, manuscript in preparation.)

3.3 Characterization of the phoE phoC Region by in Vivo Analysis

All three isolated clones had mol. wts. of 40–50 kb, and segregation analysis could assign the *Pho⁺* phenotype to a plasmid location. This was shown by growing the *Pho⁺* transformants under nonselective conditions which led to a simultaneous loss of the cloned *phoE* complementing DNA sequence and the *his3* plasmid marker.

By subcloning restriction fragments of the cosmid hybrid and subsequent testing by yeast transformation a single 5 kb *Bam*HI restriction fragment which contained *phoE*

and *phoC* complementing activity could be identified. Since complementation of a mutational defect is not sufficient evidence for the cloning of a gene, we determined the genomic location of the cloned DNA fragment. This was done by cloning *phoE* onto an integrating yeast vector with *his3* as selective marker. The transformants resulting with such a hybrid would be expected to have the whole plasmid integrated into a homologous genomic region. Since *his3* and the cloned DNA fragment were the only sequences of genomic origin, the hybrid could only integrate into one of these two regions. From size considerations (5 kb for *phoE phoC*, 1.7 kb for *his3*) it could be assumed that most transformants contained plasmid copies integrated into sequences homologous to the cloned DNA. By crossing two of the transformants obtained with suitable tester strains we could clearly establish a close linkage relationship between the cloned yeast DNA and the *phoE* locus on chromosome II with the *his3* plasmid marker located at the same site.

Since genetic data show close linkage of *phoE* and *phoC*, we tried to localize the two genes on the *Bam*HI fragment and to separate them by molecular subcloning. We partially digested the purified *Bam*HI fragment with restriction enzyme *Sau*3A and constructed a series of overlapping subclones with the high copy number 2-μ vector pJDB207 (Fig. 1). When tested for *phoC* activity only clones containing *Sau*3A fragment A were positive. In addition, a regulatory response to phosphate was only obtained in the presence of *Sau*3A fragments A and B. Further subcloning with other restriction enzymes established the location of *phoE* and *phoC* on the *Bam*HI restriction fragment (Fig. 6).

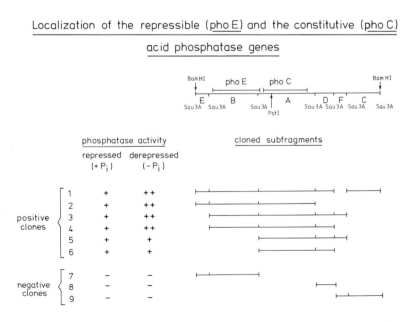

Fig. 6. Acid phosphatase activity of subfragments generated by partial digestion of a 5 kb *Bam*HI restriction fragment with restriction endonuclease *Sau*3A cloned into high copy number 2-μ vector pJDB207 (Fig. 1)

3.4 Functional Complementation of Yeast Mutations with Foreign DNA

Procedures of isolating eukaryotic genes generally depend upon the availability of naturally enriched mRNA species, most often found in specialized tissues or during specific developmental stages. A potential procedure for the cloning of eukaryotic genes which give rise to less abundant mRNAs is given by the possibility of functionally complementing specific yeast mutations with a population of genomic foreign DNA inserted into a yeast vector. Choosing this approach *Henikoff* et al. (1981) have been able to clone a *Drosophila* adenine gene. From a *Drosophila* gene library constructed with a yeast 2-μ vector a clone was obtained which was able to transform an *ade8⁻* yeast strain to *ADE8⁺*. It was convincingly shown that the expression in yeast derives from a cloned piece of *Drosophila* DNA which produces a 0.8 kb poly A-containing RNA responsible for *ADE8⁺* function. Physical characterization of the transcripts by S1 mapping (*Berk* and *Sharp* 1977) could not reveal any intervening sequences. By in situ hybridization a single site on a *Drosophila* salivary gland chromosome could be labeled which corresponds well to an area where adenosine auxotrophic mutations have been mapped. It is important to note that only one of six yeast adenine mutations could be complemented with *Drosophila* DNA and it seems unlikely that this heterologous complementation procedure is generally applicable. *Pardo* (personal communication) likewise has unsuccessfully tried to isolate the genes involved in the rudimentary locus of *Drosophila* by complementation of a yeast *ura2* mutation. One possible barrier for the expression of foreign genes in yeast could be the presence of intervening sequences. *Beggs* et al. (1980) have clearly shown that the rabbit β-globin gene is incorrectly spliced in yeast. On the other hand, yeast should have the potential for RNA splicing, since *Gallwitz* and *Sures* (1980) have found that the yeast actin gene contains an intervening sequence.

Acknowledgments. We thank all those who contributed results prior to publication. Many thanks also to *I. Mattaj* for critical comments on the manuscript and *D. Martin* for skilful technical assistance.

References

Bach ML, Lacroute F, Botstein D (1979) Evidence for transcriptional regulation of orotidine-5'-phosphate decarboxylase in yeast by hybridization of mRNA to the yeast structural gene cloned in Escherichia coli. Proc Natl Acad Sci USA 76:386–390

Beach D, Piper M, Shall S (1980) Isolation of chromosomal origins of replication in yeast. Nature 284:185–187

Beach D, Nurse P (1981) High frequency of transformation of the fusion yeast Schizosaccharomyces pombe. Nature 290:140–142

Beggs JD (1978) Transformation of yeast by a replicating hybrid plasmid. Nature 275:104–109

Beggs JD, Van den Berg J, Van Ooyen A, Weissmann C (1980) Abnormal expression of chromosomal rabbit β-globin gene in Saccharomyces cerevisiae. Nature 283:835–840

Berk AJ, Sharp PA (1977) Sizing and mapping of early adenovirus mRNAs by gel electrophoresis of S1 endonuclease-digested hybrids. Cell 12:721–732

Blanc H, Gerbaud C, Slowinski PP, Guerineau M (1979) Stable yeast transformation with chimeric plasmids using a 2 μm-circular DNA-less strain as a recipient. MGG 176:335–342

Bostian KA, Lemire JM, Cannon LE, Halvarson HO (1980) In vitro synthesis of repressible yeast acid phosphatase: identification of multiple mRNAs and products. Proc Acad Sci USA 77:4504–4508

Broach JR, Strathern JN, Hicks JB (1979) Transformation in yeast: development of a hybrid cloning vector and isolation of the CAN 1 gene. Gene 8:121–133

Broach JR, Hicks JB (1980) Replication and recombination functions associated with the yeast plasmid, 2 μ circle. Cell 21:501–508

Case ME, Schweizer M, Kushner SR, Giles NH (1979) Efficient transformation of *Neurospora crassa* by utilizing hybrid plasmid DNA. Proc Natl Acad Sci USA 76:5259–5263

Chevallier MR, Aigle M (1979) Qualitative detection of penicillinase produced by yeast strains
carrying chimeric yeast-coli plasmids. FEBS Lett 180:179–180

Clarke L, Carbon J (1976) A colony bank containing synthetic Col El hybrid plasmids representative of the entire *E. coli* genome. Cell 9:91–99

Clarke L, Carbon J (1980a) Isolation of a yeast centromere and construction of a functional small
circular chromosome. Nature 287:504–509

Clarke L, Carbon J (1980b) Isolation of the centromere-linked CDC10 gene by complementation
in yeast. Proc Natl Acad Sci USA 77:2173–2177

Cohen JD, Eccleshall TR, Needleman RB, Federoff H, Buchferer BA, Marmur J (1980) Functional
expression in yeast of the *Escherichia coli* plasmid gene coding for chloramphenicol acetyltransferase. Proc Natl Acad Sci USA 77:1078–1082

Collins J, Hohn B (1978) Cosmids: a type of plasmid gene-cloning vector that is packageable in
vitro in bacteriophage λ heads. Proc Natl Acad Sci USA 75:4242–4246

Corden J, Wasylyk B, Buchwalder A, Sassone-Corsi P, Kedinger C, Chambon P (1980) Promoter
sequences of eukaryotic protein-coding genes. Nature 209:1406–1413

Dobsen MJ, Futcher AB, Cox BS (1980a) Control of recombination within and between DNA plasmids of *Saccharomyces cerevisiae*. Curr Genetics 2:193–200

Dobsen JM, Futcher AB, Cox BS (1980b) Loss of 2-μm DNA from *Saccharomyces cerevisiae* transformed with the chimeric plasmid pJDB219. Curr Genetics 2:201–206

Duncan CH, Wilson GA, Young FE (1977) Transformation of *Bacillus subtilis* and *Escherichia coli*
by a hybrid plasmid pCDI. Gene 1:153–160

Fried HM, Warner JR (1981) Cloning of yeast gene for trichodermin resistance and ribosomal protein L3. Proc Natl Acad Sci USA 78:238–242

Gallwitz D, Sures I (1980) Structure of a split yeast gene: complete nucleotide sequence of the actin
gene in *Saccharomyces cerevisiae*. Proc Natl Acad Sci USA 77:2546–2550

Gerbaud C, Fournier P, Blanc H, Aigle M, Heslot H, Guerineau M (1979) High frequency of yeast
transformation by plasmids carrying part of entire 2-μm yeast plasmid. Gene 5:233–253

Goodman HM, Olson MV, Hall BD (1977) Nucleotide sequence of a mutant eukaryotic gene: the
yeast tyrosine-inserting ochre suppressor SUP4-o. Proc Natl Acad Sci USA 74:5453–5457

Guerry P, LeBlanc DJ, Falkow S (1973) General method for the isolation of plasmid deoxyribonucleic acid. J Bacteriol 116:1064–1066

Guerry-Kopecko P, Wickner RB (1980) Cloning of the URA1 gene of *Saccharomyces cerevisiae*. J
Bacteriol 143:1530–1533

Henikoff S, Tatchel K, Hall BD, Nasmyth KA (1981) Isolation of a gene from *Drosophila* by complementation in yeast. Nature 289:33–37

Hicks J, Fink GR (1977) Identification of chromosomal location of yeast DNA from hybrid plasmid pYeleu10. Nature 269:265–267

Hicks J, Hinnen A, Fink GR (1979a) Properties of yeast transformation. Cold Spring Harbor, Symp
Quant Biol 43:1305–1313

Hicks J, Strathern JN, Klar AJS (1979b) Transposable mating type genes in *Saccharomyces cerevisiae*. Nature 282:478–483

Hinnen A, Hicks JB, Fink GR (1978) Transformation of yeast. Proc Natl Acad Sci USA 75:
1929–1933

Hinnen A, Farabaugh PJ, Ilgen C, Fink GR (1979) Isolation of a yeast gene (his4) by transformation
of yeast. In: Axel R, Maniatis T, Fox M (eds) Eukaryotic gene regulation, vol 14. Academic
Press, New York, pp 43–51

Hohn B, Hinnen A (1980) Cloning with cosmids in *E. coli* and yeast. In: Setlow JK, Hollaender A
(eds) Genetic engineering principles and methods, vol 2. Plenum Press, New York London
169–183

Hollenberg CP, Borst P, van Bruggen EFJ (1970) Mitochondrial DNA. V. A 25 μ closed circular
duplex DNA molecule in wild-type yeast mitochondria. Structure and genetic complexity.

Biochim Biophys Acta 209:1-15

Hollenberg CP (1979) The expression in *Saccharomyces cerevisiae* of bacterial β-lactamase and other antibiotic resistance genes integrated in a 2-µm DNA vector. ICN-UCLA Symp Mol Cell Biol 15:325-338

Holmberg S, Nilsson-Tillgren T, Petersen JGL, Kielland-Brandt MC (1980) An acentric fragment of chromosome III formed by sister-strand recombination of integrated 2-micron DNA. Abstracts 10th Int Conf of Yeast Genetics and Molecular Biology, Louvain-la-Neuve Belgium, p 34

Hsiao CL, Carbon J (1979) High frequency transformation of yeast by plasmids containing the cloned ARG4 gene. Proc Natl Acad Sci USA 76:3829-3833

Ilgen C, Farabaugh PJ, Hinnen A, Walsh JM, Fink GR (1979) Transformation of yeast. In: Setlow JK, Hollaender A (eds) Genetic engineering, principles and methods, vol 1. Plenum Press, New York London, p 117

Jimenez A, Davies J (1980) Expression of transposable antibiotic resistance element in *Saccharomyces cerevisiae*; a potential selection for eukaryotic cloning vectors. Nature 287:869-871

Kielland-Brandt MC, Wilken B, Holmberg S, Petersen LJG, Nilsson-Tillgren T (1980) Genetic evidence for nuclear location of 2-micron DNA in yeast. Carlsberg Res Commun 45:119-124

Kingsman AJ, Clarke L, Mortimer RK, Carbon J (1979) Replication in Saccharomyces cerevisiae of plasmid pBR313 carrying DNA from the yeast trpl region. Gene 7:141-152

Kingsman AJ, Gimlich RL, Clarke L, Chinault AC, Carbon J (1981) Sequence variation in dispersed repetitive sequences in Saccharomyces cerevisiae. J Mol Biol 145:619-632

Klein HL, Petes TD (1981) Intrachromosomal gene conversion in yeast. Nature 289:144-148

Lauer G, Roberts TM, Klotz LC (1977) Determination of the nuclear DNA content of Saccharomyces cerevisiae and implications for the organization of DNA in yeast chromosomes. J Mol Biol 114:507-526

Mayer H (1978) Optimization of the EcoRI*-activity of the EcoRI endonuclease. FEBS Lett 90: 341-344

McNeil JB, Storms RK, Friesen JD (1980) High frequency recombination and the expression of genes cloned on chimeric yeast plasmids: identification of a fragment of 2-µm circle essential for transformation. Curr Genetics 2:17-25

Nasmyth KA, Reed SI (1980) Isolation of genes by complementation in yeast. Molecular cloning of a cell-cycle gene. Proc Natl Acad Sci USA 77:2119-2123

Nasmyth KA, Tatchel K (1980) The structure of transposable yeast mating type locus. Cell 19:753-764

Pellicer A, Robins D, Wold B, Sweet R, Jackson J, Lowy I, Roberts JM, Sim GK, Silverstein S, Axel R (1980) Altering genotype and phenotype by DNA-mediated gene transfer. Nature 209: 1414-1422

Petes TD (1980) Molecular genetics of yeast. Annu Rev Biochem 49:845-876

Ratzkin B, Carbon J (1977) Functional expression of cloned yeast DNA in *Escherichia coli*. Proc Natl Acad Sci USA 74:487-491

Royer HD, Hollenberg CP (1977) Saccharomyces 2 µm DNA. An analysis of the monomer and its multimers by electron microscopy. MGG 150:271-284

Scherer S, Davis RW (1979) Replacement of chromosome segments with altered DNA sequences constructed in vitro. Proc Natl Acad Sci USA 76:4951-4955

Scott JH, Schekman R (1980) Lyticase: endoglucamase and protease activities that act together in yeast cell lysis. J Bacteriol 142:414-423

Southern EM (1975) Detection of specific sequences among DNA fragments separated by gel electrophoresis. J Mol Biol 98:503-517

Stinchcomb DT, Struhl K, Davis RW (1979) Isolation and characterisation of a yeast chromosomal replicator. Nature 282:39-43

Stinchcomb DT, Thomas M, Kelly J, Selker E, Davis RW (1980) Eukaryotic DNA segments capable of autonomous replication in yeast. Proc Natl Acad Sci USA 77:4559-4563

Struhl K, Cameron JR, Davis RW (1976) Functional genetic expression of eukaryotic DNA in *Escherichia coli*. Proc Natl Acad Sci USA 73:1471-1475

Struhl K, Stinchcomb DT, Scherer S, Davis RW (1979) High-frequency transformation of yeast: autonomous replication of hybrid DNA molecules. Proc Natl Acad Sci USA 76:1035-1039

Toh-e A, Kakimoto S, Oshima Y (1975) Genes coding for the structure of acid phosphatases in

Saccharomyces cerevisiae. MGG 143:65–70

Tschumper G, Carbon J (1980) Sequence of a yeast DNA fragment containing a chromosomal replicator and the TRP1 gene. Gene 10:157–166

Wigler M, Sweet R, Sim GK, Wold B, Pellicer A, Lacy E, Maniatis T, Silverstein S, Axel R (1979) Transformation of mammalian cells with genes from prokaryotes and eukaryotes. Cell 16: 777–785

Williamson VM, Bennetzen J, Young ET, Nasmyth K, Hall BD (1980) Isolation of the structural gene for alcohol dehydrogenase by genetic complementation in yeast. Nature 283:214–216

Cloning with 2-μm DNA Vectors and the Expression of Foreign Genes in Saccharomyces cerevisiae

C.P. HOLLENBERG*

1 Introduction

During the last few years the yeast *Saccharomyces cerevisiae* has enjoyed ever-increasing interest among molecular biologists. This development has been particularly stimulated by the establishment of a transformation system which permits adequate analysis of gene function in yeast and, moreover, has opened the way to study the yeast cell as a host for foreign genes. *S. cerevisiae* can be used to clone genes which are not selectable in, or compatible with, bacterial systems. Another approach aims at exploiting the yeast cell for production of specific proteins. Among eukaryotes, *S. cerevisiae* has unique advantages because it is the only organism that is known to carry a high copy number, autonomously replicating, extrachromosomal DNA element, the 2-μm DNA. In many respects, especially in relation to gene cloning, the function of 2-μm DNA is comparable to that of bacterial plasmids. In this review, I will discuss the function of cloning vectors that use the 2-

---

* Institut für Mikrobiologie der Universität Düsseldorf, Universitätsstraße 1, 4000 Düsseldorf,
  Federal Republic of Germany

µm DNA replicon. To introduce this topic adequately, recent data on the structure of 2-µm DNA including the complete DNA sequence determined by *Hartley* and *Donelson* (1980) are summarized. This is of central importance in studies on functions relevant to its use as a vector.

The second part of this review deals with the expression of foreign genes in yeast. In general, it seems to be established that structural gene products from prokaryotic as well as eukaryotic organisms can be synthesized by the yeast cell if the corresponding coding sequence is introduced in an appropriate DNA context. In a few cases tested, the original DNA fragment is sufficient to permit functional expression. The observation of functional expression of a foreign gene is directly followed by questions about mode and efficiency of expression and methods to control or increase production of the desired protein. Advances in this direction require knowledge about the control of gene expression in the yeast cell, especially regulatory control sequences for transcription, processing, and transport. A brief discussion of interesting recent data on chromosomal gene regulation in yeast, therefore, will conclude this review. Mitochondrial gene expression will not be considered here because data on this semiautonomous organelle are not necessarily applicable to the other gene systems in the cell.

2 Extrachromosomal DNA in Yeast

In laboratory strains 2-µm DNA is found in approximately 50 copies per cell (*Hollenberg* et al. 1970; *Clark-Walker* 1974) and is stably maintained. The first natural strains found to lack the plasmid DNA were *S. carlsbergensis* strains (*Livingston* 1977). A more extensive screen among brewery strains showed its absence in a number of bottom yeasts (*Stewart* et al. 1980). Also some top-fermenting yeast strains have been found to lack the plasmid. In yeast species other than *Saccharomyces* 2-µm DNA has not been found. During recent screening by *Philippsen* (personal communication) using Southern blots, no 2-µm DNA sequences could be detected in *Kluyveromyces lactis, Kluyveromyces marxianus* and in about 20% of natural *Saccharomyces* isolates from all over the world.

In several yeast strains, extrachromosomal copies of *rRNA genes* (rDNA) occurring as closed circular DNA species with a monomeric circumference of 3-µm have been observed (*Meyerink* et al. 1979; *Clark-Walker* and *Azad* 1980; *Larionov* et al. 1980). The low copy number of less than five per cell in *S. cerevisiae* strongly suggests that the extrachromosomal rDNA molecules are not the result of a gene amplification process such as that observed in other lower eukaryotes. They are probably transient and not stably maintained.

The DNA plasmids recently described in species other than *Saccharomyces* are associated with a killer system (*Gunge* et al. 1980). Some strains of *K. lactis* contain two species of linear, doublestranded DNA with molecular weights of 5.4×10^6 and 8.4×10^6. The density of both these plasmids is 1.687 g/cm^3 – lower than the densities of nuclear (1.699 g/m^3) and mitochondrial (1.692 g/cm^3) DNA. The larger plasmid occurs in most *K. lactis* strains, while only those that also carry the smaller express the killer character.

Every stably maintained DNA species can, in principle, serve as a vector for the introduction of foreign genes into the yeast cell and provide features important for the maintenance or expression of inserted genes. In this respect also the mitochondrial DNA or selected fragments may prove useful.

2.1 Structure and Function of 2-μm DNA

The extrachromosomal genetic element 2-μm DNA plays an important role as a vector for yeast transformation. I will discuss here the more recent data on the structure and function of naturally occurring 2-μm DNA and refer for earlier information to some recent reviews (*Guerineau* 1979; *Petes* 1980). Although the intracellular location of 2-μm DNA has not been conclusively determined, many indirect arguments suggest close association with the chromosomal genome. 2-μm DNA has a nucleosome structure (*Nelson* and *Fangman* 1979; *Livingston* and *Hahne* 1979; *Seligy* et al. 1980) and uses for its replication at least some genes in common with the chromosomal DNA (*Petes* and *Williamson* 1975; *Livingston* and *Kupfer* 1978; *Zakian* et al. 1979). Earlier genetic (*Livingston* 1977) and physical data (*Clark-Walker* and *Miklos* 1974) indicating a cytoplasmic location could be due to the fact that DNA molecules of this size can migrate through the nuclear membrane. In heterokaryons caused by a *karl* mutation in one of the parent cells, even whole chromosomes have been shown to migrate from one nucleus to the other (*Nilsson-Tilgren* et al. 1980). Transfer of 2-μm DNA can presumably occur by a similar process, and could lead to its being scored as cytoplasmic (*Kielland-Brandt* et al. 1980). 2-μm DNA sequences do not occur integrated within nuclear or mitochondrial DNA (*Cameron* et al. 1977; *Tabak* 1977).

The plasmid has been sequenced by *Hartley* and *Donelson* (1980). The main features of the molecules are depicted in Fig. 1. To facilitate comparison, another version of a 2-μm DNA map is included. The yeast plasmid contains 61% dA + dT, essentially the same as the 60.3% dA + dT content of total chromosomal DNA (*Bernardi* et al. 1970). The full DNA sequence contains a number of previously mapped restriction sites and predicts others (Fig. 1). Several regions of 2-μm DNA are implicated in function. Both IR1 and IR2 contain extensive elements of symmetry similar to those found near the replication origins of many genomes. *Hartley* and *Donelson* (1980) suggested that the neighboring unique DNA sequences determine which of the regions of symmetry is part of the active origin. The origin of replication was more accurately localized by deletion mapping (*Broach* and *Hicks* 1980). By identifying the DNA segments capable of promoting high frequency transformation of yeast, *Broach* and *Hicks* (1980) could assign the origin of replication to a sequence extending from the middle of the inverted repeat to within 100 base pairs of the large unique region. This region corresponds in form 23 to the DNA segment from about 3600 to 4065 bp. The XbaI site at position 3945 is located close to the middle of the symmetry region from 3892–4013 that seems to constitute one end of the functional origin. In form 14 the identical sequence of IR2 with the XbaI site at position 703 is part of the origin.

For stable replication and maintenance in the cell, additional regions of 2-μm DNA are required (*Broach* and *Hicks* 1980). More recent data indicate the requirement of the DNA regions covering open reading frames Baker and Charlie for stable replication (*Broach*, personal communication). Precise function and location of these so-called REP functions have not yet been established. In the absence, rapid loss of the plasmid occurs under non-selected conditions, whereas on selective media a much lower number of plasmids per cell is present (*Broach* and *Hicks* 1980). REP could be involved in controlling replication or transmission. In the presence of endogenous 2-μm DNA, the cell can maintain a high number of *rep*° plasmids (*Broach* and *Hicks* 1980) indicative of trans-acting gene products. Recombination between endogenous 2-μm DNA and *rep*° plasmids

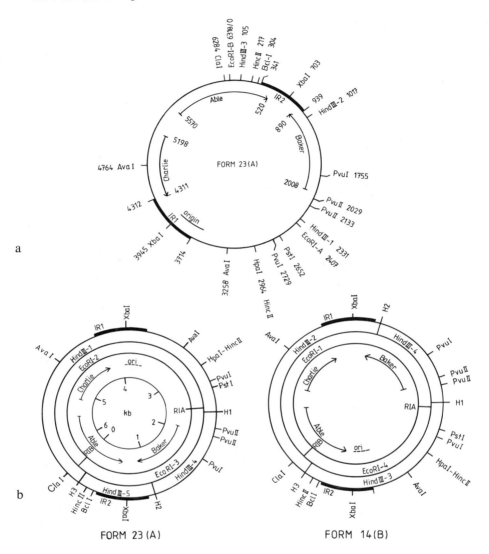

Fig. 1. a) Map of 2-μm DNA form A (form 23) showing the positions of some restriction sites and the *open reading frames* as determined by *Hartley* and *Donelson* (1980). No sites are present for the enzymes *Bam*HI, *Bgl*I, *Bgl*II, KpnI, *Sal*I, *Sma*I, *Sst*I, *Sst*II, *Xho*I. *Arrows* indicate directions of the three *large open reading frames Able, Baker,* and *Charlie* which have a coding capacity for polypeptides of mol. wt. 48625, 43235, and 33199, respectively. The inverted duplications (*thick lines*) are indicated as IR1 and IR2.

b) Maps of the two forms of 2-μm DNA, form 23 (A) and form 14 (B), showing location of the *Eco*RI and *Hind*III restriction fragments (*Holenberg* et al. 1976). The map positions on form 23 are identical to those in Fig. 1a H1, H2, and H3 refer to *Hind*III sites and RIA and RIB to *Eco*RI sites

is observed, as will be discussed further below. Plasmids that contain the total 2-µm DNA, but have the Able reading frame interrupted by integration of a bacterial DNA fragment at site RIB, are maintained in *cir°* strains (*Blanc* et al. 1979). This likewise rules out the Able reading frame for a direct function in replication.

A third function on 2-µm DNA is required for the interconversion between the two forms of 2-µm DNA, present in equal amounts in most yeast strains. The assumption that the two forms could interconvert by intramolecular recombination between the inverted repeats was verified by the first transformant analysis of *Beggs* (1978). Transformants with a recombinant plasmid containing the total 2-µm DNA, showed a rapid inversion of the DNA segment between the inverted repeats. The inversion process requires a specific coding region designated FLP (*Broach* and *Hicks* 1980), that encompasses site RIB. Integration at this site precludes interconversion only in *cir°* strains indicating that also this function can act in trans and probably involves a diffusible gene product (*Blanc* et al. 1979; *Broach* and *Hicks* 1980). Site RIB is located within the open reading frame Able from 5570 to 520, a region of the molecule which is transcribed into a 1420 base mRNA (*Broach* et al. 1979). Hence the FLP gene product is probably encoded by the Able reading frame. The interconversion seems to have no effect on the replication and transmission of plasmids containing total 2-µm DNA. Although the Able reading frame was interrupted, each form of 2-µm DNA integrated in a recombinant plasmid showed the same stability in *cir°* strains.

3 Yeast Transformation

The results obtained in transformation experiments with eukaryotes have generated the idea that any genomic DNA sequence can serve as a vector for integration of DNA into a genome. The mechanism in such cases is most probably recombination with the homologous sequence in the genome and has been shown to occur in yeast with a frequency 1–10 transformants/µg of DNA and under optimal conditions $1/10^6$ cells (*Hinnen* et al. 1978). In most cases, the inserted DNA segment is integrated at the homologous locus in the genome, as was shown for LEU2 (*Hinnen* et al. 1978) and HIS3 transformants (*Hicks* et al. 1979; *Struhl* et al. 1979). Repetitive sequences present on the same DNA fragment can also direct gene copies to other sites in the genome as was found for the LEU2 gene on the recombinant plasmid pYe-LEU10 (*Ratzkin* and *Carbon* 1977) that is thought to contain a *TY*1-17 sequence (*Kingsman* et al. 1981). This sequence is possibly responsible for integration at other loci nonhomologous to LEU2 which occurred, however, with much lower frequency (*Hicks* et al. 1979).

Certain yeast DNA fragments such as those containing the TRP1 gene (*Struhl* et al. 1979) or the ARG4 gene (*Hsia* and *Carbon* 1979) permit a much higher frequency of transformation (10^3–10^4 transformants/µg), because they are able to replicate as extrachromosomal elements. An autonomous replication sequence adjacent to the TRP1 gene, the ARS1 locus, has been analyzed (*Stinchcomb* et al. 1980) and sequenced (*Tschumper* and *Carbon* 1980). The resulting transformants, in contrast to the integration transformants, are mitotically unstable but stable enough to permit the use of these DNA sequences as transformation vectors (*Struhl* et al. 1979).

Highest transformation frequencies are obtained when the vector carries 2-µm DNA. Transformation with such vectors will be described in detail in Sect. 3.2, while the reader is referred to *Hinnen* and *Meyhack* (this volume) for more information on other types of vectors.

3.1 Procedure for Yeast Transformation

The procedure for *S. cerevisiae* transformation has been developed by *Hinnen* et al. (1978) and is described in detail by *Ilgen* et al. (1979). The method starts with protoplasts, which are transformed using a combination of calcium treatment and polyethylene glycol as used for the fusion of protoplasts (*van Solingen* and *van der Plaat* 1977). Transformed protoplasts are regenerated in agar (*van Solingen* and *van der Plaat* 1977). In the context of this review, I will make a few comments on the transformation procedure: Yeast protoplasts can be made by the use of different enzyme preparations that degrade the yeast cell wall. In our experience, both zymolyase and helicase give identical results although the second preparation is more crude. Apparently degradation of DNA during transformation by nucleases outside the protoplasts is not a problem. Also β-glucuronidase/arylsulfatase (*Kielland-Brandt* et al. 1979) and lyticase (*Struhl* et al. 1979) are used. The stage of protoplast formation is not critical. In order to determine whether the progression of protoplast formation has an influence on the transformation frequency or the rate of protoplast regeneration, we transformed yeast cells at different stages of protoplast formation. The first transformants were obtained after only 10 min of incubation with helicase, although complete lysis of the protoplasts in H_2O was reached only after 1 h. We observed no influence on the time of appearance of the first transformant colonies, usually after 4 days. In certain cases, however, it might be preferable to use cells that are only partly converted into protoplasts to reduce protoplast fusion occurring during normal transformation (*Hicks* et al. 1979).

Protoplast regeneration is performed in a rather thick layer of 3% agar using sorbitol as an osmotic stabilizer. Two modifications of this procedure can be useful under certain conditions. The agar layer of 7 ml per petri dish is rather thick and results in many transformant colonies not reaching the surface of the plate. The colonies then have to be picked individually for further analysis. If the transformation mixture is plated without the addition of regeneration agar only about 10%–20% of the transformants grow, but all can be directly replica-plated for further analysis. Also the medium in plates and regeneration agar used to grow initial transformants influences the regeneration of yeast protoplasts. We noticed this most strongly on yeast-extract-peptone-glucose plates plus the aminoglucoside G418 used to select pMP81-containing transformants (compose Sect. 3.5). The regeneration efficiency of protoplasts on this medium is only 1%–10% of that on yeast nitrogen-base medium plus glucose. For this reason we perform the selection for G418 resistant transformants on yeast nitrogen-base medium supplemented with casamino acids (Table 3).

If sorbitol is not desired as an osmotic stabilizer because the yeast species can use it as a carbon source or because the research budget is restricted, one can use 0.6 M KCl. In some cases, this leads to even better regeneration frequencies than with sorbitol, which is not always of high quality. Cotransformation has been shown to work well for yeast (*Hicks* et al. 1979; *Jimenez* and *Davis* 1980). If yeast protoplasts are transformed with a mixture of two plasmids each carrying one selectable gene, transformants that contain both selectable genes and apparently have taken up both plasmids are obtained with a high frequency. Apparently only a small percentage of protoplasts is competent for transformation and such cells take up a number of plasmid molecules. Similar observations have been made for *E. coli* (*Kretschmer* et al. 1975) and mammalian cells (*Wigler* et al. 1979). *Jimenez* and *Davis* (1980) used cotransformation with a 2-μm DNA vector carrying LEU2

Table 1. Summary of restriction fragments and identified regions on 2-µm DNA

Form	Restriction fragments				Intact genes or reading frames
	EcoRI		HindIII		
	Number	Size (bp)	Number	Size (bp)	
23 (A)	2	3912	1	4092	Replication Origin/ Charlie/IR1
	3	2406			Baker/4600/IR2
			4	1314	4600
			5	912	IR2
14 (B)	1	4073			Charlie/Baker/4600/ IR1
			2	2787	Charlie/IR1
	4	2245	3	2217	Replication Origin/IR2
			4	1314	4600

Designation of EcoRI and HindIII restriction fragments according to size. Form 23 or 14 indicate configurations comprising EcoRI fragments 2 and 3 or 1 and 4 (Hollenberg et al. 1976). The alternative designations A and B (Guerineau et al. 1976) or L and R (Livingston and Klein 1977) are in use; 4600 = the open reading frame at position 2129–2267 (Fig. 1a). Reading frame positions are derived from Hartley and Donelson (1980), origin position from Broach and Hicks (1980)

to introduce a ColEI derivative carrying a kanamycin resistance gene into S. cerevisiae. About 8% of the Leu$^+$ transformants were resistant to G418 and carried the resistance gene, as was shown enzymatically.

3.2 Transformation with 2-µm DNA Vectors

Recombinant plasmids that contain the whole 2-µm DNA plasmid can generally transform S. cerevisiae strains with frequencies as high as 10^4–10^5 transformants/µg of DNA (Beggs 1978; Hicks et al. 1979; Struhl et al. 1979). The frequency of transformants per viable cell is around 10^{-3}. The percentage of viable cells, usually 10%–20%, reflects the efficiency of protoplast regeneration (van Solingen and van der Plaat 1977; Hinnen et al. 1978). Assuming that transformed and nontransformed protoplasts regenerate with the same efficiency, the actual transformation frequency is 10^5–10^6 transformants/µg of DNA.

Not all 2-µm DNA vectors have the same stability in the transformed cell. The question of stability is closely bound up with the nature of the genetic functions on 2-µm DNA. As discussed above, functions have been distinguished in connection with replication and transmission (Broach and Hicks 1980), the possible locations of which are summarized in Table 1: (a) An origin of replication functioning in cis that is absolutely required for plasmid replication, (b) genes that function in trans that are required for stable maintenance of the recombinant plasmids. The first function is comparable with the replication origin of bacterial plasmids. The nature of the second functions have not yet been determined. It has been described as functions required for efficient use of the replication origin (Broach and Hicks 1980), but transmission of plasmids to daughter cells could also be involved.

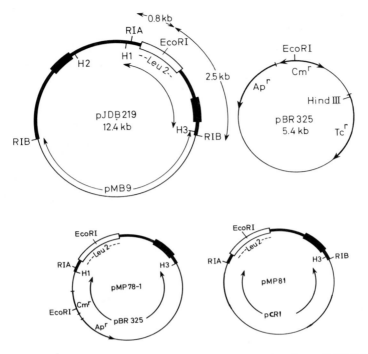

Fig. 2. To construct pMP78-1, the HindIII fragment of 3.3 kb from pJDB219 (*Beggs* 1978) was inserted into the HindIII site of pBR325. The 3.3 kb HindIII fragment carries the yeast LEU2 gene segment inserted by AT linkage into the PstI site on HindIII fragment 3 of 2-μm DNA form 14. pMP81 consists of plasmid pCRI (*Covey* et al. 1976) and a double EcoRI fragment carrying the LEU2 gene from pJDB219. The Ap^r (ampicillin resistance) and Cm^r (chloramphenicol resistance) genes are indicated (*arrows*). pCR1 carries a kanamycin resistance on Tn601. Symbols as in Fig. 1

In the absence of those so-called REP functions, the stability of the plasmid is drastically reduced, but it can be maintained in a culture under selective conditions. The difference can best be shown in a strain that has been cured for 2-μm DNA (*Erhart* and *Hollenberg* 1981). This ensures that the genome is able to support normal 2-μm DNA replication, a condition not necessarily fulfilled in naturally isolated cir^0 strains. The two plasmids used here, pJDB219 (*Beggs* 1978) and pMP78 (*Hollenberg* 1979), are depicted in Fig. 2. pMP78 is rep^0 since the only 2μm DNA fragment present is HindIII fragment 3 (cf. Fig. 1 and Table 1), which carries the replication origin and none of the open reading frames.

The cir^0 strain YT6-2-1L *leu2 his4*, derived from AH22 was transformed with plasmids pJDB219 and pMP78 and Leu$^+$ transformants were selected. Transformant colonies were inoculated into complete medium and tested for the presence of an active LEU2 gene after increasing periods of growth (Fig. 3). Clearly pMP78 is lost much faster from the cir^0 than from the Cir$^+$ strain. pJDB219 seems slightly more stable in cir^0 strains.

Under selective conditions, the difference in stability is reflected in slower growth and lower plasmid content of pMP78 in a cir^0 strain compared to a Cir$^+$ strain. The slow growth must be due to the continuous loss of pMP78 from the cir^0 strain, as can be con-

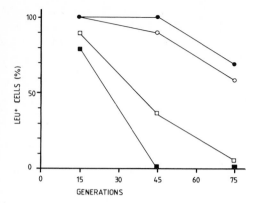

Fig. 3. Stability of different yeast transformants. Strain AH22 *leu2 his4* and the *cir°* strain YT6-2-IL *cir° leu2 his4* were transformed with plasmids pMP78 and pJDB219. Transformants were selected as described (*Erhart* and *Hollenberg* 1981) under selective conditions on minimal medium and after 15 generations transferred to rich medium (YEPD). Aliquots of the cultures were plated on selective minimal medium and on YEPD. The % Leu⁺ cells was scored as the ratio of the number of colonies on both types of media. O–O, AH22/pJDB219; ●–●, YT6-2-IL/ pJDB219; □–□, AH22/pMP78; ■–■, YT6-2-IL/pMP78

cluded from the fact that most cells of a selectively grown culture of transformant YT6-2-IL/*cir°*/pMP78 are pMP78°. The same phenomenon has been observed in a natural *cir°* strain transformed with a plasmid containing only *Eco*RI fragment 4 (*Hicks* and *Broach* 1980). It is of interest to note that the transformation frequency of pMP78 in *cir°* or Cir⁺ strains does not differ dramatically, although the stability of their transformants shows marked differences. The results show that for the stable transformation of *cir°* strains, vectors that contain the whole 2-μm DNA must be preferred at present. Though a vector like pMP78 can be maintained in such strains, the small amount of plasmid molecules in the culture can cause problems in the study of integrated DNA fragments.

The situation in Cir⁺ strains, mostly used hitherto as receptor strains for yeast transformation, is different. The endogenous 2-μm DNA supplies the Rep gene functions and possibly other gene functions so that plasmids like pMP78 can be maintained more stably. The stability of recombinant plasmids in Cir⁺ cells, however, is influenced by three other factors: a. recombination with the endogenous 2-μm DNA; b. incompatibility between identical replication systems of the recombinant and endogenous plasmids; c. copy number.

3.3 Recombination

Recombination between incoming and endogenous plasmids has been observed by many authors in cells transformed with 2-μm DNA containing vectors (*Beggs* 1978; *Storms* et al. 1979; *Gerbaud* et al. 1979; *Thomas* and *James* 1980; *McNeil* et al. 1980; *Dobson* et al. 1980a; *Erhart* and *Hollenberg* 1981). The general scheme in Fig. 4 seems to be consistent with the published data. A cointegrate of the 2-μm DNA vector and the endogenous 2-μm DNA is formed perhaps analogous to the formation of multimeric forms of 2-μm DNA described previously (*Guerineau* et al. 1976; *Royer* and *Hollenberg* 1977).

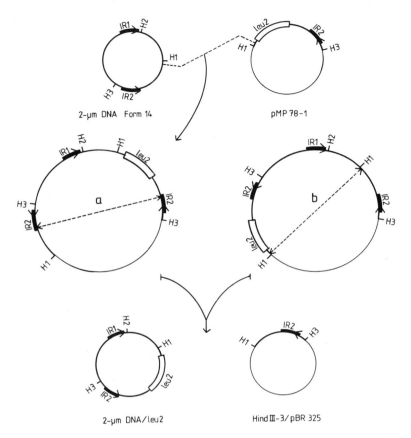

Fig. 4a, b. Formation of 2 μm DNA-LEU2 plasmids through cointegrates of pMP78 and 2 μm DNA. a) Integration occurs at a point between H1 and the end of the LEU2 segment; b) integration occurs at a point within the IR2 sequence. Dissociation points of the cointegrates are indicated (*dashed arrows*). Symbols as in Fig. 2, pBR325 segment is drawn with a *thin line*

Recombinations take place not only between the IR sequences but also between other homologous sequences in the molecules. The analysis of Leu⁺ transformants stable even under nonselective conditions showed that in most cases the LEU2 gene of pMP78 had been integrated into the endogenous 2-μm DNA. The fragment analysis of such transformants (Fig. 5) shows that both pBR325 and the normal 2-μm DNA have disappeared and that only a 2-μm DNA-LEU2 (both in forms 14 and 23) remains. The formation of such molecules most likely takes place via a cointegrate with subsequent dissociation as depicted in Fig. 4. To generate the 2-μm DNA-LEU2 molecule, at least one of the two recombination events has to take place between homologous sequences other than the IR sequences. If intramolecular recombination between the IR sequences is very frequent, the intermolecular recombinational event probably takes place outside the IR sequences, in this case in the region between site H1 and the LEU2 gene segment as shown in Fig. 4a. As this region is relatively short, the frequency might be rather low. Cointegrates that are formed via the IR sequences (Fig. 4b) can lead to the same disso-

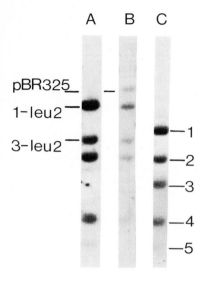

pBR325

1-leu2

3-leu2

Fig. 5. Autoradiogram of a Southern blot of HindIII digested DNA isolated from transformant strains that have a stable Leu⁺ phenotype. Yeast DNA was isolated from 2 ml overnight YEPD cultures according to the method of *Struhl* et al. (1979). Restriction digests were electrophoresed on 0.7% agarose gels and transferred to nitrocellulose filters (*Southern* 1975). Filters were hybridized with ³²P-labeled BTYP-2 DNA, a 2-μm DNA-pBR322 plasmid. *A,* YT600 8/8 contains only 2-μm DNA-LEU2 as concluded from the absence of HindIII fragments 1 and 3 and presence of fragments 1-LEU2 and 3-LEU2. *B,* YT600 8/10 contains 2-μm DNA-LEU2, but still some pMP78 persists, as concluded from the presence of the pBR325 band. 2-μm DNA is not present. *C,* control AH22 with HindIII fragments 1-5. Fragment 5 is present in all lanes, but hybridizes weakly under the conditions used

ciation products by intramolecular recombination in the region from site H1 to the LEU2 gene segment. Intramolecular inversion can explain the occurrence of LEU2 in the other form of 2-μm DNA. *Dobson* et al. (1980a) have described the occurrence of 2-μm DNA-LEU2 in pJDB219 transformed strains. The schemes in Fig. 4 predict that the formation of 2-μm DNA-LEU2 is more frequent in pJDB219 transformed strains than in strains transformed with pMP78.

In our experience the composite plasmid carrying both, 2-μm DNA and LEU2 can provide very stable transformants as, e.g., desired for biotechnology production purposes. It is most easily obtained through pMP78 by isolating stable Leu⁺ transformants lacking, e.g., the β-lactamase gene or pBR325 sequences. The results obtained with pMP78 or pJDB219, however, cannot necessarily be extended to cloning vectors carrying other selective markers. As will be discussed below, plasmid pMP78 is able to displace endogenous 2-μm DNA in *leu2* host cells, a phenomenon also observed for pJDB219 by *Dobson* et al. (1980b). As mentioned above, the occurrence of recombination with endogenous 2-μm DNA has been observed for many vectors that carry the 2-μm DNA replicon. In most cases only the presence of additional restriction fragments has been reported and no information is available on the stability of the recombinant forms or their fate.

3.4 Incompatibility Between Endogenous 2-μm DNA and Vectors Carrying its Replicon

2-μm DNA containing vectors use the same replication and transmission system as endogenous 2-μm DNA. Differences in size and DNA sequences between a recombinant plasmid and the resident 2-μm DNA in a transformed cell might, depending on the conditions, favor the transmission of only one of the plasmids and lead to loss of the other. Under nonselective conditions, i.e., no selection for a gene present on the recombinant plasmid, the natural 2-μm DNA plasmid is in most cases retained at the expense of the ar-

Table 2. Loss of 2-µm DNA induced by pMP78

Growth conditions	Properties of transformants			
	β-lactamase production (%)	BTYP-2 hybridization (%)	Leu+ (%)	cir⁰ (%)
15 generations in YNBH	85	85	90	10
36 generations in YEPD	15	20	15	75
72 generations in YEPD	10	10	10	90

S. cerevisiae strain GRF18 LEU2 HIS2 was transformed with pMP78-1 following mainly the proce-dure described by *Beggs* (1978). Transformants were grown on selective medium YNBH (yeast nitrogen base + 20 µg histidine/ml) for 15 generations. Culture samples were plated out on selective medium and on YEPD (1% yeast extract, 2% peptone and 2% glucose) to determine the fraction of cells that are *Leu+*. For colony hybridization small amounts of cells were transferred to millipore filters and grown on YEPD medium. Hybridization probes: BTYP-2 DNA, a 2-µm DNA-pBR322 plasmid (*Hollenberg* 1978), and 2-µm DNA *Hind*III fragment 4. The loss of 2-µm DNA as deter-mined with colony hybridization was confirmed by hybridizations on Southern blots (*Erhart* and *Hollenberg* 1981)

tificially constructed molecule. The recombinant plasmid is unstable under nonselective conditions, a situation reported for all vector plasmids with the 2-µm DNA replicon. Under conditions selective for the recombinant plasmid, only cells that carry at least one copy can grow and replicate their plasmids. This maintains a stable level of the recom-binant plasmid in the growing cell population, the level of which will probably depend on the number of copies of the selected gene required per cell for optimal growth. During the analysis of pMP78 transformants, we noticed that under certain conditions we obtained a large number of transformants that had lost all plasmids and had become *cir⁰* (*Erhart* and *Hollenberg* 1981). It appears that pMP78 can compete out the resident 2-µm DNA in a population of transformed cells under selective conditions. Since pMP78 does not carry the REP genes, it is not stably maintained in *cir⁰* cells, but subsequently lost. Based on these data, we developed a method for curing *S. cerevisiae* strains of 2-µm DNA and applied it to AH22 (*Hinnen* et al. 1978) and GRF18, strains frequently used as hosts for transformation. After transformation with pMP78, cells are grown selectively to displace endogenous 2-µm DNA. Plasmid pMP78 is lost during ensuing nonselective growth and up to 100% of the cells completely lack plasmids. The results of such an experiment are presented in Table 2.

It has to be stressed, however, that individual transformant colonies showed remarkable differences in stability. The majority of the transformants behaved as in Table 1 and showed after 50–70 generations of nonselective growth 70%–90% *cir⁰* cells. A low percentage of the transformants showed a total loss of plasmids within 20 generations, whereas about 5% of the transformants appeared to be completely stable within the experimental period. In order to score for the presence of pMP78 in yeast transformants, the activity of the product of the bacterial ampicillin resistance gene borne on pMP78 β-lactamase was used. This bacterial gene is expressed in *S. cerevisiae* (*Hollen-berg* 1979a) and can be detected with a simple color reaction. The expression of β-lac-tamase in yeast will be discussed in more detail below.

Table 3. Direct selection of transformants on G418

Plasmid	Carbon source	Medium + G418 (600 μg/ml)	Transformants among G418 resistant colonies (%)	Leu⁺ colonies (%)
pMP81	2% glucose	YEPD	15	22
pMP81	0.2% glucose	YNB-0.2% casaminoacids	15	12
PTY75	0.2% glucose	YNB-0.2% casaminoacids	19	–
PTY75	0.2% glucose	YNB-his-leu	20	–
PTY75	2% glucerol	YNB-0.2% casaminoacids	52	–

Strain AH22 *leu2 his4* was transformed with plasmids pMP81 and PTY75. Both plasmids contain vector pCR1 encoding a kanamycin resistance factor. PTY 75 contains in addition total 2 μm DNA (*Hollenberg* et al. 1976). For pMP81 see Fig. 2. Transformation as described in Table 2. Transformed cells were plated on different media and incubated at 34 °C for 4–7 days. Number of transformants about 10^3/μg DNA and 10^{-3}–10^{-4}/regenerated protoplast. For the detection of true PTY75 transformants 20 G418 resistant colonies were tested by colony hybridization with ^{32}P-labeled pCR1. In the case of pMP81 containing the LEU2 gene 50 colonies were analyzed with regard to the *leu2* phenotype and the presence of pCR1 sequences by colony hybridization

It has hitherto been impossible to inhibit specifically the replication or transmission of 2-μm DNA in *S. cerevisiae* and thus generate *cir*° strains. None of the methods applied for the curing of bacterial plasmids have proved successful for yeast. The method described above to cure *S. cerevisiae* for 2-μm DNA permits the generation of *cir*° derivatives of defined strains. Such defined *cir*° strains can be used as hosts for defective 2-μm DNA molecules in which to analyze functional sequences on the plasmid. Furthermore, *cir*° strains are useful hosts for recombinant plasmids carrying the 2-μm replicon, as such plasmids cannot be altered by recombination with endogenous 2-μm DNA. Finally, *cir*° strains are desirable hosts for the construction of stable transformants with 2-μm DNA vectors as discussed above.

3.5 Curing of Wild Strains of S. cerevisiae

Curing with pMP78 depends on the availability of a *leu2* mutation. It is not yet clear whether 2-μm DNA vectors carrying other structural genes like HIS3 or URA3 on 2-μm DNA can exert the same effect. In any case, curing with this type of plasmid requires auxotrophic markers that, in general, are neither present nor desirable in industrial yeast strains. To extend the method to prototrophic strains, we made use of plasmid pMP81 (Fig. 2), which contains approximately the same yeast DNA sequences as pMP78 but integrated at the *Eco*RI site of bacterial plasmid pCR1. pCR1 carries the transposable element Tn601, which contains a gene conferring resistance to kanamycin (*Hollenberg* 1979a) and gentamicin G418 (*Jimenez* and *Davis* 1980) on yeast transformants. In Table 3 we show that *S. cerevisiae* transformants can be selected on G418. The transformants, however, could not be scored directly on G418 plates since a high number of spontaneously resistant colonies grew. True transformants were, therefore, detected by colony hybridization with labeled plasmid pCR1. The number of spontaneous resistant

colonies can be influenced by raising the G418 concentration in the plates. If, however, very high concentrations are applied, the number of transformants falls off drastically. Table 3 shows further that the percentage of transformants on glycerol plates was two to three times higher than on glucose. This decrease in the number of spontaneously resistant cells is probably due to inhibition of mitochondrial biosynthesis. G418 inhibits both cytoplasmic and mitochondrial protein synthesis (*Jimenez* and *Davis* 1980).

To detect a curing effect of pMP81 or PTY75 in transformants initially selected on G418 plates, we determined the plasmid composition of such transformants after 60 generations on nonselective medium. Colony hybridization showed that PTY75 was lost in most cells whereas 2-μm DNA was still present. About 5% of the transformants lacked both plasmids and had become *cir⁰*. The effects of higher concentrations of G418 during initial selective growth on the generation of *cir⁰* cells are currently being investigated.

The curing experiments do not permit definite conclusions about the mechanism responsible. I would like to discuss one model, though other possibilities are not excluded. One can suppose that the selection for the marker gene on the recombinant plasmid leads to a high copy number with a concomitant decrease in the number of resident 2-μm DNA molecules such that *cir⁰* daughter cells are generated. Although the mechanism is attractive, it is not clear why a high copy number of the LEU2 gene on a plasmid is required, whereas only one chromosomal copy can support optimal growth in the wild-type cell. Hitherto no curing effects of recombinant plasmids carrying other selective markers have been reported. Whether a lower efficiency of expression of the LEU2 gene on pMP78 which is not associated with a *TY* element as on pYe-LEU10 (*Ratzkin* and *Carbon* 1977; *Kingsman* et al. 1981) is responsible for the difference has to await further clarification. A decrease in the efficiency of expression of the LEU2 gene on pMP78 relative to a functional chromosomal copy is conceivable. A position effect or a loss of DNA sequences during cloning could be responsible. The extrachromosomal location per se does not seem to be involved as similar vectors carrying the HIS3 gene do not appear to have a curing effect. In the case of G418 resistance, selection for a high copy number leading to segregation of the resident 2-μm DNA is easier to imagine.

3.6 Copy Number and Stability

Little information is available on the copy number of recombinant plasmids. By measuring renaturation kinetics with 2-μm DNA probes, *Gerbaud* and *Guerineau* (1980) found the number of natural 2-μm DNA molecules in several *S. cerevisiae* strains to vary from 24 to 88 copies per haploid genome according to the strain. No influence of the physiological state of the cells was observed. The data for the transformant containing plasmid G9 are not easy to evaluate since recombinant plasmid G9 does not contain the 2-μm DNA replication origin and probably recombination with endogenous 2-μm DNA had taken place in the transformants. A particular *cir⁰* mutant DR19/T7 contained about 8–10 copies of recombinant plasmid G18 comprising the URA3 gene and total 2-μm DNA, when grown selectively for the URA3 gene. The stability of this transformant, i.e., 95% Ura⁺ cells after 10 generations in nonselective medium, is similar to the stability of other plasmids with the whole 2-μm DNA in a *cir⁰* strain (*Erhart* and *Hollenberg* 1981). The copy number of G18 in those transformants was only 20% of the 2-μm DNA copy number in a corresponding Cir⁺ strain DR19/T8, a phenomenon that could be due to the hot strain or the plasmid.

For other recombinant plasmids, only copy number estimates from gels or Southern blots are available. The loss of 2-µm DNA recombinant plasmids under nonselective conditions, however, can give an indication for the copy number as well, on the simple assumption that recombinant plasmids maintained at a high copy number per cell are segregated out less frequently than those present in only a few copies. pJDB219 is such a recombinant plasmid that is relatively stable under nonselective conditions (Fig. 3). As discussed before, the copy number of recombinant plasmids is most probably influenced primarily by the nature of the selectable gene. If a cell requires only a few copies of the gene for optimal growth, the plasmid will be maintained at a low level, whereas a gene required in high number leads to higher plasmid copy numbers. A high copy number of LEU2 carrying plasmids is assumed to be necessary for the curing effect of pMP78 (*Erhart* and *Hollenberg* 1981). The extreme instability of HIS3 vectors (*Hicks* et al. 1979; *Struhl* et al. 1979), a LEU2 vector (*Hicks* et al. 1979), and URA3 vectors (*Blanc* et al. 1979) in Cir$^+$ strains could indicate low copy numbers and explain why a similar curing effect has not been observed with these markers. Certainly, additional plasmid features like size, integration sites, and DNA sequences are of importance as well. *Tho-e* et al. (1980) reported an increased stability under non-selective conditions for a plasmid similar to pMP78 but lacking the bacterial pBR325 part.

3.7 Transcription of 2-µm DNA and Recombinant Molecules Containing 2-µm DNA Sequences

Several species of polyadenylated RNA transcripts complementary to 2-µm DNA have been reported (*Broach* et al. 1979). The two major transcripts are 1325 and 1275 bases in length and map in regions compatible with the Baker (1118 bp) and Charlie (887 bp) reading frames respectively. Also the sizes are compatible with an mRNA function for those open reading frames. A very rough estimate comes to no more than 10 copies per cell for these major transcripts. The significance of other species of polyadenylated RNA mapping on both strands of the 2-µm DNA molecule remains unresolved. Only one minor RNA transcript of 2600 bases covers the Able reading frame of 1268 bp.

In yeast transformants containing pMP78 or other 2-µm DNA containing plasmids, the amount of polyA RNA transcribed from 2-µm DNA sequences was much higher than in untransformed cells (*Hollenberg* et al. 1980). Apparently functional promoter sites are present on the inserted bacterial or chromosomal DNA fragments. The presence of discrete transcripts demonstrates that not only new promoter sites but also termination sites must be present. Preliminary mapping results indicated two main transcripts covering the β-lactamase gene (*Hollenberg* et al. 1980). The determination of the direction of transcription using single-stranded DNA of different fd vectors carrying the β-lactamase gene in each orientation (*Herrmann* et al. 1980) showed that both transcripts are complementary to the non-sense strand and thus cannot represent the functional mRNA. The coding strand is transcribed into a third transcript starting from 2-µm DNA sequences and a fourth minor one originating from E. coli DNA sequences. Deletion of the bacterial promoter region leads to an inactive gene whose function can be restored by insertion of the bacterial lac promoter. This indicates that expression in yeast of the β-lactamase gene on pMP78 is dependent on bacterial DNA sequences close to or identical with the bacterial β-lactamase promoter (*Breunig* und *Hollenberg*, in preparation).

The level of URA3 enzyme activity in transformants carrying the yeast URA3 gene on a 2-µm DNA vector was 25 times higher than the wild-type level (*Chevallier* et al. 1980). These data indicate that the higher copy number of the URA3 gene can lead to an increase in translation products. In the same experiments, an 80-fold increase in RNA complementary to the URA3 region was observed, but this might in part be due to nonsense transcription as discussed above. The URA3 plasmid used in this study has much higher stability under nonselective conditions than the URA3 plasmids used by *Gerbaud* et al. (1979), demonstrating the influence of plasmid structure.

Beggs et al. (1980) reported on the expression in yeast of a rabbit β-globin gene integrated in the pMB9 part of pJDB219. The β-globin specific transcripts were about 20–40 nucleotides shorter at the 5′ end than normal globin mRNA, contained the small intron, and terminated heterogenously in an AT-rich region present in the first half of the large intron. It is not clear whether the 5′ end represents the initiation site of transcription by a yeast RNA polymerase or whether some form of processing had taken place. Clearly, correct splicing was not performed to a detectable extent, although some splicing system must be present in the yeast cell as concluded from the presence of at least one split gene, the actin gene (*Gallwitz* and *Sures* 1980; *NG* and *Abelson* 1980). Most of the globin transcripts did not contain a detectable polyA tail. The amount of β-globin RNA was estimated as 0.01% of the total yeast RNA. The presence of β-globin protein in the transformants was not mentioned.

4 Gene Expression in Yeast

Although in most cloning experiments in yeast the ultimate goal is the synthesis of functional product from the isolated gene, various possible objectives can be distinguished. If *S. cerevisiae* is only used as a host organism for the cloning of, e.g., plant genes, the functional product of the cloned gene is used to identify the clone and only detectability is required. If the aim of the cloning experiments is the overproduction of a yeast protein or production of a foreign gene product, additional optimization of the gene expression might be appropriate.

In this section I discuss the expression of foreign genes and give a short review of the present data on the expression of yeast genes. The expression of a gene involves many steps from gene activation through transcription, RNA processing, and translation to protein maturation, with transport processes at different stages. The recognition that all these events can be subjected to specific control systems initially led to the feeling that successful expression of a foreign gene would be unlikely. The results obtained in this area of research during the last few years suggest, however, that many of the processes involved in gene expression, such as protein maturation and transport, are more universal than anticipated.

With the knowledge that a number of *S. cerevisiae* genes can be functionally expressed in *E. coli* (*Struhl* et al. 1976; *Ratzkin* and *Carbon* 1977), we investigated the expression of bacterial genes in yeast. The positive results showed that prokaryotic genes can be expressed successfully in an eukaryotic organism like *Saccharomyces* (*Hollenberg* 1979). The *E. coli* genes analyzed were the resistance genes for ampicillin, chloramphenicol, and kanamycin and more recently the *ompA* gene (*Janowicz, Henning* and *Hollenberg,* in

preparation), the structural gene for the major outer membrane protein II* (*Henning* et al. 1979). This collection of genes permits us to test different aspects of the functioning of foreign gene products in yeast.

1. The *ampicillin resistance* gene codes for an enzyme, β-lactamase, which inactivates penicillins by cleaving the lactam ring. The primary gene product in *E. coli* is a pre-protein (*Sutcliffe* 1978; *Ambler* and *Scott* 1978) carrying a signal peptide that is cleaved off during secretion into the periplasmic space (*Talmadge* et al. 1980a; *Koshland* and *Botstein* 1980). The analysis of the gene products in yeast provides insight into the specificity of processing and transport mechanisms.

2. Yeast is sensitive to *chloramphenicol* and *kanamycin*. The analysis of bacterial resistance genes that code for modifying enzymes of the corresponding antibiotics shows whether such resistance genes can be used as selective markers in yeast transformation systems.

3. *Protein II**, the *ompA* gene product, is a structural protein that traverses the *E. coli* outer membrane with termini protruding at both sides (*Garten* et al. 1975; *Endermann* et al. 1978; *Datta* et al. 1977). This protein exhibits several special features such as efficient plasma membrane transport, precursor processing, and outer membrane location; all properties that can be studied in the transformed yeast cell.

4.1 Expression of an Ampicillin Resistance Gene

Although *S. cerevisiae* is not sensitive to ampicillin, the expression of a β-lactamase gene can be shown in several ways (*Hollenberg* 1979). Cell-free extracts of yeast transformants carrying plasmid pMP78 can be shown to have β-lactamase activity by a spot test on soft agar plates containing penicillin-sensitive *E. coli* cells and an inhibitory concentration of ampicillin. During subsequent incubation, ampicillin around the spot is inactivated and *E. coli* cells can grow. The same test permits the demonstration of β-lactamase in intact cells as well by incubating a streak of a transformant carrying pMP78 on the soft agar plate.

The production of β-lactamase in yeast cells can be demonstrated directly with two color reactions routinely used in our laboratory. Both methods are derived from bacterial diagnostics. The first method uses a chromogenic derivative of cephalosporin, called nitrocefin (*O'Callaghan* et al. 1972). β-lactamase converts the yellow color of the intact molecule into a red cleavage product. When nitrocefin is applied to outgrown yeast cell streaks, the color change can be observed after 15–60 min, depending on the age of the cells (*Hollenberg* 1979a). A second method is based on abstraction of iodine from an iodine-starch complex by the cleavage products of penicillin antibiotics (*Chevallier* and *Aigle* 1979). Both methods offer a cheap and simple way to check for the presence of a recombinant plasmid in a yeast clone.

The total β-lactamase activities in yeast transformant YT6-2 and *E. coli* pBR322 differ by a factor of 500, if extracts from the same quantity of cells (wet weight) are compared. The β-lactamase activity was purified from crude extracts and found to be indistinguishable by several criteria from the purified enzyme from *E. coli* (*Roggenkamp* et al. 1981). Electrophoresis in sodium dodecylsulfate (SDS) polyacrylamide gels revealed no difference in mobility between the enzyme activity from the yeast transformant and *E. coli* (Fig. 6). This means that processing of the bacterial signal peptide has occurred in

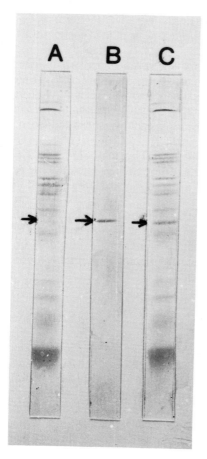

Fig. 6. SDS electrophoresis of β-lactamase from yeast transformant YT6-2 and from *E. coli*. After electrophoresis the proteins were renatured and β-lactamase activity was localized with the nitrocefin color reaction before staining with Coomassie Blue. *A*, partly purified yeast β-lactamase (100 μg); *B*, purified β-lactamase from *E. coli* (1 μg); *C*, mixture of *A* and *B*. *Arrows* indicate the positions of the nitrocefin color reaction. A gel concentration of 12.5% polyacrylamide was used

yeast since most probably the β-lactamase preprotein is also the primary translation product in yeast.

This conclusion is further supported by the evidence suggesting that the bacterial preprotein of β-lactamase has no enzymatic activity (*Roggenkamp* et al. 1981; *Kreft* et al. 1978). The possibility that the preprotein is responsible for the activity in yeast is, therefore, rendered even more unlikely. The observation that a yeast cell can process a bacterial preprotein substantiates the notion that the processing of precursor molecules, at least those steps involved in membrane transport, follows a mechanism universal to prokaryotic and eukaryotic organisms. Processing of eukaryotic preproteins in *E. coli* cells has been shown for the insulin gene products (*Talmadge* et al. 1980b).

In *E. coli*, the signal peptide is cleaved from β-lactamase during secretion into the periplasmic space (*Koshland* and *Botstein* 1980). We have not been able to detect any secretion of β-lactamase through the plasma membrane of the yeast cell and we believe that processing occurs during passage through another membrane system, presumably the endoplasmic reticulum. The secretion process in yeast (*Novick* et al. 1980) is likely to require additional recognition signals.

4.2 Expression of a Chloramphenicol Resistance Gene

Under respiratory growth conditions *S. cerevisiae* is sensitive to chloramphenicol and we have shown that yeast transformant YT6, which contains plasmid pMP78 bearing an intact chloramphenicol resistance gene, displays an increased resistance to chloramphenicol on plates (*Hollenberg* 1979a). The chloramphenicol resistance gene of pBR325 codes for an acetyltransferase which inactivates chloramphenicol by acetylation (*Shaw* 1975). The enzyme activity could also be shown in cell extracts of transformants (*Hollenberg* et al. 1980; *Cohen* et al. 1980a). *Cohen* et al. (1980b) were able to select for mutations in the recombinant plasmids of the transformants that lead to a 3- to 50-fold increase in enzyme activity. A deletion of 120 bp immediately upstream from the 5′ end of the chloramphenicol resistance gene was responsible for a 50-fold increase in enzyme activity. This deletion actually reduced expression in *E. coli*, indicating that DNA sequences in this region are functioning differently in both organisms. Transformants containing these plasmids may be directly selectable on chloramphenicol.

4.3 Expression of the Kanamycin Resistance Gene of Tn601 (Tn903)

We have shown earlier that recombinant plasmid pMP81 carrying Tn601 can confer increased resistance to kanamycin and neomycin on yeast transformants (*Hollenberg* 1979a). Not only did the pMP81 transformants acquire higher resistance, but the kanamycin-inactivating activity could be demonstrated directly in cell free extracts (*Hollenberg* et al. 1980). Clearly the yeast cell is able to express the resistance gene encoding an aminoglycoside phosphotransferase that inactivates a number of aminoglycoside antibiotics by phosphorylation (*Davis* and *Smith* 1978). The sensitivity of *S. cerevisiae* to kanamycin or neomycin is not sufficient to make a direct selection for transformants practicable.

Fortunately *Jimenez* and *Davis* (1980) have discovered an aminoglycoside antibiotic G418 that is related to gentamicin and has a strong inhibitory effect on yeast. The resistance determinant on Tn601 also inactivates G418. Yeast transformants carrying Tn601 on a 2-μm DNA vector were found to be resistant to concentrations above 1 mg/ml of G418 (*Jimenez* and *Davis* 1980). We have used G418 to directly select transformants carrying pMP81 (Table 3). At the G418 concentrations optimal for transformant selection, a high number of spontaneously resistant colonies appeared. These could be strongly reduced on plates containing glycerol as the sole carbon source. G418 is the first antibiotic that in combination with vectors carrying Tn601 can be used for direct selection of transformants of wild yeast strains. This is especially valuable because at present no yeast resistance genes that inactivate antibiotics or prevent uptake are known or available as selective markers for transformation of wild-type strains. Known resistances that alter the primary structure of ribosomal proteins are generally recessive. One such gene, encoding trichodermin resistance in *S. cerevisiae*, has been cloned and shown to code for ribosomal protein L3 (*Fried* and *Warner* 1981). A direct transformation selection for plasmids carrying the trichodermin resistance gene was not possible. Future work will show whether dominant resistances can be found or made.

4.4 Expression of the ompA Gene

The cloned *E. coli ompA* gene, encoding the outer membrane protein II* (*Henning* et al. 1979), was inserted into a 2-μm DNA vector and introduced into *S. cerevisiae* (*Janowicz, Henning,* and *Hollenberg,* in preparation). Though the recombinant plasmid carried the LEU2 gene as selective marker, the recovery of yeast transformants carrying the *ompA* gene was extremely low. The positive transformants produce the II* protein, as identified by a specific antibody reaction. The II* protein synthesized in yeast has a molecular weight very close to or identical with that of the *E. coli* II* protein. Since protein II* is synthesized as a preprotein in *E. coli*, we assume that the II* signal sequence can be cleaved off by the yeast cell. A general observation is that yeast cells carrying the *ompA* gene grow slowly.

An *E. coli* cell contains about 10^5 copies of the II* polypeptide, located in the outer membrane (*Endermann* et al. 1978). Cloned *ompA* could only be maintained on a low copy number plasmid in the *E. coli* cell, probably because a high gene dosage is lethal to the cell (*Henning* et al. 1979). Protein II* could have a similar detrimental effect in yeast if, as we assume, it invades yeast membranes. The location of protein II* in the yeast transformants is currently under investigation.

4.5 Expression of Other Foreign Genes

The *E. coli lacZ* gene encoding β-galactosidase can be functionally expressed in *S. cerevisiae* (*Panthier* et al. 1980). The gene was present on a 2-μm DNA vector and positive transformants produced an activity comparable to 30 molecules of bacterial enzyme per cell.

Positive expression of eukaryotic genes that contain introns can be problematic as demonstrated with the rabbit β-globin gene (*Beggs* et al. 1980). cDNA clones, however, preferably under the control of a yeast promoter are more likely to lead to functional expression. A cDNA copy of the avian ovalbumin gene has been successfully expressed in *S. cerevisiae* (*Mercereau-Puijalon* et al. 1980).

Direct selection of a higher eukaryotic gene in yeast, namely a *Drosophila* gene that can substitute for the yeast ADE8 gene, was accomplished by *Henikoff* et al. (1980). The random pool of recombinant plasmids containing the *Drosophila* genome was passed through *E. coli* for amplification before *S. cerevisiae* was transformed. Evidence is presented that a 0.8 kb transcript from within the *Drosophila* DNA fragment is the mRNA from the ADE8 complementary gene. More recently, expression in yeast of a cDNA for leukocyte interferon D under control of the yeast ADH1 promoter was obtained (*Hitzeman* et al., 1981). Under optimal conditions the interferon content amounted to 8% of the cell protein.

4.6 Transcription and Translation of S. cerevisiae Genes

The study of transcription and translation in yeast has provided interesting results of significance for the expression of foreign genes. Extensive transcription data are available for the mating type genes (*Klar* et al. 1981; *Nasmyth* et al. 1981) and the *iso*-1-cytochrome *c* gene (*Faye* et al. 1980) and several novel findings have emerged.

Evidence has been presented for long-distance effects that block transcription of silent mating type genes. This phenomenon is generally described as position effect and could involve chromatin structure, nucleosome phasing or DNA modification to inactivate the promoter (*Klar* et al. 1981; *Nasmyth* et al. 1981). The transcripts of the mating type genes and the *iso*-1-cytochrome *c* gene have heterogenous 5′ ends probably due to multiple initiation sites. The 5′ termini of yeast *iso*-1-cytochrome *c* transcripts have been mapped to seven different locations ranging from 29 bp to 93 base pairs upstream from the initiating ATG codon. The same transcripts were found when the gene was inserted into a 2-μm DNA plasmid, but the level was about fifteen times higher than in a wild-type strain, presumably reflecting the higher copy number. Deletion mapping suggests the existence of four different functional regions within the promoter with large effects on the level and/or starting points of the transcripts. A control region for mRNA initiation located within the coding sequence as observed for several genes transcribed by eukaryotic RNA polymerase III seems to be ruled out for the *iso*-1-cytochrome *c* gene. Deletion of the central part of the gene still yielded RNA transcripts with the same spectrum of 5′ ends as the intact gene (*Faye* et al. 1980). Sequences important for initiation of transcription encompass at least a 250 bp flanking region adjacent to the starting codon of the gene.

The effect of *TY* elements (*Cameron* et al. 1979; *Farabaugh* and *Fink* 1980; *Gafner* and *Philippsen* 1980) on gene expression in yeast is intriguing. *TY* elements, present in about 35 copies in the yeast genome, are transcribed (*Cameron* et al. 1979), occur in different families related in DNA sequence (*Kingsman* et al. 1980) and bear resemblance to *copia* and *412* of *Drosophila* and transposable elements in *E. coli* (Cold Spring Harbor Symp Quant Biol Vol 45, 1980). In several cases, *TY* elements have been observed to affect gene expression in yeast in different ways. The presence of a *TY* element adjacent to the structural gene for *iso*-2-cytochrome *c* resulted in overproduction of the gene product (*Errede* et al. 1980) whereas a *TY* element near the 5′ end of the structural gene for ADHII led to constitutive expression of the normally glucose-repressible gene (*Williamson* et al. 1981). Evidence has been presented that mating signals control expression of various yeast genes which have become overproducers due to *TY* insertion (*Errede* et al. 1980).

Negative effects on expression have been described for the HIS4 gene. Integration of *TY1* upstream from the structural gene leads to *his⁻* phenotype (*Chaleff* and *Fink* 1980; *Roeder* and *Fink* 1980). *TY* elements can be excised except for a δ-sequence. This single δ-sequence permits almost normal gene function in the case of HIS4, but blocks gene function if present near the 5′ end of the ADHII structural gene (*Ciriacy* and *Williamson* 1981).

In the case of the gene for ADHII, the *TY* insertions leading to constitutive expression are located about 100 bp upstream from the coding region and all in the same orientation, suggesting direct interaction with transcription.

Expression vectors carrying cloned promoters similar to those available for bacterial systems (*Backman* and *Ptashne* 1978; *Charnay* et al. 1978; *Guarente* et al. 1980) can be expected soon (Compare *Hitzeman* et al. 1981). The low level of 2-μm DNA transcripts suggests that its promoters are not suitable for high gene expression.

In addition to efficient transcription, good expression of a cloned gene requires that the mRNA is efficiently translated. The requirements for efficient translation in yeast have not yet been studied in detail. A ribosome binding site (*Shine* and *Dalgarno* 1975) that in bacterial mRNAs precedes the AUG codon by 4–11 bases, has not been detect-

ed in yeast mRNA. Translation studies on *iso*-1-cytochrome *c* in yeast indicate that translation starts at the AUG codon closest to the 5' end of the mRNA and can initiate with normal efficiency at any site within a region of 25 nucleotides around the wild-type AUG position (*Sherman* et al. 1980). Data on more mRNAs are required to permit a general conclusion. Effects of sequences further upstream from the AUG codon cannot be excluded. The leader sequence on *iso*-1-cytochrome *c* mRNA varies between 30 and 90 bp among the different transcripts. The most frequent mRNAs have leader sequences of 35 and 45 bases. It is not known which mRNAs are translated most efficiently.

Maximal translation of a mRNA will further depend on codon usage and product turnover, two aspects that can be determined with the help of synthetic genes.

Acknowledgments. We thank Paul Hardy for critical reading of the manuscript. This work was supported by the Deutsche Forschungsgemeinschaft.

References

Ambler RP, Scott GK (1978) Partial amino acid sequence of penicillinase coded by Escherichia coli plasmid R6K. Proc Natl Acad Sci USA 75:3732–3736

Backman K, Ptashne M (1978) Maximizing gene expression on a plasmid using recombination in vitro. Cell 13:65–71

Beggs JD (1978) Transformation of yeast by a replicating hybrid plasmid. Nature 275:104–109

Beggs JD, van den Berg J, van Ooyen A, Weissmann C (1980) Abnormal expression of chromosomal rabbit β-globin gene in Saccharomyces cerevisiae. Nature 283:835–840

Bernardi G, Faures M, Piperno G, Slonimski PP (1970) Mitochondrial DNA's from respiratory-sufficient and cytoplasmic respiratory-deficient mutant yeast. J Mol Biol 48:23–42

Blanc H, Gerbaud C, Slonimski PP, Guerineau M (1979) Stable yeast transformation with chimeric plasmids using a 2 μm-circular DNA-less strain as a recipient. MGG 176:335–342

Broach JR, Hicks JB (1980) Replication and recombination functions associated with the yeast plasmid, 2μ circle. Cell 21:501–508

Broach JR, Atkins JF, McGill C, Chow L (1979) Identification and mapping of the transcriptional and translational products of the yeast plasmid, 2μ circle. Cell 16:827–839

Cameron JR, Philippsen P, Davis RW (1977) Analysis of chromosomal integration and deletion of yeast plasmids. Nucleic Acids Res 4:1429–1448

Chaleff DT, Fink GR (1980) Genetic events associated with an insertion mutation in yeast. Cell 21:227–237

Charnay P, Perricaudet M, Galibert F, Tiollais P (1978) Bacteriophage lambda and plasmid vectors, allowing fusion of cloned genes in each of the three translational phases. Nucleic Acids Res 5:4479–4494

Chevallier MR, Aigle M (1979) Qualitative detection of penicillinase produced by yeast strains carrying chimeric yeast-coliplasmids. FEBS Lett 180:179–180

Chevallier MR, Bloch JC, Lacroute F (1980) Transcriptional and translational expression of a chimeric bacterial-yeast plasmid in yeasts. Gene 11:11–19

Ciriacy M, Williamson VM (1981) Analysis of mutations affecting Ty-mediated gene expression in Saccharomyces cerevisiae. MGG (to be published)

Clark-Walker GD, Azad AA (1980) Hybridizable sequences between cytoplasmic ribosomal RNAs and 3 micron circular DNAs of Saccharomyces cerevisiae and Torulopsis glabrata. Nucleic Acids Res 8:1009–1023

Clark-Walker GD, Miklos GL (1974) Localization and quantification of circular DNA in yeast. Eur J Biochem 41:359–365

Cohen JD, Abrams E, Eccleshall TR, Buchferer B, Marmur J (1980b) Expression of a prokaryotic gene in yeast: isolation and characterization of mutants with increased expression. Abstracts

of 10th Int Conference of Yeast Genetics and Molecular Biology, Louvain-la-Neuve, Belgium, p 127

Cohen JD, Eccleshall TR, Needleman RB, Federoff H, Buchferer BA, Marmur J (1980a) Functional expression in yeast of the Escherichia coli plasmid gene coding for chloramphenicol acetyltransferase. Proc Natl Acad Sci USA 77:1078-1082

Covey C, Richardson D, Carbon J (1976) A method for the deletion of restriction sites in bacterial plasmid deoxyribonucleic acid. MGG 145:155-158

Datta DB, Arden B, Henning U (1977) Major proteins of the Escherichia coli outer cell envelope membrane as bacteriophage receptors. J Bacteriol 131:821-829

Davis J, Smith DI (1978) Plasmid-determined resistance to antimicrobial agents. Annu Rev Microbiol 32:469-518

Dobsen MJ, Futcher AB, Cox BS (1980a) Control of recombination within and between DNA plasmids of Saccharomyces cerevisiae. Curr Genetics 2:193-200

Dobsen JM, Futcher AB, Cox BS (1980b) Loss of 2-μm DNA from Saccharomyces cerevisiae transformed with the chimeric plasmid pJDB219. Curr Genetics 2:201-206

Endermann R, Krämer C, Henning U (1978) Major outer membrane proteins of Escherichia coli K-12: evidence for protein II* being a transmembrane protein. FEBS Lett 86:21-24

Erhart E, Hollenberg CP (1981) Curing of Saccharomyces cerevisiae 2-μm DNA by transformation. Curr Genetics 3:83-89

Errede B, Cardillo TS, Sherman F, Dubois E, Deschamps J, Wiame JM (1980) Mating signals control expression of mutations resulting from insertion of a transposable repetitive element adjacent to diverse yeast genes. Cell 25:427-436

Farabaugh PJ, Fink GR (1980) Insertion of the eukaryotic transposable element Ty1 creates a 5-base pair duplication. Nature 286:352-355

Faye G, Leung DW, Tatchell K, Hall BD, Smith M (1981) Deletion mapping of sequences essential for in vivo transcription of the iso-1-cytochrome c gene. Proc Natl Acad Sci USA

Fried HM, Warner JR (1981) Cloning of yeast gene for trichodermin resistance and ribosomal protein L3. Proc Natl Acad Sci USA 78:238-242

Gafner J, Philippsen P (1980) The yeast transposon Ty1 generates duplications of target DNA on insertion. Nature 286:414-418

Gallwitz D, Sures I (1980) Structure of a split yeast gene: complete nucleotide sequence of the actin gene in Saccharomyces cerevisiae. Proc Natl Acad Sci USA 77:2546-2550

Garten W, Hindennach I, Henning U (1975) The major proteins of the Escherichia coli outer cell envelope membrane characterization of protein II* and III, comparison of all proteins. Eur J Biochem 59:215-221

Gerbaud C, Guerineau M (1980) 2 μm plasmid copy number in different yeast strains and repartition of endogenous and 2 μm chimeric plasmids in transformed strains. Curr Genetics 1:219-228

Gerbaud C, Fournier P, Blanc H, Aigle M, Heslot H, Guerineau M (1979) High frequency of yeast transformation by plasmids carrying part of entire 2-μm yeast plasmid. Gene 5:233-253

Guarente L, Lauer G, Roberts TM, Ptashne M (1980) Improved methods for maximizing expression of a cloned gene: a bacterium that synthesizes rabbit β-globin. Cell 20:543-553

Guerineau M (1979) Plasmid DNA in yeast. In: Lemke PA (ed) Viruses and plasmids in fungi. Marcel Dekker, New York, pp 539-593

Guerineau M, Grandchamp C, Slonimski PP (1976) Circular DNA of a yeast episome with two inverted repeats: Structural analysis by a restriction enzyme and electron microscopy. Proc Natl Acad Sci USA 73:3030-3034

Gunge N, Tamaru A, Ozawa F, Sakaguchi K (1981) Isolation and characterization of linear deoxyribonucleic acid plasmids from Kluyveromyces lactis and the plasmid-associated killer character. J Bacteriol 145:382-390

Hartley JL, Donelson JE (1980) Nucleotide sequence of the yeast plasmid. Nature 286:860-865

Henikoff S, Tatchel K, Hall BD, Nasmyth KA (1981) Isolation of a gene from Drosophila by complementation in yeast. Nature 289:33-37

Henning U, Royer HD, Teather RM, Hindennach I, Hollenberg CP (1979) Cloning of the structural gene (ompA) for an integral outer membrane protein of Escherichia coli K-12. Proc Natl Acad Sci USA 76:4360-4364

Herrmann R, Neugebauer K, Pirkl E, Zentgraf H, Schaller H (1980) Conversion of bacteriophage

fd into an efficient single-stranded DNA vector system. MGG 177:231–242

Hicks JB, Hinnen A, Fink GR (1979) Properties of yeast transformation. Cold Spring Harbor Symp Quant Biol 43:1305–1313

Hinnen A, Hicks JB, Fink GF (1978) Transformation of yeast. Proc Natl Acad Sci USA 75:1929–1933

Hitzeman RA, Hagie FE, Goeddel DV, Ammerer G, Hall BD (1981) Expression of a human gene in yeast. Nature (to be published)

Hollenberg CP (1978) Mapping of regions on cloned Saccharomyces cerevisiae 2-μm DNA coding for polypeptides synthesized in Escherichia coli minicells. MGG 162:23–34

Hollenberg CP (1979a) The expression in Saccharomyces cerevisiae of bacterial β-lactamase and other antibiotic resistance genes integrated in A 2-μm DNA vector. ICN-UCLA Symp Mol Cell Biol 15:325–338

Hollenberg CP (1979b) The expression of bacterial antibiotic resistance genes in the yeast Saccharomyces cerevisiae. In: Timmis KN, Pühler A (eds) Plasmids of medical, environmental and commercial importance. Elsevier/North-Holland Biomedical Press, pp 481–494

Hollenberg CP, Borst P, van Bruggen EFJ (1970) Mitochondrial DNA. V. A 25-μ closed circular duplex DNA molecule in wild-type yeast mitochondria. Structure and genetic complexity. Biochim Biophys Acta 209:1–15

Hollenberg CP, Kustermann-Kuhn B, Royer HD (1976) Synthesis of high molecular weight polypeptides in Escherichia coli minicells directed by cloned Saccharomyces cerevisiae 2-μm DNA. Gene 1:33–47

Hollenberg CP, Kustermann-Kuhn B, Mackedonski V, Erhart E (1980) The expression of bacterial antibiotic resistance genes in the yeast Saccharomyces cerevisiae. Alfred Benzon Symp 16:109–120

Hsiao CL, Carbon J (1979) High-frequency transformation of yeast by plasmids containing the cloned yeast ARG4 gene. Proc Natl Acad Sci USA 76:3829–3833

Ilgen C, Farabaugh PJ, Hinnen A, Walsh JM, Fink GR (1979) Transformation of yeast. In: Setlow JK, Hollaender A (eds) Genetic engineering principles and methods, vol 1. Plenum Press, New York London, pp 117–132

Jimenez A, Davies J (1980) Expression of a transposable antibiotic resistance element in Saccharomyces cerevisiae; a potential selection for eukaryotic cloning vectors. Nature 287:869–871

Kielland-Brandt MC, Nilsson-Tillgren T, Holmberg S, Litske Petersen JG, Svenningsen BA (1979) Transformation of yeast without the use of foreign DNA. Carlsberg Res Commun 44:77–87

Kielland-Brandt MC, Wilken B, Holmberg S, Litske Petersen JG, Nilsson-Tillgren T (1980) Genetic evidence for nuclear location of 2-micron DNA in yeast. Carlsberg Res Commun 45:119–124

Kingsman AJ, Gimlich RL, Clarke L, Chinault AC, Carbon J (1981) Sequence variation in dispersed repetitive sequences in Saccharomyces cerevisiae. J Mol Biol 145:619–632

Koshland D, Botstein D (1980) Secretion of Beta-lactamase requires the carboxy end of the protein. Cell 20:749–760

Klar AJS, Strathern JN, Broach JR, Hicks JB (1981) Regulation of transcription in expressed and unexpressed mating type cassettes of yeast. Nature 289:239–244

Kreft J, Bernhard K, Goebel W (1978) Recombinant plasmids capable of replication in B. subtilis and E. coli. MGG 162:59–67

Kretschmer PJ, Chang CY, Cohen SN (1975) Indirect selection of bacterial plasmids lacking identifiable phenotypic properties. J Bacteriol 124:225–235

Larionov VL, Grishin AV, Smirnov MN (1980) 3 μm DNA – an extrachromosomal ribosomal DNA in the yeast Saccharomyces cerevisiae. Gene 12:41–49

Livingston DM (1977) Inheritance of the 2-μm DNA plasmid from Saccharomyces. Genetics 86:73–84

Livingston DM, Hahne S (1979) Isolation of a condensed, intracellular form of the 2-μm DNA plasmid of Saccharomyces cerevisiae. Proc Natl Acad Sci USA 76:3727–3731

Livingston DM, Klein HL (1977) Deoxyribonucleic acid sequence organization of a yeast plasmid. J Bacteriol 129:472–481

Livingston DM, Kupfer DM (1978) Control of Saccharomyces cerevisiae 2-μm DNA replication by cell division cycle genes that control nuclear DNA replication. J Mol Biol 116:249–260

McNeil JB, Storms RK, Friesen JD (1980) High frequency recombination and the expression of genes cloned on chimeric yeast plasmids: identification of a fragment of 2-μm circle essential

for transformation. Curr Genetics 2:17–25

Mercereau-Puijalon O, Lacroute F, Kourilsky P (1980) Synthesis of a chicken ovalbumin-like protein in the yeast Saccharomyces cerevisiae. Gene 11:163–167

Meyerink JH, Klootwijk J, Planta RJ, van der Ende A, van Bruggen EFJ (1979) Extrachromosomal circular ribosomal DNA in the yeast Saccharomyces carlsbergensis. Nucleic Acids Res 7:69–75

Nasmyth KA, Tatchell K, Hall BD, Astell C, Smith (1981) A position effect in the control of transcription at yeast mating type loci. Nature 289:244–252

Nelson RG, Fangman WL (1979) Nucleosome organization of the yeast 2-μm DNA plasmid: a eukaryotic minichromosome. Proc Natl Acad Sci USA 76:6515–6519

Ng R, Abelson J (1980) Isolation and sequence of the gene for actin in Saccharomyces cerevisiae. Proc Natl Acad Sci USA 77:3912–3916

Nilsson-Tillgren T, Litske Petersen JG, Holmberg S, Kielland-Brandt MC (1980) Transfer of chromosome III during kar mediated cytoduction in yeast. Carlsberg Res Commun 45:113–117

Novick P, Field C, Schekman R (1980) Identification of 23 complementation groups required for post-translational events in the yeast secretory pathway. Cell 21:205–215

O'Callaghan CM, Morris A, Kirby SM, Shingler AH (1972) Novel method for detection of β-lactamases by using a chromogenic cephalosporin substrate. Antimicrobial Agents and Chemotherapy 1:283–288

Panthier JJ, Fournier P, Heslot H, Rambach A (1980) Cloned β-galactosidase gene of Escherichia coli is expressed in the yeast Saccharomyces cerevisiae. Curr Genetics 2:109–113

Petes TD (1980) Molecular genetics of yeast. Annu Rev Biochem 49:845–876

Petes TD, Williamson DH (1975) Replicating circular DNA molecules in yeast. Cell 4:249–253

Roeder GS, Fink GR (1980) DNA rearrangements associated with a transposable element in yeast. Cell 21:239–249

Ratzkin B, Carbon J (1977) Functional expression of cloned yeast DNA in Escherichia coli. Proc Natl Acad Sci USA 74:487–491

Roggenkamp R, Kustermann-Kuhn B, Hollenberg CP (1981) Expression and processing of bacterial β-lactamase in the yeast Saccharomyces cerevisiae. Proc Natl Acad Sci USA 78:4466–4470

Royer HD, Hollenberg CP (1977) Saccharomyces cerevisiae 2-μm DNA. An analysis of the monomer and its multimers by electron microscopy. Molec Gen Genet 150:271–284

Seligy VL, Thomas DY, Miki BLA (1980) Saccharomyces cerevisiae plasmid, Scp or 2 μm: Intracellular distribution, stability and nucleosomal-like packages. Nucleic Acids Res 8:3371–3391

Shaw WV (1975) Chloramphenicol acetyltransferase from chloramphenicol resistant bacteria. Meth Enzymol 43:737–755

Van Solingen P, van der Plaat JB (1977) Fusion of yeast spheroplasts. J Bacteriol 130:946–947

Sutcliffe JG (1978) pBR322 restriction map derived from the DNA sequence: accurate DNA size markers up to 4361 nucleotide pairs long. Nucl Acids Res 5:2721–2728

Stewart GG, Russell I, Panchal CJ (1980) The genetic manipulation of industrial yeast strains. Abstracts VI. Int Symp on Yeasts, London (Ontario), Canada, p 212

Stinchcomb DT, Thomas M, Kelly J, Selker E, Davis RW (1980) Eukaryotic DNA segments capable of autonomous replication in yeast. Proc Natl Acad Sci USA 77:4559–4563

Storms RK, McNeil JB, Khandekar PS, An G, Parker J, Friesen JD (1979) Chimeric plasmids for cloning of deoxyribonucleic acid sequences in Saccharomyces cerevisiae. J Bacteriol 140:73–82

Struhl K, Cameron JR, Davis RW (1976) Functional genetic expression of eukaryotic DNA in Escherichia coli. Proc Natl Acad Sci USA 73:1471–1475

Struhl K, Stinchcomb DT, Scherer S, Davis RW (1979) High-frequency transformation of yeast: autonomous replication of hybrid DNA molecules. Proc Natl Acad Sci USA 76:1035–1039

Tabak HF (1977) Absence of 2 μm DNA sequences in Saccharomyces cerevisiae Y 379-5D. FEBS Letters 84:67–70

Talmadge K, Stahl S, Gilbert W (1980a) Eukaryotic signal sequence transports insulin antigen in Escherichia coli. Proc Natl Acad Sci USA 77:3369–3373

Talmadge K, Kaufman J, Gilbert W (1980b) Bacteria mature preproinsulin to proinsulin. Proc Natl Acad Sci USA 77:3988–3992

Sherman F, Stewart JW, Schweingruber AM (1980) Mutants of yeast initiating translation of iso-1-cytochrome c within a region spanning 37 nucleotides. Cell 20:215–222

Shine J, Dalgarno L (1975) Determinant of cistron specificity in bacterial ribosomes. Nature 254: 34–38

Southern EM (1975) Detection of specific sequences among DNA fragments separated by gel electrophoresis. J Mol Biol 98:503–517

Thomas DY, James AP (1980) Transformation of Saccharomyces cerevisiae with plasmids containing fragments of yeast 2μ DNA and a suppressor tRNA gene. Curr Genetics 2:9–16

Toh-e A, Guerry-Kopecko P, Wickner RB (1980) A stable plasmid carrying the yeast LEU2 gene and containing only yeast deoxyribonucleic acid. J Bacteriol 141:413–416

Tschumper G, Carbon J (1980) Sequence of a yeast DNA fragment containing a chromosomal replicator and the TRP1 gene. Gene 10:157–166

Wigler M, Sweet R, Sim GK, Wold B, Pellicer A, Lacy E, Maniatis T, Silverstein S, Axel R (1979) Transformation of mammalian cells with genes from prokaryotes and eukaryotes. Cell 16:777–785

Zakian VA, Brewer BJ, Fangman WL (1979) Replication of each copy of the yeast 2 micron DNA plasmid occurs during the S phase. Cell 17:923–934

Williamson VM, Young ET, Ciriacy M (1981) Transposable elements associated with constitutive expression of yeast alcohol dehydrogenase II. Cell 23:605–614

Selectable Markers for the Transfer of Genes into Mammalian Cells

F. Colbère-Garapin*, A. Garapin**, and P. Kourilsky**

1 Introduction

The introduction of purified DNA into mammalian cells is an important approach to the understanding of gene expression and regulation, as well as, to engineering cell lines endowed with new properties. A variety of procedures can be used, including precipitation with calcium phosphate (*Graham* and *Van der Eb* 1973), microinjection (*Capecchi* 1980 an references therein), fusion of the cells with liposomes (*Wong* et al. 1980; *Fraley* et al. 1980), microcells (*Smiley* et al. 1978), or bacterial protoplasts (*Schaffner* 1980). Following such manipulations, some of the exogenous DNA either remains transiently within the living cells (usually for a few days) or becomes stably associated with the dividing cells (usually as the result of its integration within the host chromosomes). Since the latter event is rather rare, it is desirable and, in many instances, essential, to select for it, which requires selectable markers. In this survey, we focus mainly on the Herpes simplex virus type I thymidine kinase gene (referred to as Herpes TK gene), which has been used in many of the gene transfer experiments performed to date.

2 The Herpes TK Gene

Thymidine kinase is found in all growing mammalian cells tested so far. There are at least two enzymatic activities, one (F) found in the cytosol and the other (A) found mostly in

---

* Unité de Virologie Médicale
** Groupe de Biologie Moléculaire du Gène, E.R. C.N.R.S. 201 and S.C.N. I.N.S.E.R.M. 20
   Institut Pasteur, 28 rue du Dr. Roux, 75724 Paris Cédex 15, France

mitochondria, although it is probably coded for by nuclear DNA (*Kit* et al. 1974). Cells deficient in the cytosol activity (F) can be obtained after repeated passage in a bromodeoxyuridine (BUdR) containing medium. Some of these mutants (denoted TK⁻ cells although they retain A activity) but not all, are very stable. For instance, the widely used fibroblastic TK⁻ mouse L cells (clone 1D) (*Kit* et al. 1963) have never been reported to revert. TK⁺ cells can be selected out of a TK⁻ population after growth in HAT medium containing hypoxanthine, aminopterin, and thymidine, where TK⁻ cells die in a few days (*Szybalska* and *Szybalski*, 1962).

Cells infected by certain DNA viruses show an increase in thymidine kinase activity. In some cases (papova- and adenoviruses) it is due to induction of cellular F activity. In others (vaccinia and a number of Herpes viruses), it corresponds to the synthesis of a virus-coded enzyme. The Herpes simplex virus type 1 enzyme (Herpes TK) has been shown to differ from cellular enzymes by biochemical, serological, and genetic methods. Herpes TK is more efficient in phosphorylating iododeoxycytidine, which allows an easy measurement of the viral enzymatic activity in the presence of the cellular TK (s) (*Summers* and *Summers* 1977). It also phosphorylates acycloguanosine, a drug which is not toxic for normal cells (which cannot utilize it) but will inhibit replication upon phosphorylation in virus-infected cells (*Crumpacker* et al. 1980 and references therein).

Complementation of TK⁻ cells by Herpes TK in cells infected by irradiated virus was first demonstrated by *Munyon* et al. (1971). Since the viral gene could be isolated much more easily than a cellular TK gene, it was an early candidate to serve as a selective marker. *Wigler* et al. (1977) localized the Herpes simplex virus type I TK gene within a specific 3.4 kb *Bam*H1 restriction fragment of the viral genome. We, and others, have cloned this fragment (or a larger *Kpn*I fragment) in a bacterial plasmid (*Colbère-Garapin* et al. 1979; *Wilkie* et al. 1979; *Enquist* et al. 1979; *McKnight* and *Gavis* 1980). Specifically, we ligated the purified 3.4-kb *Bam*H1 fragment with *Bam*H1 cleaved pBR322 DNA. Its insertion inactivated the plasmid tetracycline resistance gene, such that recombinants were scored as ampicillin resistant and tetracycline sensitive. After cloning in *Escherichia coli*, the Herpes TK gene carried in this pFG5 plasmid transformed TK⁻ mouse L cells (clone 1D) with about the same efficiency as the viral gene not cloned in *E. coli* (2–20 colonies per 10^{10} gene equivalents using the calcium technique). This suggested that mouse cells do not severely restrict DNA grown in bacteria, contrary to bacterial cells, which usually restrict foreign DNA.

The number of transformants was proportional to the amount of input DNA, indicating that one TK gene is sufficient to induce transformation. Linearized plasmid DNA transformed mouse cells two to three times more efficiently than supercoiled DNA (*Colbère-Garapin* et al. 1979).

3 Structure of the TK Gene

To map the TK gene within the cloned 3.4 kb *Bam*H1 fragment, we built a detailed restriction map of the latter and correlated it with the transforming activity of the gene carried in pFG5 after cleavage with a variety of enzymes. All transforming activity was lost upon digestion by many enzymes, but total or partial (1%–10%) activity was retained upon digestion by *Pvu*II or EcoRI and *Sma*I, respectively. The 2-kb *Pvu*II fragment was

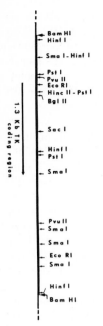

Fig. 1. Restriction map of the *Bam*HI herpes DNA fragment carrying the TK gene. The TK coding region and the orientation of the gene are shown

recloned in the *Pvu*II site of pBR322 to yield pAGO, a plasmid as efficient as pFG5 in the transformation of mouse cells, but smaller and more convenient since it carries both ampicillin and tetracyclin resistance genes and several unique restriction sites useful for cloning. Most of the TK coding sequence could thus be positioned in a 1.3-kb *Pvu*II-*Sma*I segment (*Colbère-Garapin* et al. 1979; Fig. 1).

The orientation of the TK gene was determined by three point crosses with viruses carrying mutations in the TK and DNA polymerase genes (*Smiley* et al. 1980). It was confirmed by our studies on the expression of the Herpes TK gene in *E. coli*. The *Eco*RI fragment containing the TK coding sequence was fused to the very beginning of the *E. coli lac* Z gene, and the recombinant molecules were introduced into thymidine kinase deficient *E.coli* strain. Bona fide complementation was observed only when the *Eco*RI TK fragment was oriented such that the *Bgl*II and *Hinc*II sites were close to the *lac* promoter, thus indicating the polarity of the gene (*Garapin* et al. 1981).

In these experiments, we encountered the paradoxical situation that functional TK was produced under *lac* promoter control in each of the three reading frames starting in the *lac* Z gene. The same TK-like polypeptides, in the range of 40–42 000 daltons, were immunoprecipitated from *E. coli* extracts in all three cases. We sequenced the corresponding region of the TK gene and found stop signals in all three phases close to the *Eco* RI site used for fusion with *lac* Z. Thus, the TK polypeptide was not fused to the first amino acids of β-galactosidase; instead, reinitiation of translation took place in Herpes DNA (*Garapin* et al. 1981).

Complementary to our partial sequence, the complete sequence of the Herpes TK gene was determined (*McKnight* 1980; *Wagner* et al. 1981). In the first 200 base pairs located at the 5' end of the 2 kb *Pvu*II fragment, there are two AT-rich sequences, known

as TATA boxes, similar to those found at the 5' end of other eukaryotic genes, where they have been shown to be involved in promoting transcription (see, for example, *Mathis* and *Chambon* 1981). The sequence CATATTAA closest to the *Hinc*II site at 5' of the gene is the one likely to be used *in vivo* for several reasons: 1. It lies at the expected distance (some 30 nucleotides) from the site where the 5' end of TK mRNA has been mapped; and 2. The other TATA box is deleted in DNA segments engineered by *McKnight* and *Gavis* (1980) which retain full TK coding capacity upon injection into *Xenopus laevis* oocytes. The sequence GGCGAATTC found 85 nucleotides upstream of the mRNA start is similar to a consensus sequence found in many eukaryotic genes in the same position (*Benoist* et al. 1980). In the TK gene, this sequence encompasses the *Eco*RI site at 5' of the gene. The finding that the *Eco*RI-cleaved TK gene transforms with poor efficiency points to a functional significance of this sequence in vivo. From nucleotide sequencing, the first AUG in the mRNA sequence occurs 107 nucleotides from its 5' end. It is preceded (29 nucleotides upstream) by a sequence showing some complementarity with the 3' end of 18 S ribosomal RNA, which may function as a possible ribosome binding site (*Hagenbüchle* et al. 1978). It is followed by an open reading frame coding for a polypeptide chain of 376 residues, the expected size of the viral enzyme. Downstream of the stop codon, the sequence AATAAAA is found twice. It has been suggested that this sequence signals the polyadenylation site (*Proudfoot* and *Brownlee* 1976), which has been mapped by *McKnight* (1980) some 16 nucleotides downstream of the first heptamer. *Sma*I cuts within the coding sequence but only a few triplets before the stop codon, making it understandable that the *Sma*I-digested TK gene retains some transforming activity (*Colbère-Garapin* et al. 1979). Finally, several lines of evidence, including DNA sequencing, S 1 mapping experiments (*McKnight* 1980; *Wagner* et al. 1981), and, more indirectly, expression of TK in *E. coli* (*Garapin* et al. 1981) (which is known not to carry out RNA splicing, *Mercereau-Puijalon* and *Kourilsky* 1979) converge to demonstrate that there is no intervening sequence in the TK gene. This agrees with the size of the TK mRNA (*Cremer* et al. 1978).

4 Transformation of TK⁻ Cells with the Herpes TK Gene

TK⁺ selection in HAT medium was in widespread use in somatic cell genetics well before the cloned Herpes TK gene was available (*Szybalska* and *Szybalski* 1962; *Littlefield* 1963). Selection based upon the transfer of the cellular chromosomal TK gene have been particularly useful in cell fusion experiments and the transfer of chromosomes — the latter being either isolated as metaphase chromosomes (*Klobutcher* and *Ruddle* 1979; *Scangos* et al. 1979) or present in microcells obtained by treatment of normal cells with colcemid and cytochalasin B (*Smiley* et al. 1978) (for review, see *McBride* and *Peterson* 1980). The procedure used to select TK⁺ cells from TK⁻ cells transfected with Herpes TK DNA is essentially the same, with minor modifications.

Herpes TK enzyme differs from the cellular human and murine enzymes in several respects. These differences were exploited by *Mercola* et al. (1980) to devise a selection procedure in which TK⁺ cells which acquire the Herpes TK gene can be isolated in a modified HAT medium.

DNA transfer can be accomplished by various means. The most broadly used has been the calcium precipitation technique first described by *Graham* and *van der Eb* (1973) and modified by *Wigler* et al. (1977, 1979b). In this technique, only a small proportion of

the cells receive and stably express the Herpes TK gene (10^{-6}%–1%of the cell population). Similarly, only a tiny fraction of the DNA molecules added (10^{-8} or less) finish as active genes in the selected TK$^+$ cells. These figures are somewhat increased when transfection is carried out in the presence of helper DNA (see below).

The low proportion of cells and DNA molecules which participate in a successful transformation event raises questions as to whether particular cells or DNA molecules are selected in this process. Although this cannot be ruled out presently it should be noted that other procedures, particularly microinjection, yield much higher transformation efficiencies. Purified Herpes TK gene can be injected into the cytoplasm and the nucleus of individual mammalian cells, even as small as TK$^-$ mouse L cells. In a series of informative experiments, *Capecchi* (1980) showed that transformation could take place in 20% of microinjected cells provided that: 1. DNA was injected into the nucleus, rather than into the cytoplasm where much lower efficiencies were observed, and 2. a viral sequence isolated from the SV40 genome (see below) was present adjacent to TK. Carrier DNA was unnecessary and most transformants were stable.

These data are compatible with the simple model that, in the stable transformation of a TK$^-$ cell into TK$^+$ with the Herpes TK gene, the DNA must successfully pass through three major steps:

1. The DNA has to enter the cell. Unless it is microinjected into the nucleus, it reaches the cytoplasm.

2. It must then transit from the cytoplasm to the nucleus.

3. In the nucleus, stable transformation will usually result from integration of the DNA into the chromosomes. This is not sufficient per se to guarantee expression since: a) gene activity will be destroyed if the gene is split in the course of integration; and b) gene activity may depend on the chromosomal site of integration. Therefore, the DNA must become integrated in a configuration adequate for expression at the appropriate level of selection.

This simple formalization should not obscure additional complexities. It is conceivable, for instance, that some ways of delivering DNA to the cells, or the physical state of the DNA, may activate DNA repair mechanisms which could favor integration. Thus, the above steps may include several distinct events and not be independent of each other.

Parameters Involved in the Transformation of TK$^-$ Cells with the Herpes TK Gene

Maximum transformation requires an intact TK gene. Genes truncated by *Eco*RI (in the GGCGAATTC sequence 85 nucleotides upstream of the mRNA start) or *Sma*I (a few triplets before the stop codon in 3′) exhibit low transformation efficiencies. The reasons for this behavior have not been fully explored yet and raise interesting possibilities. For instance, integration could take place in sites which compensate for the defect; or multiple copies may become integrated in order to balance lower gene expression or enzyme activity, etc.

Transformation varies drastically with the recipient cell. Of the various cell types used to date, clone 1D of TK$^-$ mouse L cells appears the most susceptible to transformation. Efficiencies in the order of 10^4 colonies per µg of DNA per 10^6 cells have been reported with the calcium phosphate technique (*Pellicer* et al. 1978, 1980a; *Wold* et al. 1979). Other

TK ⁻ cell lines, such as the human 143 BUdR line of Huebner and Croce (*Bachetti* and *Graham*, 1977) or mouse teratocarcinoma cells (*Pellicer* et al. 1980a; *Linnenbach* et al. 1980), are much less permissive (about 1 colony per 1–5 μg of DNA per 10^6 cells). TK^+ mouse bone marrow cells transformed with the Herpes TK gene in a modified HAT medium (*Mercola* et al. 1980) were also poorly susceptible. Whether such drastic differences reflect differences in cell permeability, activity of nucleolytic enzymes and recombination systems, or any other events involved in the three steps defined above is largely unknown at present.

Transformation efficiency is greatly influenced by the way in which DNA is delivered to the cells. As mentioned above, microinjection is very efficient, but is not adapted to the treatment of large cell populations. By contrast, the calcium phosphate technique is suitable for the transfection of cell populations, but lower efficiencies are obtained. Other methods have been tried, such as trapping of the DNA within liposomes (*Wong* et al. 1980; *Fraley* et al. 1980). Fusion of mammalian cells with bacterial protoplasts containing amplified plasmids might prove to be a convenient tool (*Schaffner* 1980). It must be emphasized that many of these methods have not been carefully calibrated yet (number of DNA molecules delivered per cell, etc.) nor evaluated with regard to their subsequent biological impact.

The addition of carrier DNA (usually eukaryotic DNA) increases transformation efficiency. It amounts, however, to the cotransfer of large quantities of DNA with unspecified structure and function and results in a loss of definition of transformation systems, which has initially been underestimated (see Sect. 5).

Questions addressed to the role of carrier DNA can be formulated in a more general way: Do nonspecific or specific DNA sequences influence transformation? It is indeed conceivable that specific sequences may promote integration, replication, or integration of an (adjacent) TK gene. For example, some repeated sequences could stimulate integration by promoting homologous recombination with, or transposition into, the chromosomes of the recipient cells.

We mentioned above the finding that a portion of the SV40 genome, when built into the same molecule as the TK gene, enhances transformation of TK⁻ into TK^+ cells dramatically. *Capecchi* (1980) showed that the stimulation is not due to the ability of the SV40 genome to replicate or to induce neoplastic transformation. It could be due to the presence, in the viral genome, of an integration- or expression-promoting signal. In this respect, there is increasing evidence that at least one copy of the two 72-bp repeat sequences close to the SV40 origin of replication is essential in *cis* for the expression of early and late SV40 genes or genes inserted in the molecule (see review by *Gruss* and *Khoury*, this volume). Whether the stimulation of TK transformation is due to the enhanced expression of the TK gene caused by this sequence is not clear.

These experiments represent the first well-documented example of a *cis*-acting DNA sequence influencing transformation, a line of research likely to be productive in the near future. *Herpes saimiri* repetitive DNA (*Fleckenstein* 1979) has been included into derivatives of pAGO (Colbère-Garapin et al. 1981). The corresponding plasmids yielded slightly higher transformation efficiency than pAGO; however, TK activity in extracts of transformed cells was on the average five times higher (our unpublished results).

5 Cotransfer of Genes Along with the Herpes TK DNA

One of the first genes which was cotransferred and expressed along with the Herpes TK marker was the Herpes simplex virus N103 gene: upon transfection of TK⁻ cells with sheared HSVI genomes, a few TK⁺ colonies acquired the ability to complement thermosensitive HSVI mutants defective in N103. TK⁻ revertants lost this ability (*Minson* et al. 1978). In these experiments, the two genes were probably carried by the same DNA molecule. *Wigler* et al. (1979b) were first to demonstrate the cotransfer of unlinked DNA sequences. They showed that, in cells cotransfected with Herpes TK DNA and plasmid pBR322, or phage ØX174, or rabbit β-globin gene DNA in excess of TK DNA, most of the TK⁺ transformants had acquired the additional DNA sequence.

An alternative procedure consists in presenting the cell with gene sequences previously ligated in vitro with marker DNA, or cloned in the same recombinant molecule. These various methods may not be as different as they look because it is now believed that exogenous DNA, at least when introduced into cells by the calcium phosphate technique in the presence of carrier DNA, ligates in vivo to form large concatenates prior to integration (*Perucho* et al. 1980a, b). This conclusion derives from the demonstration that, even though sequences were not ligated prior to transfer, they are genetically linked in transformants, since TK⁻ revertants usually lost the additional sequence. Moreover, plasmid DNA was shown to be embodied within carrier DNA sequences in the transformed cells (*Perucho* et al. 1980b). These observations raise questions about the actual function of carrier DNA in the transformation process. It could provide replication origins such that the concatenate structure could be replicated and persist in a semi-stable episomic state for a number of generations. This would increase its probability of being eventually integrated. Concatenates with carrier DNA could also be a better substrate for integration. Any of these hypotheses would be consistent with the observation that TK⁺ transformants obtained in this way often revert to TK⁻ with rather high frequency.

Robins et al. (1981) have been able to map human growth hormone genes to metaphase chromosomes of transformed rat cells by in situ hybridization. Although limited in sensitivity (single copy genes would hardly be seen) this technique has provided useful information. In cell lines where multiple copies of the gene were integrated, the gene copies were found to reside in a single locus. This is most easily explained by assuming that they were ligated in vivo prior to integration. Integration occurred in different chromosomes in different cell clones. In several instances, chromosomal rearrangements were observed at the site of integration. These data confirm previous chromosome transfer experiments (*Donner* et al. 1977; *Smiley* et al. 1978) in demonstrating that integration of exogenous DNA takes place into host chromosomes and further suggest that integration is random. Herpes-specific antigens have been found to be associated with cellular chromosomes in TK-transformed cells (*Kit* et al. 1979, 1980).

The experiments described by *Robins* et al. (1981) make it clear that the fate of a gene transfected along with carrier DNA in the recipient cell can not be fully eludidated without some understanding of the role played by the latter. Thus, the many Southern blot experiments showing that transfected TK DNA is found associated with high molecular weight DNA in a nonunique configuration do not necessarily imply that it has become integrated in chromosomal DNA in a nonunique site. Although this is

apparently the correct conclusion, additional work with and without carrier DNA is needed before it can be firmly established that integration is random, or takes place in preferred sites and whether or not the choice of sites is influenced by the presence of carrier DNA.

6 Expression of Cotransferred Genes

It is useful to distinguish between two situations, depending upon whether or not expression of the cotransferred gene is analyzed in cells where it is normally expressed. Thus, SV40, adenovirus, and Herpes genes are normally expressed in cells where their products are detected upon viral infection. In contrast, globin, ovalbumin, or growth hormone genes do not belong to this category, when they are introduced into cells (such as fibroblasts) which are not properly differentiated to promote their expression.

Hanahan et al. (1980) have cotransferred the SV40 early region along with Herpes TK into mouse TK⁻ cells and analyzed the expression of SV40 large T antigen by indirect immunofluorescence in TK⁺ transformants. Unselected expression of large T showed great variations, both between cell lines (about half of the lines were positive) and between individual cells of a cloned line. When the early region was connected to the TK marker in a recombinant plasmid, heterogeneous expression of T was still observed. These observations were interpreted to mean that unselected expression of the SV40 early region may depend on the site of integration in various cell lines and, in a given one, on some presently nonunderstood regulatory mechanism.

Grodzicker and *Klessig* (1980) have cotransformed TK⁻ human cells with Herpes TK and adenovirus 2 genes. TK⁺ transformants were then screened for their ability to complement superinfecting adenovirus mutants. Several such lines were obtained. Again, they were not completely stable for the expression of adenovirus genes. The reduction in expression in some TK⁺ subclones was not related to loss or detectable rearrangement of viral DNA.

A correlation between gene activity and its state of methylation has been reported by several authors. It has been shown that the methylation pattern of a gene can be replicated and thereby inherited, albeit with limited fidelity (*Wigler* et al. 1981). It would thus be conceivable that heterogeneity in expression is related to switches in the methylation pattern of the gene. However, experiments reported by *Pellicer* et al. (1980a) on the Herpes TK gene and by *Grodzicker* and *Klessing* (1980) on adenovirus genes do not presently support this hypothesis.

A remarkable result was obtained by *Dubois* et al. (1980) on cotransformation of TK⁻ mouse L cells with the Herpes TK gene and a tandem repeat of the Hepatitis B genome carried in separate plasmids. All of the 15 TK⁺ transformed clones examined produced Hepatitis B surface antigen. Moreover, the latter was excreted into the cell culture medium in the form of 22-nm particles resembling those found in human serum and not containing viral DNA. In two lines, as many as $2-4. \times 10^6$ molecules of surface antigen were produced per cell per 24 h. Other Hepatitis B antigens (HBc and HBe or DNA polymerase) were not detected. This suggests that selective expression of some viral genes takes place in the murine cells.

Genes whose expression is restricted to given differentiated cell types may show some expression upon transformation into other cell types. *Mantei* et al. (1979) transfect-

ed TK⁻ mouse L cells with ligated Herpes TK and rabbit chromosomal β-globin sequences. Most TK⁺ clones contained several copies of the β-globin gene and expressed up to 2000 copies of globin mRNA per cell. In similar experiments, *Wold* et al. (1979) found much lower expression (a few copies of mRNA per cell). Moreover, 45 nucleotides were missing at the 5' end of the globin transcript. In experiments carried out with the chicken ovalbumin gene, *Lai* et al. (1980) isolated mouse fibroblasts producing immunoprecipitable ovalbumin-like material. Expression varied between independant clones, one clone producing up to 100 000 molecules of ovalbumin-like protein per cell. *Breathnach* et al. (1980) however, found that in independently generated cell lines, transcription of the ovalbumin gene within mouse fibroblasts yielded a transcript 650 nucleotides longer at its 5' terminus than native chicken ovalbumin mRNA.

From these and several other experiments (see, for example, *Mulligan* et al. 1979; *Hamer* and *Leder* 1979; *Hsiung* et al. 1980; *Anderson* et al. 1980, etc.), a few general conclusions can be drawn. First, RNA splicing takes place accurately even when the gene acquired by the cell originates from a different animal species. Second, expression is very heterogenous. Finally, transcription is, in at least some instances, improperly initiated.

Why should genes whose expression is restricted to specifically differentiated cell types be expressed in other types of cells at all? This can be explained in a variety of ways, but raises questions as to whether the regulation of gene expression may be properly reproduced with transformed genes.

In untransformed TK⁺ cells, thymidine kinase activity is absent in resting cells until a sharp induction of activity takes place at the beginning of the S phase. In TK⁻ cells transformed with Herpes virus, thymidine kinase is present throughout the cell cycle. *Schlosser* et al. (1981) have examined the expression of transformed Herpes TK or cellular TK genes in synchronized populations of TK⁻ mouse L cells. They found that the Herpes TK gene is expressed throughout the cell cycle while cellular genes of various origin show the expected variations. These experiments suggest that the expression of the transformed genes is properly regulated. However, nothing is known about the regulatory mechanisms involved, which could be posttranscriptional rather than transcriptional.

Evidently, regulation of gene expression would best be studied in cells which normally express the gene. This implies that genes should be introduced into all sorts of cell types, particularly cells at the appropriate stage of differentiation. This has prompted attempts to transform various undifferentiated and differentiated cell types. For this purpose dominant selective markers should be particularly useful (see next section). An interesting approach is to introduce genes into multipotent mouse teratocarcinoma stem cells, which can differentiate in vitro to yield several cell types. *Pellicer* et al. (1980b) mutagenized a teratocarcinoma cell line and selected a TK⁻ derivative by growth in BUdR-containing medium. They next showed that these cells could be transformed to TK⁺ by the Herpes TK gene, albeit at low efficiency. TK⁺ transformants could be grown as tumors in the animals and usually maintained an active TK gene. A human β-globin gene was cotransferred in two out of ten clones, but no data on its expression have been reported. Linnenbach et al. (1980) have cotransferred SV40 DNA along with Herpes TK into F9 TK⁻ stem cells. TK⁺ transformants retained antigenic characteristics of stem cells, showing that they had not differentiated in the course of transformation by DNA. SV40 T antigen was not initially expressed but appeared as stem cells treated by retinoic acid acquired differentiated traits — as expected since SV40 early functions are not expressed in undifferentiated stem cells.

A still more general approach would be to introduce genes into embryonic cells and then to generate animals, every cell of which, including all differentiated cell types, would carry the integrated gene. Experiments along these lines are presently under way (*Gordon* et al. 1980).

In general, doubts may be raised, however, as to whether all kinds of gene regulation will be easily reproduced with transformed genes. If some types of regulation depend on the chromosomal gene locus (i.e., a larger DNA region surrounding a gene or a group of genes), they will not be mimicked unless the latter is reconstructed around the gene to be transformed, or the exogenous gene is integrated at the appropriate locus, rather than randomly. Site-specific integration may, therefore, be an important challenge for the future and require tools which are not yet available.

7 Other Selectable Markers

Total cellular DNA can be used as donor in gene transfer experiments. For example, TK^- mouse L cells can be transformed with human DNA and TK^+ transformants can be obtained which express the human enzyme (*Wigler* et al. 1978). Similar experiments have been carried out with cellular DNA from TK^- cells transformed with the Herpes TK gene and with other genes coding for adenine phosphoribosyltransferase (APRT) (*Wigler* et al. 1979a), hypoxanthine-guanine phosphoribosyltransferase (HGPRT) (*Graf* et al. 1979; *Lester* et al. 1980), and dihydrofolate reductase (dHFR) (*Lewis* et al. 1980; *Wigler* et al. 1980).

Experimental designs which permit the isolation, from total cellular DNA, of the genes responsible for transformation have been described by *Perucho* et al. (1980a) and *Lowy* et al. (1980). Basically, they consist in tagging the donor DNA at random by in vitro recombination with a marker sequence such as plasmid DNA. The marker sequence is then used to recover the adjacent material from successfully transformed cells. In one protocol, this is achieved by rescuing biologically active plasmid from the DNA of the transformed cells, as exemplified by the isolation of the chicken thymidine kinase gene by *Perucho* et al. (1980a). In another device, a library of genomic DNA of transformed cells is constructed in a phage vector and screened with a plasmid probe by in situ hybridization. The APRT hamster gene was isolated in this way (*Lowy* et al. 1980).

To permit the transformation of any cell type, without requirement for previous mutation to allow selection, a search for dominant selectable markers has been undertaken. Along this line, three bacterial genes have recently been coupled to eukaryotic expression signals in order to serve as selectable markers in mammalian cells. *Mulligan* and *Berg* (1980) have isolated the *E. coli* gene (Ecogpt) coding for xanthine-guanine phosphoribosyl-transferase. The gene was fused to SV40 expression signals (which include a promoter, an intron to provide RNA splicing sites and a polyadenylation site). The *E. coli* gene is thus expressed in mammalian cells, where it can compensate for the human Lesch-Nyhan defect in HGPRT. Moreover, under certain growth conditions, this marker permits selection of cells expressing the Ecogpt gene over a background of normal cells (*Mulligan* and *Berg* 1981). It can thus be used as a dominant selective marker. Similarly, a bacterial gene coding for dihydrofolate reductase has been linked to SV40 molecules along the same general principles. The bacterial gene, when expressed

in mouse L cells, confers for resistance to high concentrations of methotrexate and thereby permits selection of transformed cells (*O'Hare* et al. 1981).

Following observations of *Jimenez* and *Davies* (1980) on yeast, we have shown that an antibiotic (G418) is toxic for mammalian cells. Bacteria harboring the Tn5 transposon and synthesizing the enzyme aminoglycoside 3' phosphotransferase type II (APH (3')-II) phosphorylate the antibiotic and become resistant to it. By attaching the Tn5 APH(3')-II gene onto the Herpes TK promoter and polyadenylation site, we promoted expression of the enzyme in mammalian cells, such that the TK-APH(3')-II hybrid gene could serve as a dominant selective marker. It is noteworthy that there was no need to introduce RNA-splicing signals in the hybrid gene to promote its expression. All cell lines tested were sensitive to G418 and in all of them, the dominant marker was expressed and permitted selection of transformed cells (*Colbère-Garapin* et al. 1981).

As mentioned above, there is a mean of selecting normal mouse TK$^+$ cells transformed with the Herpes TK gene in a modified HAT medium. This procedure was developed by *Mercola* et al. (1980) for the transformation of mouse bone marrow stem cells. These studies, as well as that of *Cline* et al. (1980) with a cellular dihydrofolate reductase gene, illustrate the use of dominant markers for gene transfer in intact animals. In the experiments of *Mercola* et al. (1980), the TK^{++} bone marrow cells (i.e., possessing both the cellular and the Herpes genes), selected in the presence of methotrexate (10^{-4} M), were reimplanted in marrow-depleted mouse recipients. To provide the TK^{++} cells with a selective advantage and favor their proliferation, the mouse recipients were injected with methotrexate. After some time, it was observed that the spleens of these animals contained dividing hematopoeitic cells harboring the Herpes TK gene marker. This remarkable result suggests that gene replacements may have some future therapeutic value in humans but many unkowns have yet to be solved.

8 Gene Transfer and Animal Cell Genetics

Gene transfer methods appear to have an increasing potential for the study of numerous aspects of gene and cell functions. Better tools are expected to emerge in the near future, particularly new vectors for the transfer of genes and the amplification of transformed genes. In murine cells treated with methotrexate, the cellular gene coding for dihydrofolate reductase is amplified (*Alt* et al. 1978; *Nunberg* et al. 1978). In cotransformation experiments, sequences linked to this gene are also amplified upon selection with methotrexate (*Perucho* et al. 1980b). DNA sequences integrated in viral vectors can be replicated many times within the host cell, causing overexpression of a cloned gene (see *Gruss* and *Khoury*, this volume). However, no stable episomal vector analogous to bacterial plasmid vectors has been reported to date.

Although gene transfer opens obvious prospects in animal cell genetics, it is important to emphasize that the tools and basic knowledge underlying their use are still primitive. As an example, utilization of all the selectable markers reviewed above (TK, APRT, HGPRT, dHFR) except the Tn5 transposon derived marker, is based upon a manipulation of the nucleotide synthesis pathway. It may be feared that in the present state of the art, selective forces are not always properly evaluated. (It is conceivable, for instance, that overproduction of some of these enzymes may be deleterious to the host cell.) The stability of the transformed cells must be examined with caution.

Some transformed lines are very unstable once the selective pressure is removed and even the more stable ones are often less stable than untransformed lines (see, for example, *Pellicer* et al. 1980a and our unpublished observations). The role of carrier DNA and of the way of introducing DNA into the cells deserves more attention (*Capecchi* 1980). Also, as reviewed above, expression of the transformed gene often shows heterogeneity of variations. Finally, the mechanism of integration is unknown and it may be significant that chromosomal rearrangements have been observed at, or near, the locus of integration (*Robins* et al. 1981). The question of the relative extent of homologous versus illegitimate recombination of exogenously added sequences in the cell chromosomes may turn out to be a central one in developing means of integrating genes into defined chromosomal loci where their expression would be properly regulated. In conclusion, much additional work is needed before the mechanisms of gene transfer are fully understood and mastered.

References

Alt FW, Kellemo RE, Bertino JR, Schimcke RT (1978) J Biol Chem 253:1357–1370
Anderson WF, Killos L, Sanders-Haigh L, Kretschmer PJ, Diacumakos EG (1980) Proc Natl Acad Sci USA 77:5399–5403
Bacchetti S, Graham FL (1977) Proc Natl Acad Sci USA 74:1590–1594
Benoist C, O'Hare K, Breathnach R, Chambon P (1980) Nucleic Acids Res 8:127–142
Breathnach R, Mantei N, Chambon P (1980) Proc Natl Acad Sci USA 77:740–744
Capecchi M (1980) Cell 22:479–488
Cline MJ, Stang H, Mercola K, Morse L, Ruprecht R, Browne J, Salser W (1980) Nature 284: 422–426
Colbère-Garapin F, Chousterman S, Horodniceanu F, Kourilsky P, Garapin AC (1979) Proc Natl Acad Sci USA 76:3755–3759
Colbère-Garapin F, Horodniceanu F, Kourilsky P, Garapin AC (1981) J Mol Biol 150:1–14
Cremer K, Bodemer M, Summers WC (1978) Nucleic Acids Res 5:2333–2344
Crumpacker CS, Chartrand P, Subak-Sharpe J, Wilkie N (1980) Virology 105:171–184
Donner L, Dubbs DR, Kit S (1977) Int J Cancer 20:256–267
Dubois MF, Pourcel C, Rousset S, Chany C, Tiollais P (1980) Proc Natl Acad Sci USA 77:4549–4553
Enquist L, Vande Woude G, Wagner M, Smiley J, Summers WC (1979) Gene 7:335–342
Fleckenstein B (1979) Biochim Biophys Acta 560:301–342
Fraley R, Subramani S, Bery P, Papahadjopoulos D (1980) J Biol Chem 255:10431–10435
Garapin AC, Colbère-Garapin F, Cohen-Solal M, Horodniceanu F, Kourilsky P (1981) Proc Natl Acad Sci USA 78:815–819
Gordon JW, Scangos GA, Plotkin DJ, Barbosa JA, Ruddle FH (1980) Proc Natl Acad Sci USA 77:7380–7384
Graf LH, Urlaub G, Chasin LA (1979) Somatic Cell Genet 5:1031–1044
Graham FL, Van der Eb AJ (1973) Virology 52:456–467
Grodzicker T, Klessig DF (1980) Cell 21:453–463
Hagenbüchle O, Santer M, Steitz JA, Mans R (1978) Cell 13:551–563
Hamer DH, Leder P (1979) Nature 281:35–40
Hanahan D, Lane D, Lipsich L, Wigler M, Botchan M (1980) Cell 21:127–139
Hsiung N, Warrick H, De Riel JK, Tuan D, Forget BG, Skoultchi A, Kucherlapati R (1980) Proc Natl Acad Sci USA 77:4852–4856
Jimenez A, Davies J (1980) Nature 287:869–871
Kit S, Dubbs D, Piekarski L, Hsu T (1963) Exp Cell Res 31:297–312
Kit S, Leung WC, Jorgensen GN, Dubbs DR (1974) Int J Cancer 14:598–610
Kit S, Qavi H, Hazen M, Dubbs DR (1979) Int J Cancer 23:721–732
Kit S, Otsuka H, Qavi H, Trkula D, Dubbs DR (1980) Virology 105:103–122

Klobutcher LA, Ruddle FH (1979) Nature 280:657–660
Lai EC, Woo SLC, Bordelon-Riser ME, Fraser TH, O'Malley BW (1980) Proc Natl Acad Sci USA 77:244–248
Lester SC, le Van SK, Steglick C, De Mars R (1980) Somatic Cell Genet 6:241–249
Lewis WH, Srinivasan PR, Stokoe N, Siminovitch L (1980) Somatic Cell Genet 3:333–347
Linnenbach A, Huebner K, Croce CM (1980) Proc Natl Acad Sci USA 77:4875–4879
Littlefield J (1963) Proc Natl Acad Sci USA 50:568–573
Lowy I, Pellicer A, Jackson JF, Sim G-K, Silverstein S, Axel R (1980) Cell 22:817–823
Mantei N, Boll W, Weissmann C (1979) Nature 281:40–46
Mathis DJ, Chambon P (1981) Nature 290:310–315
McBride OW, Peterson JL (1980) Annu Rev Genet 14:321–345
McKnight SL (1980) Nucleic Acids Res 8:5949–5964
McKnight SL, Gavis E (1980) Nucleic Acids Res 8:5931–5947
Mercereau-Puijalon O, Kourilsky P (1979) Nature 279:647–649
Mercola KE, Stang HD, Browne J, Salser W, Cline MJ (1980) Science 208:1033–1035
Minson AC, Wildy P, Buchan A, Darby G (1978) Cell 13:581–587
Mulligan RC, Howard BH, Berg P (1979) Nature 277:108–114
Mulligan RC, Berg P (1980) Science 209:1422–1427
Mulligan RC, Berg P (1981) Proc Natl Acad Sci USA 78:2072–2076
Munyon W, Kraiselburd E, Davis D, Mann J (1971) J Virol 7:813–820
Nunberg H, Kaufman RJ, Schimcke RT, Urlaub G, Chasin LA (1978) Proc Natl Acad Sci USA 75:5553–5556
O'Hare K, Benoist C, Breathnach R (1981) Proc Natl Acad Sci USA 78:1527–1531
Pellicer A, Wigler M, Axel R, Silverstein S (1978) Cell 14:133–141
Pellicer A, Robins D, Wold B et al. (1980a) Science 209:1414–1422
Pellicer A, Wagner EF, Karch AE et al. (1980b) Proc Natl Acad Sci USA 77:2098–2102
Perucho M, Hanahan D, Lipsich L, Wigler M (1980a) Nature 285:207–210
Perucho M, Hanahan D, Wigler M (1980b) Cell 22:309–317
Proudfoot NJ, Brownlee GG (1976) Nature 263:211–214
Robins DM, Ripley S, Henderson AS, Axel R (1981) Cell 23:29–39
Scangos G, Huttner K, Silverstein S, Ruddle F (1979) Proc Natl Acad Sci USA 76:3987–3990
Schaffner W (1980) Proc Natl Acad Sci USA 77:2163–2167
Schlosser CA, Steglich C, De Wet JR, Scheffler IE (1981) Proc Natl Acad Sci USA 78:1119–1123
Smiley J, Steege DA, Juricek DK, Summers WP, Ruddle F (1978) Cell 15:455–468
Smiley J, Wagner H, Summers WP, Summers WC (1980) Virology 102:83–93
Summers WC, Summers WP (1977) J Virol 24:314–318
Szybalska EH, Szybalski W (1962) Proc Natl Acad Sci USA 48:2026–2034
Wagner MJ, Sharp JA, Summers WC (1981) Proc Natl Acad Sci USA 78:1441–1445
Wigler M, Silverstein S, Leu LS, Pellicer A, Cheng YC, Axel R (1977) Cell 11:223–232
Wigler M, Pellicer A, Silverstein S, Axel R (1978) Cell 14:725–731
Wigler M, Pellicer A, Silverstein S et al. (1979a) Proc Natl Acad Sci USA 76:1373–1376
Wigler M, Sweet R, Sim GK et al. (1979b) Cell 16:777–785
Wigler M, Perucho M, Kurtz D, Dana S, Pellicer A, Axel R, Silverstein S (1980) Proc Natl Acad Sci USA 77:3567–3570
Wigler M, Levy D, Perucho M (1981) Cell 24:33–40
Wilkie NM, Clements JB, Boll W, Mantei N, Lonsdale D, Weissman C (1979) Nucleic Acids Res 7:859–877
Wold B, Wigler M, Lacy E, Maniatis T, Silverstein S, Axel R (1979) Proc Natl Acad Sci USA 76:5684–5688
Wong TK, Nicolau C, Hofschneider PH (1980) Gene 10:87–94

Gene Transfer into Mammalian Cells: Use of Viral Vectors to Investigate Regulatory Signals for the Expression of Eukaryotic Genes

PETER GRUSS* AND GEORGE KHOURY*

1 Introduction

Selective isolation of eukaryotic genes, utilizing the recently developed cloning procedures, offered molecular biologists the experimental expertise necessary to approach fundamental questions concerning regulatory events in eukaryotic cells. Initially, a comparative analysis of DNA sequences of numerous cloned genetic segments (*Konkel* et al. 1978; *Breathnach* et al. 1978; *Seif* et al. 1979; *Dawid* and *Wahli* 1979; *Benoist* et al. 1980; see also *Abelson* and *Butz* 1980) has helped to elucidate certain nucleotide signals that may be involved in the control of gene expression. The ultimate characterization of these processing signals, however, would appear to require the introduction of the cloned genes into a functional assay system. At present, this is best achieved by translocating the genes back into eukaryotic cells. Several approaches have been elaborated to reinsert DNA or RNA into eukaryotic cells. These include the mechanical microinjection of RNA or DNA, in which a large number of molecules is administered to a small number of cells (*Mertz* and *Gurdon* 1977; *Kressman* et al. 1978; *Mueller* et al. 1978), liposome fusion (*Dimitradis,* 1978; *Ostro* et al., 1978) and erythrocyte fusion (*Rechsteiner* 1978) with target cells or direct DNA transfection in the presence of facilitating agents (*Graham* and *van der Eb* 1978; *Wigler* et al. 1977). For a number of reasons each of these methods is limited

* Laboratory of Molecular Virology, National Cancer Institute, Bethesda, Maryland 20205

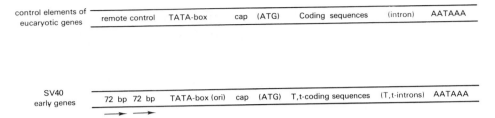

Fig. 1. The control elements involved in the regulation of eukaryotic gene transcription. *Above*, the generalized elements for a representative eukaryotic gene, *below*, the specific elements involved in the regulation of early SV40 gene expression

in efficiency. Alternatively, the DNA of interest can be inserted into a self-contained replicon, frequently employing an animal virus genome. It is not surprising that the best studied animal virus, simian virus 40 (SV40), has been employed most frequently as the eukaryotic viral vector. In these experiments, SV40 has been covalently linked to the gene of interest and infected into permissive or nonpermissive cells (*Goff* and *Berg* 1976; *Ganem* et al. 1976; *Hamer* et al. 1977a, b; *Upcroft* et al. 1978; *Mulligan* et al. 1979; *Gruss* and *Khoury* 1978). Besides SV40, a variety of virus-derived vector systems including polyoma, papilloma, adenovirus, and murine sarcoma viruses have been recently developed (*Solnik* 1981; *Howard* et al. 1981; *Sarver* et al. 1981; *G. Vande Woude*, personal communication). The choice of a particular viral vector depends on the specific experimental goals. As a vector in permissive cells increases the number of templates after infection up to 100 000 molecules per cell due to replication of the recombinants (*Hamer* 1980), the amplification of templates results accordingly in the synthesis of large amounts of transcriptional and translational products. This unique advantage of a viral vector has facilitated a number of experiments concerned with identification of genetic elements required for the expression of eukaryotic genes.

As demonstrated below and schematically outlined in Fig. 1, eukaryotic gene expression requires a number of discrete genetic elements. Like parts in a puzzle, these elements can also be assembled from heterologous sources. In order to distinguish these discrete elements, and examine their potential interactions, a number of observations have been made using viral genomes and mutants of these viruses. Some of these observations and experiments are described below using SV40 as an example. It is the interaction of such genetic elements then which serves as a basis for constructing viral recombinant DNA molecules.

2 The Anatomy of the SV40 Genome

A clear understanding of the anatomy of the SV40 genome and the temporal sequence of expression in permissive and nonpermissive cells is essential to experiments employing this agent as a viral vector for the introduction of genes into eukaryotic cells. The extensive accumulated data (for a detailed review, see *Tooze* 1980) with this viral genome has led to its frequent use as the vector of choice. In permissive cells (African green monkey kidney (AGMK), the viral life cycle is expressed in two distinct phases. The early and late

SV40 coding sequences are situated in opposite halves of the viral genome and transcribed in opposite directions on the antiparallel DNA strands. The predominant SV40 expression prior to viral DNA replication in permissive cells is that of the early viral gene region. After the onset of DNA replication, the early genes continue to be expressed but late transcription and translation products are 10–20 times as abundant. In nonpermissive cells little, if any, functional late gene products appear. A number of genetic elements have been described which are essential to the expression of mRNA from either the SV40 genome or its recombinants and to the continued propagation of these complementing viral populations.

2.1 The Origin for Viral DNA Replication

This origin is a *cis*-essential gene function that must be present in all molecules which require amplification or continued propagation. This region of the SV40 DNA has been rather well defined on the basis of the minimal retained sequences in evolutionary variants of SV40 genomes which retain only the replication function (*Gutai* and *Nathans* 1978; *Subramanian* and *Shenk* 1978).

2.2 The Promoters for the Early and Late Viral Transcripts

Until recently, the definition of a promoter for transcription, as well as the location of promoters within the SV40 genome was obscure. Based on a number of recent studies, we have come to consider at least three elements essential to a promoter (Fig. 1). These include the cap site, or 5' end of a nascent transcript which appears to be generated de novo rather than through a processing event. About 25–30 nucleotides upstream from the cap site resides the Goldberg-Hogness (G-H) box, also called the TATAA sequence. The best data to date indicate that the G-H box provides a measuring function in determining both the position and the efficiency of use of a particular 5' end (*Ghosh* et al. 1981; *Mathis* and *Chambon* 1981). In addition, a set of repeated elements located more than 100 nucleotides upstream from the 5' ends of the early nascent transcripts has recently been shown to play an essential role in the initiation of all early transcriptional events from SV40.

Studies which demonstrate the role of these repeated nucleotide sequences in early SV40 transcription were based on the work of *Benoist* and *Chambon* (1981) who analyzed the biological properties of a set of deletion mutants of the early region of SV40 which had been inserted into a plasmid vector; coincident with these studies, we reported findings based on two viral mutants of SV40 which will be described briefly below (*Gruss* et al. 1981).

The function of the SV40 72 base-pair repeated elements, situated approximately 100 nucleotides to the late side of the TATAA box (see Fig. 2) was studied through the production of two deletion mutants. These mutants were created using the restriction enzyme which cuts uniquely within each of the 72 base-pair repeats. Complete digestion with *Sph*I generated an SV40 mutant precisely missing one 72 base-pair repeat. Upon transfection of this mutant (*dl*-2355) into AGMK cells, a viable productive infection was obtained in the absence of a helper virus. This observation indicates that the absence of

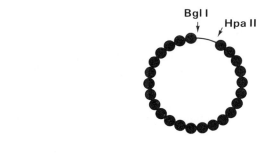

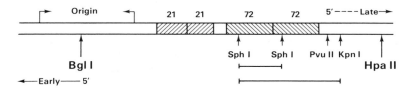

Fig. 2. The early SV40 genetic control region. *Below,* a segment of SV40 DNA near the origin of replication from the *Bgl*I site (center of the replication palindrome) to the *Hpa*II site on the late side of the origin. Included within this region are the 5′ ends of the major late and early SV40 mRNAs as well as the 21 base-pair and 72 base-pair repeating units. *Above,* the SV40 minichromosomes indicating the correspondence of the nucleosome-free area to the repeated segments

one repeated unit is consistent with the viability of the virus; nevertheless, the deleted molecule, *dl*-2355, may not replicate as efficiently as does the wild type SV40 DNA.

Production of a second mutant, *dl*-2356, resulted from cleavage with both *Sph*I and *Kpn*I restriction nucleases. This mutant deletes 122 of the 144 base pairs present in the two major repeats; additional nucleotides removed beyond the second repeat in the region between *Kpn*I and *Sph*I are from a dispensable region of the SV40 genome to the late side of the repeats (*Barkan* and *Mertz* 1981). After transfection of this extended deletion mutant, *dl*-2356 into AGMK cells, no early viral RNA, early proteins or viral replication was observed (*Gruss* et al. 1981). Furthermore, the only plaques produced in complementation with either tsA or tsB resulted from revertants of the temperature-sensitive mutants. This result confirms that at least one copy of the 72 base-pair element is a *cis*-essential structure required for the expression of both early and late SV40 genes.

In Fig. 2 an expanded segment near the origin and to the late side of the SV40 genome is presented in linear form at the bottom. The *Bgl*I restriction enzyme cleaves within the center of the palindrome essential to the origin of DNA replication. Two sets of repeated elements to the late side of the origin are demonstrated by hatched boxes, and the sequences deleted in mutants *dl*-2355 and 2356 are indicated by the bars below the figure. Above Fig. 2 is a schematic representation of the mini-chromosome of SV40. The dark circles, representing nucleosomes, have been shown by a number of investigators to be equally distributed around the genome with the exception of the region included by the repeated elements (*Waldeck* et al. 1978; *Scott* and *Wigmore* 1978; *Varshawsky* et al. 1979; *Saragosti* et al. 1980). Some of these investigators have speculated that

there is a direct correlation between the absence of the nucleosomes from this region and its ability to serve as an essential element in the promoter for SV40 early transcription.

The location of the late SV40 promoter is somewhat obscure. Clearly there is an extensive heterogeneity in the 5' ends of late SV40 transcripts; this is even more pronounced in molecules which have been altered either by deletions and/or insertions, including most late recombinants of SV40. What does seem to be clear is that there is no well-defined G-H box for the late region transcripts, and the position of the upstream controlling element (? repeats) may well overlap the early control region. Whether this late controlling element will involve the 72 base-pair repeats, the preceding 21 base-pair repeats or even the repeated element GGGCGGXG, which is found multiple times in the region of the SV40 replication origin and repeat region (*Fiers* et al. 1978; *Reddy* et al. 1978), remains to be determined.

2.3 Splice Sites

Splice junctions for the two early SV40 transcripts have been well defined. There are two separate donor sequences and a common acceptor sequence which unite to form the large T and small t mRNAs. Splice sites in the late region of the SV40 genome are more variable; although there is one predominant late splice junction for the 16S mRNA, there are at least three frequently used 19S splice junctions. For a more detailed analysis of these splice sites, the reader is referred to the analytical studies by *Weissmann* and his colleagues (*Ghosh* et al. 1978a, b). Significance of splicing events has been shown to depend on the individual transcript.

In an attempt to determine the function of the intron, we have generated a deletion mutant lacking precisely the intron of 16S mRNA (*Gruss* et al. 1979). The procedure involves the replacement of a portion of the late genomic segment of SV40 by a reverse transcription product of 16S mRNA (Fig. 3). This SV40 deletion mutant produced no stable late 16S transcript, which suggested a defect in the posttranscriptional processing of the viral RNA. Subsequently, we used this mutant as the recipient for the insertion of an isolated mouse β^{maj} globin intron. The β^{maj} globin intron (I) was inserted in both the sense and the antisense orientation relative to the late SV40 transcription unit, at a site which differs from the location of the original 16S RNA intron. Cloned recombinants harboring the inserted intron in the sense direction give rise to stable mRNA. Thus, while the polarity of the intron is critical to its function, the species of origin and the relative position in the mRNA are not. These observations suggested that introns represent functional elements in the generation of certain mRNAs (*Gruss* and *Khoury* 1981).

On the other hand, the recent detection of a number of eukaryotic genes which are uninterrupted by intervening sequences (*Alestrom* et al. 1980; *Lawn* et al. 1981; *Nagata* et al. 1981; *Schaffner* et al. 1976; *McKnight* 1980; *Wagner* et al. 1981), as well as certain late mutants of SV40 which appear to produce unspliced 19S late mRNAs (*Ghosh* et al. 1981), have suggested that alternative mechanisms must exist for the expression of nonspliced transcripts.

In an attempt to shed further light on this question, we have employed the SV40 rat insulin recombinant molecules which retain a single rat insulin intervening sequence in the late region of the genome. While all of the late recombinant transcripts appeared to use the retained rat insulin intervening sequence as a splite site, it was unclear whether or

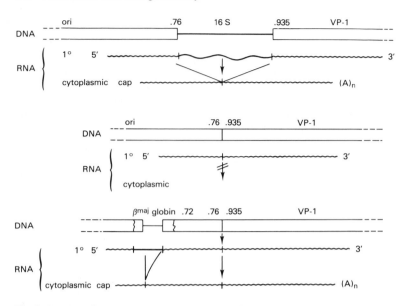

Fig. 3. Outline of experiments demonstrating the role of splicing in the generation of the stable 16S SV40 RNA. *Top*, the wild-type SV40 16S mRNA template, the primary transcription product, and the mature spliced 16S cytoplasmic mRNA, *center*, the SV40 mutant from which the 16S intervening sequence was precisely removed (as indicated this mutant failed to give rise to stable cytoplasmic mRNA); *bottom*, experiment in which the mouse β^{maj} globin small intron was inserted into this intron minus mutant (*dl*-2350). This recombinant molecule produced stable cytoplasmic mRNA which employed the mouse β^{maj} globin splice sequence

not this particular splicing event was obligatory for the production of stable mRNA. Mutants of these recombinant molecules were constructed in which the single rat intervening sequence as well as certain flanking sequences were removed (*Gruss* et al. PNAS, in press). This resulted in the fusion of the coding region from the rat chromosomal preproinsulin I gene to the 5' noncoding region of the late SV40 genes. While this reconstruction deleted potential 5' ends of the rat insulin transcripts, the genomic region of SV40 which contains sequences coding for the most abundant 5' late mRNA ends was preserved and was situated in the immediate vicinity of the new junction between the SV40 vector sequences and those of the rat insulin insert. This new recombinant molecule also deletes the first five codons of the preproinsulin coding sequence and replaces them "in frame" with the first three codons of the SV40 agnoprotein (*Jay* et al. 1981). Both RNA and protein analyses revealed that the new mutant recombinant molecules were capable of generating stable hybrid colinear transcripts lacking splice junctions. Nevertheless, it is still conceivable that this latter class of RNA molecules might appear in greater abundance if a splice junction were located at some position within the transcript.

Studies based on the insertion of a number of recombinant DNAs into the early region of papovavirus genomes seem to indicate the requirement of a splice junction for efficient expression (*B. Howard* and *P. Berg; R. Breathnach* and *P. Chambon,* personal communications). It is at present unclear what factors determine the requirement of a splicing process in stabilization of certain transcripts.

2.4 Polyadenylation Sites

Since polyadenylation events appear to be crucial to the expression of stable mRNA molecules, it is important that the vectors be constructed to contain such a signal (either from the viral or the recombinant DNA). The poly(A) signals in SV40, as is the case for poly(A) sites in most eukaryotic mRNA molecules, can be recognized by a hexanucleotide sequence AATAAA, which is usually situated 10–15 nucleotides before the poly(A) addition site (*Proudfoot* and *Brownlee* 1976). These signals have been located for both early and late regions of the SV40 genome and analyzed in detail (*Fitzgerald* and *Shenk* 1980).

2.5 Methylation Events

Internal methylation events are characteristic not only of eukaryotic mRNAs in general but of SV40 and other animal viruses as well. In part because of the difficulty in locating these sites, the significance of such modifications in RNA is not yet understood.

2.6 Translational Initiator Codons

The codons which initiate translation of SV40 polypeptides have been precisely positioned. In many cases, the proximal AUG serves as the principle site for initiation of protein synthesis. Thus, it is important to consider these sites relative to the position of the recombinant DNA species, especially if the latter is to be expressed at the protein level. For example, inserting a recombinant molecule downstream from a strong SV40 initiator may lead to the production of either a fusion polypeptide (in frame insertion) or lack of expression (an alternate reading frame).

3 Construction of Recombinant Molecules

As indicated in Sect. 2, the only indispensable region of the SV40 genome is the origin for viral DNA replication. This *cis*-required function must be present for a genome to be propagated. For use as a vector, however, it has been beneficial to retain, in addition to the origin, either the early or late set of SV40 genes. Such a recombinant molecule can be introduced into a permissive cell in the presence of a complementing helper virus. For example, a recombinant in the late region of an SV40 vector can be inoculated into African green monkey kidney cells together with an early SV40 mutant (either a temperature-sensitive or a deletion mutant). A productive infection in this system requires the presence of both viral genomes in the same cell, insuring the retention of the recombinant molecule in the virus population. As pointed out in Sect. 1, one of the advantages of SV40 as a vector system relates to the amplification of the genome and its gene products. In order for the recombinant molecules to be retained in a viral stock, however, they must be encapsidated within the SV40 coat proteins. This places an additional restriction on the size of the recombinant genome; it must be between 70% and 105% the length of the wild-type genome. Thus, recombinants in either the early or the late region of SV40 are effectively limited in size to less than 2500 base pairs. For many purposes this represents the most serious limitation to SV40 as a cloning vector.

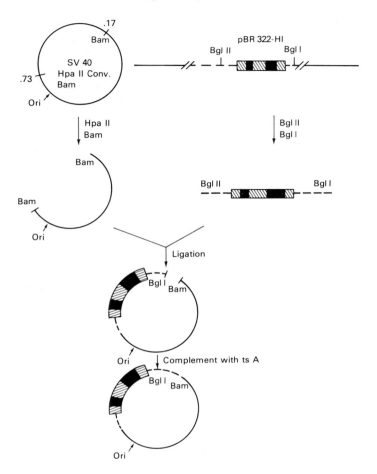

Fig. 4. Construction of an SV40-human preproinsulin recombinant molecule. The human pre-proinsulin sequences have been inserted into the late region of an SV40 vector extending from .73 to .17 map units in a clockwise direction. This late defective recombinant was complemented with an early temperature-sensitive SV40 mutant, *tsA*. (See text for details)

4 Late Region Vectors

The insertion of recombinant DNA molecules within the late region of the SV40 genome, as implied in Sect. 3, is particularly advantageous in productive (lytic) infections of AGMK cells. Amplification of the templates in the presence of a functional T-antigen supplied by the recombinant molecule leads to a high copy number per cell (*Hamer* et al. 1980). Subsequent to the viral DNA replication, the late SV40 promoters are particularly active and can lead to the expression of large numbers of transcripts from the recombinant region of these molecules. The result of such infections is, in part, the production of abundant RNA and protein molecules specific for the recombinant portion of the genome. A disadvantage to such reconstructed molecules is the potential use of the SV40

late promoter, in addition to a putative recombinant promoter. This obstacle can be partially overcome by inserting the recombinant gene within the late region in the opposite orientation. Such a construction does not benefit from the activity of the late SV40 promoter and can still be subject to some read-through transcription from the early SV40 promoter. An example of a late SV40 vector is shown in Fig. 4. In this experiment (carried out in collaboration with *A. Ullrich*), the human insulin gene (*Bell* et al. 1980; *Sures* et al. 1980) was substituted into the late region of an SV40 vector between the *Hpa*II and *Bam*HI sites (in map positions 0.725 to 0.14 clockwise). In this particular SV40 vector, the *Hpa*II site had been converted to a *Bam*HI site to facilitate the insertion of molecules containing *Bam*HI junctions at both extremities. This construct is in many ways analogous to the SV40-rat preproinsulin I gene recombinant molecules previously described (*Gruss* and *Khoury* 1981). In those studies, the complete rat preproinsulin gene I has been cloned into an SV40 vector. The initiation of stable transcripts of the insulin insert is as efficient as the production of late SV40 mRNA. Analysis of these transcripts indicated that the rat preproinsulin gene nucleotide signals involved in RNA splicing and poly(A) addition are used. The generation of a class of mRNA molecules similar in size to authentic rat insulinoma mRNA species suggests that the initiation signal of the insulin insert is also functional. In the beta cells of the pancreas the primary translation product is a preprohormone (preproinsulin) which is processed stepwise into proinsulin and finally to the mature insulin (*Lomedico* et al. 1977; *Tager* and *Steiner* 1974). The processing of preproinsulin to proinsulin involves the removal of the amino terminal 24 amino acids of the preprohormone. In monkey kidney cells (the permissive host for SV40) only the first processing step occurs, thus yielding large quantities of proinsulin. The fact that most of the proinsulin accumulates in the tissue culture medium indicates that this protein is secreted. The maturation from proinsulin to insulin, however, has not been observed in monkey kidney cells. This processing step presumably requires the eukaryotic background of the highly differentiated pancreatic beta cells.

5 Early SV40 Vectors

A number of SV40 vectors have been constructed, particularly by *Berg* and his co-workers (*Howard* et al. 1981; *Mulligan* et al. 1980; *Southern* and *Berg*, personal communication). These vectors, which for the most part retain the early SV40 promoter, have the advantage of being expressed in virtually all cells infected by the recombinant molecules, as is the case for the early SV40 genomes. As best we can tell, transcripts harbored in the early SV40 region will be preferentially expressed in nonpermissive viral cell interactions, particularly if the recombinant genome survives the abortive (the early nonpermissive) phase of this cycle, which is followed by integration into the cellular chromosome. This type of vector, when introduced into nonpermissive cells has the potential advantage, upon integration, of continued expression in the absence of cell lysis.

6 Nonrestrictive Vectors

If one is not concerned with reinfection of secondary cells, or amplification of a stock of encapsidated viral genomes, there are relatively few restrictions on size and composition

of the recombinant molecules. Thus, for example, the recombinant molecule could contain two early regions of the SV40 genome in tandem, one encoding the SV40 T-antigen, and the second containing the recombinant genome of interest. In addition, such a vector may contain sequences capable of being replicated in a prokaryotic system (e.g., pBR322 sequences minus the so-called poison sequences, which would otherwise interfere with amplification in a eukaryotic cell). This type of recombinant molecule can be purified in a prokaryotic system and can subsequently be introduced by transfection into eukaryotic cells. Assuming enough cells can be infected, the expression of the recombinant gene of interest could be studied at the molecular level. In an alternative strategy, the vector containing the origin of SV40 replication and the gene of interest as well as its prokaryotic amplifying sequences can be introduced into AGMK cells which constitutively produce T-antigen (Cos-cells; *Gluzman* et al. 1980). This allows for the amplification of the transfected DNA molecules and their subsequent expression, without the requirements for helper virus or size considerations.

Recently, *Howard* et al. have constructed a vector which can be amplified in both prokaryotic and eukaryotic cells. pSV2 contains the pBR322 sequences, including the origin for DNA replication and the ampicillin resistance cistron as well as the SV40 origin for DNA replication. This vector also contains the SV40 small-t splice site as well as the early SV40 polyadenylation site and the potential for inserting eukaryotic gene segments between the SV40 origin and these RNA processing signals. In further modifications of this plasmid, *Berg* and his co-workers have also inserted genes which allow for a dominant selection in various eukaryotic cell systems (*Mulligan* and *Berg* 1980; *Southern* and *Berg*, personal communication). For continued expression in eukaryotic cells this set of vectors appears to require integration into the host cell chromosome. Thus, efficient excision and rescue from the "transformed" eukaryotic cell is difficult. A potential solution to this problem is exemplified by recent studies which employ papilloma viruses as a vector system.

6.1 Bovine Papilloma Virus as a Eukaryotic Cloning Vector

It has recently been demonstrated that the bovine papilloma virus (BPV) can transform mouse cells while remaining in an unintegrated, episomal state (*Law* et al. 1980). This suggested the potential of using BPV as a cloning vector, which not only has the advantage of containing its own selectable marker (viz cell transformation), but also a potential means for recovering the DNA of interest from the transformed cell by selecting the episomal fraction of DNA. Additional advantages of BPV as a cloning vector include the absence of a size restriction for encapsidation and the general amplification of the vector in 20–200 copies per transformed cell. In recent studies it was demonstrated by *Sarver* et al., that the 69% transforming fragment of BPV could serve as an effective cloning vehicle for the rat I preproinsulin gene (*Sarver* et al. 1981). These studies indicated that transcription was initiated within the rat preproinsulin portion of the molecule, presumably utilizing the preproinsulin promoter signals. This property should be of advantage in future studies directed at the analysis of eukaryotic promoters. The production and secretion of rat proinsulin into the tissue culture medium suggested that this viral cell system may also serve as the source of protein encoded by recombinant molecules in eukaryotic cells. As such, the polypeptides produced by such a system have the potential advantage of under-

going eukaryotic posttranslational modifications. Present studies are directed toward the insertion of selectable markers into BPV cloning vectors so that they might be expressed in a wide variety of eukaryotic cells.

References

Alestrom P, Akusjarvi G, Perricaudet M, Mathews MB, Klessig DG, Petterson U (1980) Cell 19: 671–681

Abelson J, Butz D (eds) (1980) Science 209: No. 4463

Barkan A, Mertz J (1981) J Virol 37:730–737

Bell GI, Pictet RL, Rutter WJ, Cordell B, Tischer E, Goodman HM (1980) Nature 284:26–32

Benoist C, Chambon P (1981) Nature 290:304–310

Benoist C, O'Hare K, Breathnach R, Chambon P (1980) Nucleic Acids Res 8:127–142

Breathnach R, Benoist C, O'Hare K, Gannon F, Chambon P (1978) Proc Natl Acad Sci USA 75: 4853–4857

Cordell B, Bell G, Tischer E, DeNoto FM, Ullrich A, Pictet R, Rutter WJ, Goodman HM (1979) Cell 18:533–543

Dawid IB, Wahli W (1979) Develop Biol 69:305–328

Dimitriadis GJ (1978) Nature 274:923–924

Fiers W, Contreras R, Haegeman G, Rogiers R, Van de Voorde A, Van Heuverswyn H, Van Herreweghe J, Volckaeart G, Ysebaert M (1978) Nature 273:113–120

Fitzgerald M, Shenk T (1981) Cell 24:251–260

Ganem D, Nussbaum AL, Davoli D, Fareed GC (1976) Cell 7:349–359

Ghosh PK, Reddy VB, Swinscoe J, Lebowitz P, Weissman SM (1978a) J Mol Biol 126:813–846

Ghosh PK, Reddy VB, Swinscoe J, Choudary P, Lebowitz P, Weissman SM (1978b) J Biol Chem 253:3643–3647

Ghosh PK, Lebowitz P, Frisque FJ, Gluzman Y (1981) Proc Natl Acad Sci USA 78:100–104

Gluzman Y, Frisque RJ, Sambrook J (1980) Cold Spring Harbor Symp Quant Biol 44:293–300

Goff SP, Berg P (1976) Cell 9:695–705

Graham FL, van der Eb AJ (1973) Virology 52:456–467

Gruss P, Khoury G (1980) Nature 286:634–637

Gruss P, Khoury G (1981a) Proc Natl Acad Sci USA 78:133–137

Gruss P, Dhar R, Khoury G (1981b) Proc Natl Acad Sci USA 78:943–947

Gruss P, Lai C-J, Dhar R, Khoury G (1979) Proc Natl Acad Sci USA 76:4317–4321

Gutai MW, Nathans D (1978) J Mol Biol 126:259–279

Hamer DH (1980) In: Setlow J, Hollaender A (eds) Genetic engineering, vol 2. Plenum Press, N.Y., pp 83–101

Hamer DH, Leder P (1979a) Nature 281:35–40

Hamer DH, Davoli D, Thomas CA Jr, Fareed GC (1977) J Mol Biol 112:155–182

Hamer DH, Smith KD, Boyer SH, Leder P (1979b) Cell 17:725–735

Howard B, Southern P, Mulligan R, Berg P (1981) J Mol Applied Genetics (in press)

Jay G, Nomura S, Anderson CW, Khoury G (1981) Nature 291:346–349

Konkel DA, Tilghman SM, Leder P (1978) Cell 15:1125–1132

Kressman A, Clarkson SG, Pirrotta V, Birnstiel M (1978) Proc Natl Acad Sci USA 75:1156–1180

Law M-F, Lowy DR, Dvoretzky J, Howley PM (1981) Proc Natl Acad USA 18:2727–2731

Lawn RM, Adelman J, Franke AE, Houck CM, Gross M, Najarian R, Goeddel DV (1981) Nucleic Acids Res 9:1045–1052

Lomedico PT, Chan SJ, Steiner DF, Saunders GF (1977) J Biol Chem 252:7971–7978

Mathis DJ, Chambon P (1981) Nature 290:310–315

Mertz JE, Gurdon JB (1977) Proc Natl Acad Sci USA 74:1502–1506

McKnight SL (1980) Nucleic Acids Res 8:5949–5966

Mueller C, Graessman A, Graessman M (1978) Cell 15:579–585

Mulligan RC, Berg P (1980) Science 209:1422–1427

Mulligan RC, Howard BJ, Berg P (1979) Nature 277:108–114

Nagata S, Mantei N, Weissmann C (1980) Nature 287:401–408

Ostro MJ, Giacomoni D, Lavelle D, Paxton W, Dray S (1978) Nature 279:921–923
Proudfoot NJ, Brownlee GG (1976) Nature 263:211–214
Rechsteiner M (1978) Natl Cancer Inst Monogr 48:57–64
Reddy VB, Thimmappaya B, Dhar R, Subramanian KN, Zain BS, Ghosh PK, Celma ML, Weiss-man SM (1978) Science 200:494–502
Saragasti S, Moyne G, Yaniv M (1980) Cell 20:65–73
Sarver N, Gruss P, Law M-F, Khoury G, Howley P (1981) J Mol Cell Biol 1:486–496
Schaffner W, Gross K, Telford J, Birnstiel M (1976) Cell 8:471–478
Scott WA, Wigmore DJ (1978) Cell 15:1511–1518
Seif I, Khoury G, Dhar R (1979) Nucleic Acids Res 6:3387–3398
Solnick D (1981) Cell 24:135–143
Subramanian KN, Shenk T (1978) Nucleic Acids Res 5:3635–3642
Sures I, Goeddel DV, Gray A, Ullrich A (1980) Science 208:57–59
Tager HS, Steiner DF (1974) Annu Rev Bioch 43:509–538
Tooze J (ed) (1980) DNA tumor viruses, molecular biology of tumor viruses, part 2. Cold Spring Harbor Laboratory, Cold Spring Harbor, N.Y.
Upcroft P, Skolnik H, Upcroft JA, Solomon D, Khoury G, Hamer DH, Fareed GH (1978) Proc Natl Acad Sci USA 75:2117–2121
Varshawsky A, Sundin O, Bohn M (1979) Cell 16:453–466
Wagner MJ, Sharp JA, Summers W (1981) Proc Natl Acad Sci USA 18:1441–1445
Waldeck W, Föhring B, Chowdhury K, Gruss P, Sauer G (1978) Proc Natl Acad Sci USA 75: 5964–5968
Wigler M, Silverstein S, Lee L-S, Pellicer A, Cheng Y-C, Axel R (1977) Cell 11:223–232

Liposomes: The Development of a New Carrier System for Introducing Nucleic Acids into Plant and Animal Cells

ROBERT FRALEY* AND DEMETRIOS PAPAHADJOPOULOS**

1 Introduction

DNA-mediated transfer of genes (transformation) was originally discovered in bacteria [1] and has been utilized extensively as a tool for genetic studies in prokaryotes. More recently, transformation systems have been developed for mammalian cells [2, 3] and yeast [4], extending this powerful genetic method to eukaryotic cells. A variety of protocols and techniques have been used to facilitate the introduction of nucleic acids into eukaryotic cells; these include cell : cell fusion [5] and the uptake of isolated nuclei, microcells [6], or whole chromosomes [2] by cells. Purified nucleic acids can be introduced into cells by formation of coprecipitates with calcium phosphate [7] or polycations [8] and subsequent uptake by endocytosis. Treatment of cells with cryoprotectants and polyalcohols [4, 9] has also been used to facilitate nucleic acid uptake. Nucleic acids can also be introduced directly into cells by microinjection with small capillary needles [10, 11] or red blood cell ghosts [12].

Recently, we have shown that RNA [13] and DNA [14,15] molecules can be encapsulated within the aqueous interior of phospholipid vesicles (liposomes, for review see 85).

* Monsanto Company St. Louis, Missouri 63166 and Department of Biology, Washington University, St. Louis, Missouri 63130
** Cancer Research Institute and Department of Pharmacology, University of California Medical Center, San Francisco, California 94143

Subsequent fusion of liposomes with the plasma membrane or their uptake by endocytosis results in intracellular delivery of the encapsulated nucleic acid. The liposome-mediated delivery of nucleic acids to a variety of cell types, including bacterial [14], mammalian [15], and plant [16] cells, has been demonstrated; in some cases, the efficiency of liposome-mediated delivery is higher than that reported using other methods for introducing nucleic acids into cells [13, 16–19]. Because the uptake of liposomal contents (by fusion and/or endocytosis) bypasses the normal uptake mechanisms for molecules (receptor-mediated endocytosis, specific transport systems, etc.) and overcomes the surface restrictions imposed by the cell membrane, liposomes can be used to introduce macromolecules that normally cannot enter cells. As a result, cells which are outside the host range of a particular virus can be infected by the liposome encapsulated virus [20] and cells that are resistant to particular toxins can be killed by the liposome-entrapped toxins [21].

Other advantages of liposomes as a carrier for introducing RNA and DNA molecules into cells are their relatively simple preparation, low toxicity, and ability to protect encapsulated nucleic acids from nuclease digestion. Also important is the flexibility of liposomes as a delivery system: purified genes [22], whole chromosomes [17], and even isolated nuclei [23] can be encapsulated and introduced into cells. In addition, liposome-cell interactions can be modified and optimized for each cell system by varying incubation conditions, liposome composition, and by targeting liposomes to cells with antibodies and lectins.

As indicated above, an obvious application of liposome-mediated delivery of nucleic acids will be to maximize expression in existing transformation and transfection systems. Other likely applications of this technique include: transformation of cell types which do not respond to existing methods for introducing nucleic acids into cells, development of liposomes as carriers for delivering nucleic acids to cells in vivo, and use of the cellular expression of liposome-encapsulated nucleic acids as a sensitive and unambiguous probe for monitoring liposome-cell interactions.

In this review we will discuss these potential applications, describe methods for encapsulating nucleic acids in liposomes and consider the use of liposomes for introducing nucleic acids into both plant and animal cells.

2 Liposome Preparation

A variety of different liposome preparations have been used to encapsulate nucleic acids (Table 1). The simplest preparation yields multilamellar vesicles (MLV), which are made by the mechanical shaking of aqueous dispersions of phospholipids [24]. Extended sonication of MLV converts them into small unilamellar vesicles [SUV, 25]. Although there has been a report on the use of SUV to encapsulate DNA fragments [22], in general, SUV are not useful as carriers for nucleic acids due to their small size (diameter, 200–250 Å), low efficiency of entrapment (0.1–0.5 µl/µmol lipid) and the requirement for extensive sonication (which may shear large nucleic acids). On the other hand, MLV entrap 10–20 times more material than SUV (5–10 µl/µmol lipid) and their preparation avoids the sonication step. The principal disadvantages of MLV as a carrier are their size heterogeneity (0.1–10 µm in diameter), which is an important factor in determining liposome tissue distribution in vivo, and the multilamellar nature of each vesicle, which reduces the efficien-

Table 1. Nucleic acid encapsulation in liposomes and delivery to cells

Nucleic acid encapsulation	Liposome composition	Liposome preparation	Recipient cell type	Assay for delivery
DNA				
1) Metaphase chromosomes	PC : Chol (7 : 2)	MLV	HGPRT⁻ A9	Trans-formation to HGPRT⁺
2) Mouse and pMB9 DNA	PC	MLV	–	–
3) E. coli, pBR322 DNA	PC, PC : Chol (7 : 2)	MLV	Cowpea	Uptake of radioactive DNA
4) pB2322 DNA	PC : SA (9 : 1)	LUV	Tobacco	Uptake of radioactive DNA
5) pBR322 DNA	PC : PG (10 : 1)	LUV	E. coli	Transforma-tion to tet^R
6) Phage T7 DNA	PS	LUV	–	–
7) β-lactamase gene (pBR322)	PC : PS (10 : 1)	SUV	HeLa, chicken embryo cells	β-lactamase activity
8) SV40 DNA	PS : Chol (1 : 1), PS, PC, PG, PA, PC : SA (9 : 1) azolectin, PC : Chol : DCP (7 : 2 : 1)	LUV LUV	AGMK, HeLa, L-cells	Plaque assay, T-antigen assay
9) Chicken red blood cell nuclei	PS : PI : PE (4 : 1.5 : 1)	LUV	CHO (proline⁻)	Transforma-tion to proline⁺
10) B. subtilis	PC : SS (6 : 1), PC : SA : SS (3 : 2 : 1), PC : DCP : SS (3 : 2 : 1)	MLV, LUV	Carrot, tobacco, protoplasts	Uptake of radioactive DNA (auto-radiography)
11) Ti plasmid DNA	PS : Chol (1 : 1)	LUV	Tobacco protoplasts	Transforma-tion to hor-mone-inde-pendent growth

	Reference
1) Tenfold improvement in transformation frequency	17
2) First report demonstrating the encapsulation of free DNA in liposomes	27
3) Uptake of encapsulated radioactive DNA was enhanced by PEG treatment	26
4) Demonstration that biochemical techniques do not allow quantitative estimate of DNA uptake	50
5) DNA was delivered to bacteria by a process that was resistant to DNase	14
6) Efficiency of encapsulation by Ca^{2+}-EDTA chelation LUV is dependent on DNA size	44
7) Prokaryotic gene is expressed in animal cells	22
8) (PS-Chol) LUV are most efficient carrier, efficiency of delivery increased by glycerol treatment and dependent on vesicle binding/leakage	15, 35, 60, 64, 70, 74
9) Transformation frequency was higher than obtained with isolated chromosomes	23
10) Positively charged MLV associated with protoplasts to greatest extent	53, 54
11) First demonstration of the incorporation and expression of free DNA by plant cells	16

Table 1. (Continued)

Nucleic acid encapsulation	Liposome composition	Liposome preparation	Recipient cell type	Assay for delivery
RNA				
1) Rabbit globin mRNA	PS	LUV	–	–
2) *E. coli* RNA	PC : Chol : DCP (7 : 2 :1)	LUV	–	–
3) Polio RNA	PS	LUV	HeLa, CHO, L-cells	Infectious center or plaque assay
4) Rabbit globin mRNA	PS	LUV	Mouse spleen lymphocytes	Globin synthesis
5) Rabbit globin mRNA	PS : Chol : DCP (7 : 2 : 1)	LUV	Human epithelial carcinoma	Globin synthesis
6) *E. coli* RNA	PC : Chol : DCP (various ratios)	LUV	Carrot proto-plasts, rabbit spleen lympo-cytes, Hep 2 cells	Uptake of radioactive RNA
7) Rat uterine mRNA	PS : PC	LUV	Communi-cation-defec-tive mouse cells	Restoration of cell: cell coupling
8) CPMV RNA	PS : Chol (1 : 1), PC, PC : SA (9 : 1)	LUV	Cowpea protoplasts	ELISA assay for detecting viral pro-teins
9) TMV RNA	PS : Chol (1 : 1), PC, PC : SA (9 : 1)	LUV	Tobacco protoplasts	Radio-immune assay for detecting viral pro-tein

	Reference
1) Encapsulated RNA was resistant to RNase digestion, reisolated RNA was still biologically active	34
2) Encapsulated RNA was resistant to RNase digestion	33
3) 90% of cells could be infected, more efficient than other transfection techniques; cell-induced leakage of vesicle contents	20
4) Detection of globinlike protein in SDS gels following purification by immune adsorbent chromatography	36
5) Same as in 4)	36
6) % RNA uptake was dependent on liposome and plasma membrane fluidity	51, 52
7) Restoration of coupling was dependent on protein synthesis and could be eliminated by pretreating mRNA with RNase	43
8) PS : Chol most efficient carrier, infection higher than obtained by other methods	18
9) PS : Chol most efficient carrier, delivery could be enhanced by poly-alcohol treatments and cholesterol	19

MLV, multilamellar vesicles; SUV, small, unilamellar vesicles; LUV, large, unilamellar vesicles; PC, phosphatidylcholine; PG, phosphatidylglycerol; PI, phosphatidylinositol; SA, stearylamine; PA, phosphatidic acid; PE, phosphatidylethanolamine; DCP, dicetyl phosphate; Chol, cholesterol; SS, β-sitosterol; PS, phosphatidylserine

cy for trapping macromolecules and possibly for intracellular delivery [13]. In addition, it has been reported that encapsulation in MLV can damage nucleic acids and may not completely protect them from nuclease digestion [26, 27].

The development of procedures for making large unilamellar vesicles [LUV, for review see 28, 29], has permitted the encapsulation of nucleic acids at high efficiency. Such vesicles range in diameter from 0.1–0.6 μm and encapsulate 10^3–10^4 times as much volume as a typical SUV. These methods, which include Ca^{2+}-ethylenediaminetetra-acetic acid (EDTA) chelation [30], ether injection [31], and reverse-phase evaporation [32], have been used to encapsulate RNA and DNA molecules; the encapsulated nucleic acids are completely resistant to nuclease digestion and retain full biological activity [13–15, 33, 34].

The reverse-phase evaporation procedure [32] is particularly useful for the encapsulation of nucleic acids since DNA molecules of 50 kb (or larger) can be encapsulation of intact. In contrast, the efficiency of encapsulation by the Ca^{2+}-EDTA chelation procedure is decreased for DNA molecules larger than a few kb [44]. No size limits have yet been determined for vesicles prepared by the ether injection method. The high encapsulation efficiency of the reverse-phase evaporation procedure (30%–60%, together with the small sample volumes required (0.05–0.1 ml), combine to minimize the quantities of nucleic acid needed to obtain a sufficiently high copy number per vesicle. Achieving a high copy number per liposome is important, since "empty" liposomes will compete with DNA-containing liposomes for binding sites on the cell surface [15]. Also, because liposome binding to cells is saturable, the extent of intracellular delivery is proportional to the copy number per vesicle [15]. The other advantage of this procedure is that liposomes can be prepared from virtually all phospholipids and their mixtures and it is easy to maintain sample sterility. For a detailed description of the reverse-phase evaporation procedure and its use for nucleic acid encapsulation see 35.

Following the encapsulation of nucleic acids in liposomes, it may not be necessary to separate unencapsulated RNA and DNA molecules from liposomes; however, for specific cases where this is necessary (e.g., labeled DNA uptake is being determined or the efficiency of encapsulation is being measured) several methods have been used. Separation can be achieved by first digesting the preparation with nucleases to degrade unencapsulated nucleic acids and then separating the liposomes (and entrapped RNA and DNA) from nucleotide fragments by molecular sieve chromatography [13, 33, 34]. While this method works satisfactorily, it is time consuming, the sample is obtained in diluted form, and in addition, it is difficult to perform under sterile conditions. Recently, a method has been developed using discontinuous polymer gradients, which separates liposomes from free nucleic acids on the basis of their large density differences [15, 35]. This procedure is much quicker, more amenable to multiple samples, and is easier for maintaining sterile preparations.

The encapsulated nucleic acids can be stored under an inert atmosphere (nitrogen or argon) at 4 °C until used. Liposome-encapsulated DNA preparations simian virus 40 (SV 40), Herpes Simplex Virus thymidine kinase (HSV TK), etc. have been stored under these conditions for 3–4 months without detectable loss of activity [15, 18]. Entrapped tobacco mosaic virus (TMV) [19] and cowpea mosaic virus (CPMV) [18] RNA preparations have been stored for periods of up to 3–4 weeks with only small (<10%) decrease in infectivity. As a result of the stability of the encapsulated nucleic acids, a single liposome preparation can be used to perform a number of experiments.

3 Delivery of Nucleic Acids to Mammalian Cells

The demonstration by *Wilson* et al. [20] that polio virus could be encapsulated in LUV prepared by the Ca^{2+}-EDTA chelation method provided the first indication that liposomes might be useful for introducing large macromolecules such as nucleic acids into cells. The infectivity of the liposome-entrapped polio virus, assayed on permissive HeLa cells, was shown to be largely unaffected by the presence of neutralizing antisera. This indicates that the virus was entrapped (and protected) within the aqueous interior of liposomes and that liposomes must interact directly with cells in order to introduce their contents, since infection initiated by naked virus particles was eliminated in the presence of antisera to the virus.

Further evidence for a liposome-mediated delivery event came from experiments demonstrating that the liposome-encapsulated polio virus was also capable of infecting nonprimate cells (CHO cells), which lack a polio virus receptor and are normally resistant to infection. Presumably, by fusing directly with cell membrane or by being taken up by the cells by endocytosis, liposomes were able to introduce the virus directly into the cell cytoplasm — bypassing the normal receptor-mediated pathway for viral infection.

3.1 RNA Delivery

Following the experiments by *Wilson* et al. [20], a number of reports demonstrated that various purified RNA molecules could be encapsulated in liposomes and that following removal of the vesicle lipid by detergent lysis or solvent extraction, the RNA remained biologically active [33, 34]. That the liposome-encapsulated RNA molecule could be delivered to cells and translated was shown concurrently by two laboratories [36, 37]. Rabbit globin mRNA was entrapped in negatively charged LUV and incubated with Hep-2 cells [36] and mouse spleen lymphocytes [37]. Following incubation, a globinlike protein could be identified on gels containing sodium dodecyl sulfate (SDS) after immunoprecipitation of whole cell lysates. Control experiments with free mRNA and empty liposomes did not result in the appearance of the polypeptide, providing evidence that liposomes promoted the delivery of their contents to cells directly. It seems likely that further refinement of such a liposome-mediated delivery system for introducing mRNA into mammalian cells will facilitate studies on protein synthesis in a variety of host cells and would complement existing cell-free or frog oocyte systems for studying translation.

A quantitative study of liposome-cell interactions was carried out in series of experiments [16, 38, 39] using the expression of liposome-encapsulated polio RNA as an assay for intracellular delivery. It was shown using a variety of cell lines (HeLa, L, CHO, etc.) that virtually all the cells in the population (85%–100%) could be infected following incubation with liposome-encapsulated polio RNA. In contrast, only 40%–60% of the cells could be infected using the diethylaminoethyl (DEAE)-dextran and dimethyl sulfoxide (DMSO) treatment for introducing viral RNA into cells. The infectivity of the liposome-encapsulated polio RNA (10^4–10^5 plaque-forming units (pfu) µg) was also much higher (10^1–10^2-fold) than that obtained with the other commonly used transfection techniques.

By determining the amounts of cell-associated RNA and vesicle lipid under various conditions, it was determined that approximately 10% of the added lipid and only 1% of

the entrapped nucleic acid became cell associated [39]. This disparity in binding was shown to be due to the accelerated leakage of encapsulated material upon interaction of liposomes with cells. A number of recent studies have also confirmed the occurrence of cell-induced vesicle leakage [40, 42]. From the infectivity data, it was calculated that approximately 10 000 LUV (each containing one polio virus RNA molecule) must be added per cell in order to ensure infection. From the binding data, it was calculated that 10% of the LUV will become cell associated (1000 liposomes/cell) and that 90% of these lose their contents (100 RNA molecules/cell). These figures (1 pfu/100 RNA molecules) are comparable to those determined for infection by the whole virus (1 pfu/100 virus particles).

In later sections we will describe more recent experiments showing that the leakage of material from the liposomes can be minimized and the intracellular delivery of contents can be enhanced by at least 1000-fold.

Recently, mRNA or polysomes isolated from rat uterine tissue that is actively synthesizing gap junction proteins (proteins which are involved in the formation of intracellular gap junctions and responsible for cell : cell communication) have been incorporated into LUV [43]. Mouse cell lines which are defective in intercellular communication were able to restore cell : cell communication temporarily (24 h), after treatment with the liposome-encapsulated RNA as assayed by an increase in the incidence of electrical coupling between cell pairs. Treatment of cells with cycloheximide after their incubation with liposomes eliminated coupling. In addition, cells incubated with empty liposomes or with polysomes that had been pretreated with RNAse before liposome encapsulation did not increase the frequency of cell : cell coupling. The above results suggest that the introduction of liposome-encapsulated mRNA coding for gap junction proteins (or other regulatory proteins) into communication-defective cells can temporarily establish junctional communication.

3.2 DNA Delivery

The above demonstrations of entrapment of biologically active RNA in liposomes and their delivery and subsequent expression in cells suggested that liposomes might also be useful as carriers for introducing DNA into cells. The development of an efficient liposome-mediated DNA delivery system would be important and might help to facilitate genetic studies in those cell types which lack conventional means of genetic exchange or where current techniques for introducing nucleic acids into cells have not been successful. In addition, DNA delivery and its expression in cells can be used (in the same manner as polio RNA) as a sensitive and discriminating probe for studying liposome-cell interactions and for determining those conditions which favor the increased uptake of encapsulated materials by cells.

The first demonstration that liposomes would, in fact, be useful for introducing DNA into cells came from a study [17] in which metaphase chromosomes isolated from a hypoxanthine guanine phosphoribosyl transkenase (HGPRT$^+$) mouse/human hybrid cell line were entrapped in liposomes (phosphatidylcholine (PC) : cholesterol (Chol)) and used to transform HGPRT$^-$ A-9 cells. The frequency of transformation obtained with the liposome-encapsulated chromosomes (10^{-5}) was at least ten-fold higher than that obtained using naked chromosomes and the resulting transformants were shown to be

stable and to express HGPRT activity of human origin. Recently another study has extended this approach to the entrapment of whole nuclei in liposomes [23]. Nuclei isolated from chicken red blood cells were encapsulated in LUV phosphatidylserine (PS) formed by the Ca^{2+}-EDTA chelation method and were incubated with proline-requiring (Pro⁻) CHO cells. Pro⁺ cells arose at a frequency (3.7–6.2×10^{-5}) approximately 10^2-fold higher than the spontaneous reversion rate; the frequency of transformation was shown to be dependent on the amount of added liposomes.

Several studies demonstrated that purified DNA molecules such as λDNA [27], phage T7 DNA (44), and pBR322 DNA [14] could be entrapped in liposomes. The encapsulated DNAs were resistant to DNase digestion; isolation and analysis of the entrapped molecules revealed that liposome encapsulation did not significantly alter the physical properties of the DNA, although some nicking could be detected [14]. It was shown [14] that pBR322 DNA entrapped in negatively charged LUV could be used to transform competent *Escherichia coli* cells to tetracycline resistance. The frequency of transformation by the liposome-encapsulated pBR322 DNA was not reduced by the presence of high levels of DNAse during the incubation with cells. These results demonstrated that liposome-entrapped DNA molecules retained biological activity and also indicated that liposome-mediated DNA delivery was not limited to mammalian cells and would be applicable to a variety of cell types.

Recently, a small restriction fragment (875 bp) isolated from pBR322 and containing the β-lactamase gene was entrapped in negatively–charged SUV [22]. The incubation of these liposomes with either HeLa or chicken embryo fibroblast cells resulted in the appearance of detectable levels of β-lactamase activity in cell extracts. No enzyme activity could be detected in cell extracts from cells treated with empty liposomes or empty liposomes plus the β-lactamase gene. This report constitutes one of the first demonstrations of the expression of a prokaryotic gene in eukaryotic cells. However, it is not clear what the efficiency of the process was in this system and as discussed elsewhere in this review, estimates of intracellular delivery based on uptake of radiolabeled material can be misleading.

4 Delivery of Nucleic Acids to Plant Cells

Liposome-mediated delivery might be expected to offer several advantages over other methods for introducing nucleic acids into plant protoplasts. These include the protection of encapsulated RNA and DNA molecules from degradation by nucleases which are reported to be present in plant cell culture medium [45, 46] and the absence of a requirement for treatment with polycations which can be toxic to plant cells. Moreover, because of the larger size (30–70 µm diameter) and greater surface area of a typical protoplast in comparison to a mammalian cell, it is possible that liposome binding and delivery may be more efficient with plant protoplasts than with mammalian cells. Finally, since the plasma membrane of protoplasts is freely accessible following removal of the cell wall, protoplasts may be more prone to fuse with liposomes (especially following treatment with "fusogens") than mammalian cells, which are covered by an extensive glycocalyx.

Although the possible use of liposomes for introducing nucleic acids into plant protoplasts has been suggested frequently, there are only a few reports describing the interac-

tion of liposomes with plant protoplasts. *Nagata* et al. [47] have shown that the incubation of positively charged MLV prepared from the synthetic phospholipid, 1,2-0-dipentade-cylmethylidene-glycerol-3-phosphoryl-(n-ethyl-amino)-ethanolamine, with tobacco protoplasts results in a high level of protoplast fusion. In another study [48], positively charged MLV (egg PC : stearylamine (SA); 9 : 1) containing fluorescein diacetate were shown to be taken up by tomato protoplasts. Uptake in this case was measured by the accumulation of intracellular fluorescence following liposome-cell incubation. While this result is encouraging, it must be viewed cautiously because of the many problems associated with the use of fluorescent probes to monitor vesicle fusion [40], particularly when the probe used (fluorescein diacetate) is actively taken up by plant cells [49].

There have been several reports which describe the interaction of nucleic acid containing liposomes with various protoplast species. The incubation of tobacco and cowpea protoplasts with radioactively labeled pBR322 DNA entrapped in neutral MLV (PC or PC-Chol) was found to result in the association of significant amounts of radioactivity with the protoplasts [26]. This occurred despite the fact that large fraction of the liposomes (approximately 70%) released their contents during the short incubation with cells. The amount of radioactive DNA in association with the protoplasts could be enhanced five- to six-fold by including polyethylene glycol (PEG) during the incubation. Under these conditions up to 3% of the added DNA in liposomes was found associated with nuclei following cell fractionation. It was later shown that positively charged LUV could be used to increase the proportion of radioactive DNA that becomes associated with nuclei and with chromatin [50]. However, this study also provided evidence indicating that isolated nuclei incorporate exogenous DNA into chromatin, a result which seriously challenges the validity of using the uptake of radioactively labeled DNA into protoplast material or as an assay for monitoring DNA transfer. In similar studies by *Ostro* et al. [51, 52], radioactive *E. coli* RNA entrapped in negatively charged LUV was shown to become associated with carrot protoplasts. The extent of intracellular RNA delivery was assessed by determining the amount of degraded RNA that was found in association with the cells (see discussion below). It was determined using this assay that the percentage of intracellular RNA delivery by liposomes was decreased by including cholesterol in the liposome preparation; however, it could be increased by the addition of lysolecithin. More recently, *Rollo* et al. [53, 54] have compared the ability of differently charged MLV and LUV preparations to introduce radioactively labeled *Bacillus subtilis* DNA into carrot protoplasts. Positively charged MLV (PC : SA; 10 : 1) were found to associate with cells to the greatest extent and to promote the highest level of DNA transfer (measured by autoradiographic analysis of sectioned protoplasts). The positively charged liposomes were also shown to be significantly toxic to both carrot and tobacco protoplasts (incubation of 100–200 nmol/10^6 protoplasts reduced plating efficiency by approximately 50%).

An unambiguous interpretation of the above studies demonstrating liposome-mediated nucleic acid transfer to plant protoplasts is difficult. The determination of radioactivity associated with isolated nuclei as proof of cellular uptake of nucleic acids has been previously criticized because it has been shown that isolated nuclei bind added DNA [55, 56]. In addition, it has been suggested that damaged or broken protoplasts in the preparation may be responsible for the bulk of the apparent DNA uptake [57]. In any case, the use of radioactive markers entrapped in liposomes to follow liposomal delivery to cells is ambiguous and has also been criticized [58, 59] since this approach cannot dis-

tinguish between true delivery events and the simple adhesion of liposomes to the cell surface or the leakage of vesicle contents and uptake by other mechanisms. Despite many attempts to remove "bound" liposomes by washing (or other treatments) our experience indicates that removal is both difficult and variable. The assay developed by *Ostro* et al. [51, 52] to monitor the uptake of labeled RNA into cells rests on the assumptions that: 1. liposome-cell interactions are nonleaky and 2. that measurements of cell-associated radioactivity are analogous to measurements of intracellular delivery. Our present understanding of liposome-cell interaction indicates that these assumptions are incorrect since a large fraction of the liposomes which interact with cells remain absorbed to the cell surface [41, 60] and actually leak their contents at much elevated rates [26, 40, 42]. In Sect. 5.2 we will discuss our own experiments with plant protoplasts and methods to increase the efficiency of delivery of liposome-encapsulated nucleic acids [16, 18, 19].

5 Increasing the Efficiency of Liposome-Mediated Delivery

The results of the studies discussed in the preceding sections have shown that incubation of liposome-encapsulated nucleic acids with cells can result in delivery and expression of the entrapped RNA and DNA molecules. While these studies have clearly indicated that liposomes may be useful as a carrier for introducing nucleic acids into a variety of cell types, with a few exceptions there has been a lack of systematic investigation to determine the parameters (vesicle lipid composition, incubation conditions, treatment with fusogens, etc.) which maximize enhanced liposomal delivery. As a result, the limitations and potential usefulness of liposome-mediated delivery as a method for transforming cells or studying protein synthesis have not been clearly established.

In order to explore more fully the potential of liposomes for introducing nucleic acids into eukaryotic cells, we have undertaken an extensive investigation of the parameters that are likely to enhance the expression of liposome-encapsulated nucleic acids.

5.1 Mammalian Cells

Studies using liposomes to deliver nucleic acids to mammalian cells have utilized a wide variety of liposome compositions (Table 1); however, negatively charged liposomes have been shown to be the most successful [13, 23, 34, 35]. Our own results with liposome-encapsulated SV40 DNA have clearly confirmed the superiority of negatively charged liposomes, particularly LUV (PS), as carriers.

SV40 DNA was selected for encapsulation in liposomes and delivery to cells because of the availability of sensitive plaque assays for monitoring virus production in permissive cell lines, such as African Green Monkey Kidney (AGMK) cells. In cell lines which are nonpermissive for virus replication (HeLa, L-cells), the expression of SV40 DNA can be followed by fluorescent antibody assays directed against specific viral proteins [61]. Another important reason for developing a liposome-mediated delivery system for SV40 DNA is its increasing use as a vector for introducing cloned genes into mammalian cells [62, 63].

Table 2. Effect of vesicle lipid composition on binding, leakage and delivery[a]

Vesicle lipid composition	Infectivity[b] (pfµ/µg DNA)	Infectivity[c] (pfµ/µg DNA)	Cell-associat-ed[d] vesicle lipid	Cell-associat-ed[e] vesicle contents	Leakage[f] (% contents retained)
PC	1.1×10^2	4.9×10^2	0.51	0.14	28
PC : SA (9 : 1)	50	1×10^2	0.57	0.5	87
PS	1.5×10^3	2.3×10^4	2.7	1.16	43
PG	40	1.7×10^3	2.9	0.26	8.8
PA	60	1.9×10^3	–	–	–
PC : Chol : DCP (7 : 2 : 1)	1.5×10^2	2×10^4	–	–	–
Azolectin	5.5×10^2	1.1×10^4	–	–	–
PS : Chol (1 : 1)	1.8×10^3	3.4×10^4	2.5	2.11	84.7
PG : Chol (1 : 1)	1×10^3	2.8×10^4	2.6	1.27	49.2
PA : Chol (1 : 1)	1.1×10^3	2.6×10^4	–	–	–

[a] Liposomes were prepared by the reverse phase evaporation technique starting with 10 µmol of phospholipid and 10 µg SV40 DNA. The vesicle preparations were sized and sterilized by passage through a 0.4 µM Unipore filter and the vesicles were separated from unencapsulated DNA by floatation on Ficoll gradients. Incubation of cells with liposome-encapsulated SV40 DNA was carried out as described previously (*Fraley* et al., [15]). 0.2 ml of Tris-buffered saline (TBS) containing between 0.1–10 µg of SV40 DNA and 0.5–500 nmol of phospholipid was added to washed cell monolayers and the plates were incubated for 30 min at 37 °C. Following the incubation, the monolayers were overlaid with 1.0 ml of TBS containing 25%(v/v) glycerol for 4 min at 25 °C. The monolayers were then washed twice with TBS and overlaid with agar medium as described by *Mertz* and *Berg* [75]. Liposomes used for cell binding were prepared as above except that either ^{3}H-DPPC (50 µCi) or ^{3}H inulin were included in the preparations as radioactive labels for vesicle lipid or aqueous interior, respectively. Incubations of radioactive liposomes with AGMK cells were performed as described above. The cell monolayers were subsequently washed 3 times with 3 ml of TBS, removed from the tissue culture plate by trypsinization, and transferred to scintillation vials for determination of cell-associated radioactivity
[b] Infectivity of liposome-encapsulated SV40 DNA in the absence of glycerol treatment
[c] Infectivity of liposome-encapsulated SV40 DNA following treatment with 25% (v/v) glycerol
[d] Amount of cell-associated vesicle lipid measured after incubation with 10 nmol of vesicle lipid/ plate (nmol/10^6 cells)
[e] Amount of cell associated vesicle contents (^{3}H inulin) measured as in d (nmol lipid vesicle equivalents/10^6 cells)
[f] Determined from the amounts of cell-associated vesicle lipid and vesicle contents

Liposome-cell incubations were performed as described in the footnote to Table 2 and infectivity was measured as pfu/µg DNA. The incubation of free SV40 DNA or free DNA plus empty liposomes with AGMK cells does not result in detectable infectivity [15]. A summary of the infectivity of SV40 DNA encapsulated in various liposome preparations is shown in Table 2. Vesicles made from egg PC, a neutral phospholipid are at least ten-fold less efficient than negatively charged PS vesicles in promoting DNA delivery. Liposomes positively charged by the inclusion of 10 mol% stearylamine in PC vesicles are even less efficient. The same pattern of delivery by these liposome preparations has been observed in other cell types (L-cells, HeLa cells, mouse embryo fibroblasts) as well.

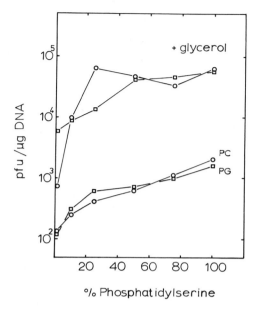

Fig. 1. Effect of phosphatidylserine on lipo-some-mediated SV40 DNA delivery to AGMK cells. Liposomes were prepared as described in footnote to Table 2. Cell incu-bations and treatment with glycerol (25% v/v) were performed as described in reference 15. Infectivity of SV40 DNA (0.1–10 ng added/plate) encapsulated in vesicles composed of a mixture of PS and PC (O---O) and PS and PG (□---□)

5.1.1 Liposome Delivery in Relation to Binding and Cell-Induced Leakage

Differences in the efficiency of delivery also exist among various negatively charged LUV preparations (Table 2). The infectivity of SV40 DNA entrapped in PS liposomes is higher than in PG, PA, azolectin, and PC:Chol:dicetylphosphate liposomes. In liposome preparations composed of mixtures of PS with PC (Fig. 1), the infectivity of the encapsulated SV40 DNA was shown to be proportional to the percentage of PS in the liposomal membrane. In a series of experiments in which the amounts of cell-associated vesicle lipid and vesicle contents were determined following the incubation of AGMK cells with various liposome preparations [64], it was found that PS vesicles exhibited higher affinity for cells and were more resistant to cell-induced leakage than either PS, PC-SA, or PG liposomes. The superiority of PS vesicles as a carrier can be partly under-stood in terms of the increased availability of encapsulated SV40 DNA molecules for intracellular delivery.

Including cholesterol in PS vesicle preparation further reduces the extent of vesicle leakage and results in slightly higher levels of infectivity (Table 2). This effect of choles-terol is consistent with its known role in reducing the fluidity of liposomal membranes [65, 66] and decreasing the extent of protein- or serum-induced leakage of vesicle con-tents [67, 68, 69]. The strong correlation between the extent of vesicle content leakage and the efficiency of SV40 DNA delivery is clearly shown in Fig. 2 for PG vesicles (which are more susceptible to cell-induced leakage) containing increasing mol% cholesterol. By reducing vesicle content leakage upon the interaction of liposomes with cells, choles-terol presumably increases the amount of encapsulated SV40 DNA that remains available for intracellular delivery.

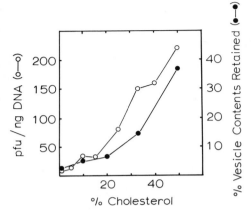

Fig. 2. Effect of colesterol on the leakage and delivery of SV40 DNA by PG liposomes. Liposomes composed of PG and the mol% cholesterol shown in the Fig. and either ^{3}H inulin or SV40 DNA, were prepared as described in the footnote to Table 2. The amount of cell-associated vesicles contents following the glycerol treatment were determined after extensive washing of the cells. A correlation between SV40 DNA infectivity and vesicles leakage also was observed in the absence of the glycerol treatment (data not shown). Infectivity of SV40 DNA encapsulated in vesicles composed of mixtures of PG and cholesterol following glycerol treatment (O). Retention of PG vesicles contents in the presence of increasing amounts of cholesterol (●)

5.1.2 Methodology for Enhancing Intracellular Delivery

Although the SV40 DNA encapsulated in LUV (PS-Chol) was at least two orders of magnitude more infectious than free SV40 DNA, the delivery via liposomes was still much less efficient than can be obtained with either the calcium phosphate (1×10^5 pfu/µg) or DEAE dextran (5×10^6 pfu/µg) methods (see below). A large fraction of the liposomes which interact with cells remain adsorbed to the cell membrane [41, 60]; therefore, incubation conditions or treatments which could facilitate cellular uptake from this "pool" would greatly increase the efficiency of delivery. A variety of agents or incubation conditions (Table 3) were tested to determine their effectiveness in enhancing the infectivity of liposome-encapsulated SV40 DNA. The most successful of these treatments was washing cells with DMSO, PEG, or glycerol following the incubation of liposomes with cells. It was found that increasing the glycerol concentration markedly increased infectivity and that following a 25% glycerol wash, the infectivity of the liposome encapsulated SV40 DNA was $2–3 \times 10^5$ pfu/µg [15, 64]. An additional two- to five fold increase in infectivity is obtained if the incubations are performed in media without serum rather than buffer [70]. Under these conditions, the efficiency of liposome-mediated SV40 DNA delivery is comparable to or greater than can be obtained with the calcium phosphate method (Table 4), although it is still less efficient than the DEAE dextran procedure.

The glycerol treatment has no effect on liposome binding, but instead, based on observations under the microscope and metabolic inhibitor studies, it appears to stimulate the uptake of cell-associated liposomes by a process resembling endocytosis [15, 64, 70]. The glycerol treatment has been shown to increase the delivery of liposomal contents into a variety of mammalian cells [60] and an interesting feature of the glycerol treatment is that it is relatively specific for negatively charged liposomes [15, 64]. This observation, together with liposome-binding data indicating that PC vesicles cannot compete with PS vesicles for binding to AGMK cells, has led to the proposal that negatively charged liposomes may interact at sites on the cell surface which are not effective for binding neutral vesicles. The lack of enhancement of delivery with neutral PC vesicles by glycerol and other polyalcohols explains previous negative results [17] using

Table 3. Effects of various treatments on the infectivity of liposome-encapsulated DNA

Incubation conditions	Infectivity (pfμ/ng DNA)	Enhancement factor
PS-Chol LUV (TBS buffer)	1.8	–
+ 3 mM Mg^{2+}	1.8	0
+ 1.5 mM Ca^{2+}	1.7	0
+ 4% DMSO	1.8	0
RNase	4.3	2.4
low salt	1.4	0
Postincubation treatments		
44% PEG (90 s)	18	10
25% DMSO (4 min)	6.0	3.3
pH 4.5 shock (4 min)	2.4	1.3
1% bromohexane (4 min)	1.8	0
15% glycerol (4 min)	34	18.8
25% glycerol (10 min)	400	220
Preincubation treatments		
chloroquine	1.8	0
chloroquine (+ 25% glycerol)	2 000	1 100
NH$_4$Cl	1.8	0
NH$_4$Cl (+ 25% glycerol)	800	440

Liposomes were composed of PS : Chol (1 : 1) and contained 0.35 µg SV40 DNA/µmol phospholipid. The effect of including Mg^{2+} (3 mM), Ca^{2+} (1.5 mM), or DMSO (4% v/v) during the incubation period; pretreating the liposomes with RNase (10 µg/nmol phospholipid); or replacing the 0.15 M NaCl in the buffer with 0.3 M sorbitol was examined using plaque assays. Postincubation treatments were performed by adding 1.0 ml of TBS buffer containing either glycerol (15%, 25% v/v), PEG (44% w/v), DMSO (25% v/v), bromohexane (1% v/v), or 10 mM sodium acetate (final pH = 4.5) directly to the monolayers for the period of time indicated in parentheses. The solutions were then removed and the monolayers were washed with TBS buffer. Loss of cell viability was not observed following any of these treatments. Pretreatment of AGMK cells with chloroquine (100 µm) or NH$_4$Cl (10 mM) was for 1 h in medium without serum prior to incubation with liposomes. Glycerol was then added and incubated for an additional 30 min before washing the cells and applying the agar medium [64]

PEG to increase the delivery of metaphase chromosomes encapsulated in MLV (PC).

The infectivity of SV40 DNA in LUV (PS/Chol) is observed to saturate at higher liposome concentrations (50–100 nmol lipid/5×10^6 AGMK cells, Fig. 3) and this appears to coincide with the saturation of the available liposome binding sites on the cell surface [15, 64]. In order to increase the percentage of cells that can be infected by liposome-encapsulated SV40 DNA, it is necessary to increase the number of SV40 DNA molecules per liposome. The effect of altering the DNA content of LUV (PS-Chol) on infectivity is illustrated in Fig. 3. Over the range of DNA/liposome ratios investigated (0.002–10 DNA copies/liposome), the specific infectivity of the liposomes (pfu/nmol lipid) increases with increasing DNA/liposome ratios (*insert*, Fig. 3).

Table 4. Comparison of SV40 DNA infectivity using different methods[a]

Method of infection	Infectivity (pfµ/ng DNA)	% Cells infected[b]
PS-Chol LUV	1.8	0.001
+ 25% glycerol	400	5
+ chloroquine (25% glycerol)	2 000	20–30
DEAE-dextran[c]	5 000	0.5–1.0
Calcium phosphate[c]	100	5–10

[a] CV-1P cells, an established cell line of AGMK cells, were cultured as described by *Mertz* and *Berg* [75], using *Dulbecco's* modified Eagle's medium supplemented with 5% bovine newborn serum. Viral DNA was obtained from CV-1 cells infected at low multiplicity with wild-type SV40 virus. The viral DNA was extracted and purified by CsCl-ethidium bromide centrifugation. Plaque assays using purified viral DNA were performed as described by *Mertz* and *Berg*[75]. The CV-1P monolayers were washed twice with TBS (*Kimura* and *Dulbecco* 1972) and incubated for 15 min at room temperature with 0.2 ml TBS containing DNA (0.01–0.1 ng) and 500 µg/ml DEAE-dextran. After washing the monolayers twice more with TBS, they were overlaid with agar medium as previously described (*Mertz* and *Berg* 1975). Incubations with liposomes were performed exactly as described in foot note to Table 3. Transformation with calcium phosphate was as described in [3]
[b] Determined by T-antigen assay according to *Mertz* and *Berg* [75]
[c] Performed as described in [3]

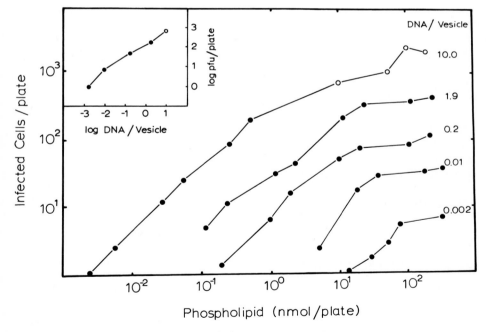

Fig. 3. Infectivity of different concentrations of encapsulated SV 40 DNA. Liposomes were prepared from PS : Chol (1:1) and contained different amounts of encapsulated SV 40 DNA (0.004, 0.0027, 0.48 and 4.2 µg DNA/µmol phospholipid). The different vesicle preparations, referred to by their DNA/vesicle ratio, were added to cells at a range of lipid concentrations (0.1–500 nmol/plate) and infectivity was enhanced by treating the cells with glycerol (20% v/v). Calculation of

5.1.3 Enhanced Delivery by Preventing Degradation or Altering Intracellular Processing

It is likely that those liposome-entrapped DNA molecules which succeed in entering the cell by fusion or endocytosis are quickly processed or degraded by the cell's lysosomal system [64, 70–72]. In order to take full advantage of liposome-mediated delivery, the intracellular degradation of newly introduced RNA and DNA molecules must be prevented. Attempts to encapsulate high concentrations of carrier DNA or nucleotides along with SV40 DNA in liposomes in order to compete with cellular nucleases have been unsuccessful (unpublished observations). Similarly, it has been shown that coinjection of carrier DNA along with the sample DNA in microinjection experiments does not increase the efficiency of transformation [10, 11]. However, pretreatment of cells with lysosomotropic agents (NH_4, chloroquine, dibucaine, etc.) results in an additional five- to ten-fold enhancement in SV40 DNA infectivity following the glycerol treatment [64, 70]. Presumably by interfering with normal lysosomal function(s), these agents increase the possibility that an incoming DNA molecule will escape lysosomal degradation and will eventually reach the cell nucleus and be expressed. An alternative interpretation is that these agents actually increase the trafficking of newly introduced molecules to the nucleus. Treatment of bovine corneal endothelial cells with chloroquine results in a 25-fold increase in the accumulation of epidermal growth factor in the nucleus [73].

The combination of chloroquine and glycerol treatments increases the infectivity of liposome-encapsulated SV40 DNA 1000-fold over that obtained by simply incubating AGMK cells with LUV (PS-Chol). This treatment produces values of infectivity that are comparable to those obtained with the DEAE-dextran method (Table 4). Under these same conditions (chloroquine and glycerol treatments) up to 30% of the cell population can be infected, as monitored by T-antigen production (Table 4); this value is substantially higher than can be obtained with the other techniques. Such a high efficiency of infection should permit a more detailed biochemical analysis of transcription and translation in this system.

It should be emphasized that although SV40 DNA expression has been used to determine optimal conditions for liposome-cell interactions, these conditions are also applicable for the delivery of other liposome-encapsulated nucleic acids or macromolecules. The incubation of LUV (PS-Chol) containing the HSV-TK gene with TK$^-$ L-cells (in conjunction with the glycerol and chloroquine treatments) results in the appearance of TK$^+$ transformants (unpublished observations) at frequencies comparable to those obtained with the calcium phosphate procedure [3].

5.2 Plant Cells

Based on our experience and success with the expression of liposome-encapsulated polio RNA [16, 38, 39] and SV40 DNA [14, 35, 60, 64, 70, 73] in mammalian cells, we have

---

the number of DNA molecules/vesicle assumed (10): a phospholipid surface area of 72 $Å^2$ condensed to 58 $Å^2$/molecule in the presence of cholesterol; a cholesterol surface area of 38 $Å^2$; and a uniform population of 0.4 µM diameter vesicles. Under these conditions there would be 4.5×10^{11} vesicles/µmol phospholipid. Calculation of the number of vesicles/cell assumed 5×10^5 cells/plate. Insert, data used are replotted for a lipid concentration of 10 nmol phosphilipid added per plate. Curve with *open circles* was determined by T-antigen assay

Table 5. Effect of vesicle lipid composition on infectivity of encapsulated TMV and CPMV RNA[a]

Vesicle lipid[b] composition	TMV RNA infectivity (ng virus/10^5 protoplasts)	CPMV RNA infectivity (ng virus/10^6 protoplasts)
PC	0	2
PC-SA	0	0
PS	430	271
Control[c]	5 500	781

[a] Tobacco (Xanthi) protoplasts were isolated from suspension cultures as described by *Fraley* et al. [19] and were resuspended at a final cell density of 2×10^6 cells/ml in Tris-buffered mannitol (TBM). Liposomes (200–300 nmol phospholipid) were added to 0.5 ml of protoplasts, incubated for 5 min, and 4.5 ml of TBM (containing 10% w/v polyvinyl alcohol) was added to cell suspension and incubated for 20 min. Protoplasts were washed, resuspended in medium (10^5 cells/ml) and allowed to incubate for 48 h before 100 μl aliquots (in duplicate) were removed for determination of virus levels by radioimmune assay. Cowpea protoplast isolation and incubation with liposomes were performed as described by *Kiefer* et al. [18]
[b] 200–300 nmol lipid, containing 5 μg viral RNA was used for the incubations
[c] Infection with whole TMV virus was performed according to [76], CPMV infection by the method of *Kiefer* and *Bruening* (unpublished)

adapted a similar strategy for the delivery of liposome-encapsulated TMV and CPMV RNA into plant protoplasts [18, 19]. There have been numerous studies demonstrating the infection of protoplasts by plant viruses and viral nucleic acids [for review, see 76]. Methods for monitoring infection include: local lesion assays on plants, examination of infected cells under the microscope, fluorescent antibody assays, and various radio- (RIA) or enzyme-linked (ELISA) immune assays. The latter methods (RIA and ELISA) offer the advantages of being rapid, sensitive, and quantitative and therefore we have used them to detect the infection of protoplasts by liposome-encapsulated viral RNA. The radioimmune assay used for TMV can easily detect 1–3 ng of virus in cell lysates. A sensitive enzyme-linked immune assay has been developed for measuring CPMV virus by our collaborators, Dr. *M. Kiefer* and Dr. *G. Bruening*. The conditions for protoplast isolation and liposome incubation are described in the footnote to Table 5.

Initial attempts to detect virus production following incubation of liposome-encapsulted TMV and CPMV RNAs with tobacco and cowpea protoplasts, respectively, were unsuccessful. Binding studies (Table 5) indicated that substantial amounts of liposomes, particularly positively charged vesicles, were becoming cell associated. The level of PS vesicle binding to tobacco protoplasts was also high (approximately tenfold higher than observed with an equivalent number of AGMK cells), but in all cases delivery was undetectable.

Based on the results with mammalian cells, various agents (glycerol, DMSO, ethylene glycol, PEG, Ficoll, polyvinyl alcohol) were tested for their ability to stimulate liposome uptake by the protoplasts. PEG was found to work effectively with cowpea protoplasts [18] and polyvinyl alcohol increased liposome infectivity with tobacco protoplasts [19]. Surprisingly, LUV (PS-Chol) were by far the most effective carrier for introducing RNA into both types of protoplasts (Table 5). As was observed with mam-

malian cells, the efficiency of delivery could be substantially enhanced by including cholesterol in the liposome preparations. This latter observation is in direct conflict with that of *Ostro* et al. [51, 52] indicating that cholesterol actually reduces the estent of intracellular delivery to protoplasts. It should also be noted that despite the high level of binding exhibited by positively charged LUV to both tobacco and cowpea protoplasts, delivery could not be detected (Table 5). This observation supports the idea that negatively charged liposomes interact at some unique site on the cell surface that facilitates their uptake by cells. This result also clearly illustrates the importance of establishing a biological assay for monitoring delivery, since simple binding data would lead to the erroneous conclusion that positively charged liposomes are the most efficient carrier.

The infectivity of the liposome-encapsulated TMV and CPMV RNA is 5–50 times higher than that with other methods currently used to infect plant protoplasts. The mechanism by which polyalcohols stimulate liposome-mediated delivery to plant cells is not clear and could involve liposome-cell fusion or an endocytosislike process. There have been several reports describing the uptake of large particles (virus, organelles, bacteria) by protoplasts [77, 78] and current efforts are being directed at understanding this process. Now that conditions have been established for introducing liposome-encapsulated nucleic acids to plant protoplasts, it is possible to extend this approach to the delivery of DNA molecules such as Cauliflower Mosaic Virus DNA [79] or Ti plasmid DNA [16] into cells. In order to demonstrate the biological expression of liposome-encapsulated DNA, we have entrapped Ti plasmid (tTi-A6) DNA in LUV (PS-Chol) and incubated them with tobacco protoplasts in the presence of PEG [16]. The "transformed" cells obtained from these experiments were selected on the basis of phytohormone-independent growth and were shown to synthesize octopine and to contain Ti DNA sequences. The frequency of transformation was very much reduced following treatment with free Ti DNA and empty liposomes, or if PEG was omitted from the incubations. This study demonstrates for the first time the incorporation and expression of foreign genes in plant cells by transformation with free DNA. Liposome-mediated delivery should prove very useful in future studies on plant cell transformation and the molecular biology of Crown Gall disease.

6 Future Prospects

As pointed out throughout this review, the expression of liposome-encapsulated nucleic acids provides a sensitive and unambiguous assay for monitoring the delivery of liposomal contents into cells. Since the incubation of even high levels of free RNA or DNA with cells does not result in their expression [13, 14, 16, 22], problems resulting from the leakage of encapsulated materials from liposomes and their subsequent uptake by cells are eliminated. Whenever possible this type of biological assay should also be used for determining optimal conditions for the liposomal delivery of other types of molecules into cells, since assays based strictly on liposome binding or the uptake of labeled contents can give misleading results. Also, in many cases, assays based on liposome binding and uptake measurements are not applicable, as in the case of determining the effects of inhibitors or altered cellular metabolism on the intracellular processing of newly introduced molecules. The latter is likely to represent an important and exciting subject for future studies on liposome-cell interactions.

In many cases the efficiency of liposome-mediated nucleic acid delivery has been shown to be comparable to or greater than that of currently used methods for infecting or transforming cells [15–19]. Further refinement and optimization of the method will undoubtably lead to its more widespread use as a delivery system. Ultimately, we hope that liposome-mediated delivery can be used as a tool by molecular biologists for introducing nucleic acids quantitatively into cells. Increasing the efficiency of RNA and DNA delivery to cells is important in that it would allow for direct biochemical analysis of certain transcriptional and translational processes than can now only be studied with great difficulty or in artificial cell-free systems. In this respect, it may be possible eventually to use liposome-mediated delivery to cells in culture in a manner analogous to frog oocyte microinjection [80] for studying gene expression. The ability to transform cells at high efficiency with DNA would minimize problems with the screening and selection of putative transformants.

Beside the possibility for increasing the efficiency of transformation and transfection in existing genetic systems, liposome-mediated delivery offers the exciting possibility of extending recombinant DNA technology to cell types which do not respond to the other available methods for introducing nucleic acids into cells. An excellent example of this is the use of liposome-encapsulated Ti plasmid DNA to transform plant protoplasts [16]. It seems likely that liposome-mediated delivery will prove useful with other cell types that can be protoplasted, including bacteria, fungi, and algae.

Certainly a future area for liposome-mediated delivery is the introduction of genes in vivo. Current research has shown that it is possible to isolate cells, transform them with DNA in vitro, and then reintroduce the transformed cell into the host animal [81]. Liposomes may offer the possibility of introducing genes directly into specific cells or tissues which are accessible to liposomes injected in vivo. The recent development of methods for conjugating antibodies to liposomes [82–84] in order to "target" them to specific cells is a necessary first step in this direction. The feasibility of in vivo liposome-mediated delivery could be initially tested, by the use of encapsulated viral nucleic acids, and by monitoring the production of antiviral antibodies and the occurrence of tumors or death in experimental animals. Delivery to specific tissues could be determined by detection of viral antigens, unique isozymes, or by sensitive in situ hybridization techniques.

In summary new liposome methodology, developed during the last few years, has made possible their use as an efficient carrier system for nucleic acids [85]. Liposome morphology, lipid composition, and incubation conditions with cells are very important parameters, which can affect the efficiency of DNA delivery and expression by more than 10^3-fold. Better understanding of the mechanism of liposome-cell interactions can be expected to increase further their usefulness as a carrier system. Liposomes targeted to specific cells [82] and containing either cytotoxic molecules or nucleic acids could play an important role in future medical treatments.

Acknowledgements. We wish to thank R. Straubinger for his contribution and help in preparing this manuscript. This work was supported by a grant from the National Cancer Institute (CA25526) and a postdoctoral fellowship (R.T.F.) from the Jane Coffin Childs Memorial Fund for Medical Research.

References

1. Avery OT, Macleod CM, McCarty M (1944) J Exp Med 79:137–158
2. McBride O, Athwall R (1977) In Vitro 12:777–786
2a.Spandidos D, Siminovitch L (1977) Proc Natl Acad Sci USA 74:2943–2947
3. Wigler M, Silverstein S, Lee L, Pelecer A, Cheng Y, Axel R (1977) Cell 11:223–232
4. Hinnen A, Hicks J, Fink G (1978) Proc Natl Acad Sci USA 75:1929–1933
5. Ringertz N, Savage R In: Cell Hybrids. cademic Press, New York, pp 1–366
6. Fournier R, Ruddle F (1977) Proc Natl Acad Sci USA 74:319–323
7. Graham F, Van der Eb A (1973) Virology 52:456–460
8. McCutchen J, Pagano J (1968) J Nat Cancer Inst 41:351–357
9. Stow N, Wilkie N (1976) J Gen Virol 33:446–458
10. Capecchi M (1980) Cell 22:479–488
11. Anderson W, Killos L, Sanders-Haigh L, Kretschmet P, Diacumakos E (1980) Proc Natl Acad Sci USA 77:5399–5403
12. Straus S, Raskas H (1980) J Gen Virol 48:241–245
13. Wilson T, Papahadjopoulos D, Taber R (1979) Cell 17:77–84
14. Fraley R, Fornari C, Kaplan S (1979) Proc Natl Acad Sci 76:3348–3352
15. Fraley R, Subramani S, Berg P, Papahadjopoulos D (1980) J Biol Chem 255:10431–10435
16. Dellaporta S, Giles K, Fraley R, Papahadjopoulos D, Powell A, Thomashow M, Nester E, Gordon M (to be published)
17. Mukherjee A, Orloff S, Butler J, Triche T, Lalley P, Schulman J (1978) Proc Natl Acad Sci USA 75:1361–1365
18. Kiefer M, Fraley R, Papahadjopoulos D, Bruening G (to be published)
19. Fraley R, Dellaporta S, Papahadjopoulos D (to be published)
20. Wilson T, Papahadjopoulos D, Taber R (1977) Proc Natl Acad Sci 74:3471–3475
21. Nicolson G, Poste G (1978) J Supramol Struct 8:235–245
22. Wong T-K, Nicolau C, Hofschneider P (1980) Gene 10:87–94
23. Kondorosi E, Duda E (1980) FEBS Letts 120:37–40
24. Bangham A, Hill M, Miller N (1974) Methods Membrane Biol 1:1–68
25. Papahadjopoulos D, Watkins J (1967) Biochem Biophys Acta 135:639–652
25a. Huang C (1969) Biochemistry 8:344–352
26. Lurquin P (1979) Nucleic Acids Res 6:3773–3784
27. Hoffman R, Margolis L, Bergelson L (1978) FEBS Lett 93:365–368
28. Szoka F, Papahadjopoulos D (1980) Ann Rev Biophys Bioeng 9:467–508
29. Deamer D, Uster P In: Baserga R, Croce C, Roueza G (eds) Introduction of Macromolecules into viable mammalian cell. Alan R Liss, New York, pp 205–220
30. Papahadjopoulos D, Vail W (1978) Ann NY Acad Sci 308:259–266
31. Deamer D, Bangham A (1976) Biochim Biophys Acta 394:483–491
32. Szoka F, Papahadjopoulos D (1978) Proc Natl Acad Sci 75:4194–4198
33. Ostro M, Giacomoni D, Dray S (1977) Biochem Biophys Res Comm 76:836–842
34. Dimitriadis G (1978) FEBS Lett 86:289–293
35. Papahadjopoulos D, Fraley R, Heath R, Straubinger R (to be published) In: Baker P (ed) Trends in life sciences, Elsevier/North Holland, Amsterdam
36. Ostro M, Giacomoni D, Lavelle D, Paxton W, Dray S (1978) Nature 274:921–923
37. Dimitriadis G (1978) Nature 274:923–924
38. Papahadjopoulos D, Wilson T, Taber R (1980) In: Celis J, Graessman A, Loyter A (eds) Transfer of cell constituents into eukaryotic cells. Plenum Press, New York, pp 155–192
39. Papahadjopoulos D, Wilson T, Taber R (1980) In Vitro 16:49–54
40. Szoka F, Jacobson K, Papahadjopoulos D (1979) Biochim Biophys Acta 551:295–303
41. Szoka F, Jacobson K, Derzko Z, Papahadjopoulos D (1980) Biochim Biophys Acta 600:1–18
42. Van Renswoude J, Hoekstra D (1981) Biochemistry 20:540–546
43. Dahl G, Azarnia R, Werner R (1981) Nature 289:683–685
44. Mannino R, Allebach E, Strohl W (1979) FEBS Lett 101:229–232
45. Lurquin P, Kado C (1977) M G G 154:113–121
46. Hughes B, White F, Smith M (1977) FEBS Lett 79:80–84
47. Nagata T, Eibl H, Melchers G (1979) Z Naturforsch C34:460–462

48. Cassells A (1978) Nature 275:760
49. Lurquin P (1976) Planta 128:213–216
50. Lurquin P (1981) Plant Sci Lett 21:31–40
51. Ostro M, Lavelle D, Paxton W, Matthew B, Giacomoni D (1980) Arch Biochem Biophys 201:392–402
52. Matthew B, Dray S, Widholm J, Ostro M (1979) Planta 145:37–44
53. Rollo F, Sala F, Cella R, Parisi B (1980) In: Sala F, Parisi B, Cella R, Ciferri O (eds) Plant cell cultures: results and perspectives. Elsevier/North Holland, New York, pp 237–246
54. Rollo F, Galli M, Parisi B (1981) Plant Sci Lett 20:347–354
55. Kleinhofs A, Behki RAM (1977) Anu Rev Genet 11:79–101
56. Ohyama K, Pelcher L, Horn D (1977) Plant Physiol 60:98–103
57. Hughes B, White F, Smith M (1978) Plant Sci Lett 11:199–206
58. Poste G (1980) In: Gregoriadis G, Allison A (eds) Use of liposomes in biology and medicine. Wiley, New York, pp 101–151
59. Pagano R, Weinstein J (1978) Anu Rev Biophys Bioeng 7:435–468
60. Rule G, Fraley R, Straubinger R, Papahadjopoulos D (to be published)
61. Robb J, Martin R (1970) Virology 41:751–760
62. Hamer D, Smith K, Boyer S, Leder P (1979) Cell 17:725–736
63. Mulligan R, Berg P (1980) Nature 209:1422–1427
64. Fraley R, Straubinger R, Rule G, Springer L, Papahadjopoulos D (1981) (submitted for publication)
65. Oldfield E, Chapman D (1972) FEBS Lett 23:285–297
66. Papahadjopoulos D, Jacobson K, Nir S, Isac R (1973) Biochem Biophys Acta 311:330–348
67. Papahadjopoulos D, Cowden M, Kiwelberg H (1973) Biochem Biophys Acta 330:8–26
68. Kirby C, Clarke J, Gregoriadis G (1980) FEBS Lett 111:324–328
69. Allen T, Cleland L (1980) Biochem Biophys Acta 597:418–426
70. Straubinger R, Fraley R, Papahadjopoulos D (to be published)
71. Hoekstra D, Tomasini R, Scherphof G (1978) Biochem Biophys Acta 542:456–469
72. Hernandez-Yago J, Knecht E, Martinez-Ramon A, Grisolia S (1980) Cell Tissue Res 205:303–309
73. Savion N, Vlodavsky I, Gospodarowicz D (1981) J Biol Chem 256:1149–1154
74. Papahadjopoulos D, Fraley R, Heath T (1980) In: Tom B, Six H (eds) Liposomes and immunology, Elsevier/North Holland, New York, pp 151–164
75. Mertz J, Berg P (1974) Virology 62:112–124
76. Takebe I (1977) In: Fraenkel-Conrat H, Wagner R (eds) Comprehensive Virology 11. Plenum Press, New York, pp 237–283
77. Suzuki M, Takebe I, Kajita S, Honda Y, Matsui C (1977) Exp Cell Res 105:127–135
78. Cocking E (1970) Int Rev Cytol 28:89–112
79. Shepherd R (1979) Anu Rev Plant Physiol 30:405–423
80. Gurdon J, DeRobertis E, Partington G (1976) Nature 260:116–120
81. Mercola K, Stang H, Browne J, Salser W, Cline M (1980) Science 208:1033–1035
82. Heath T, Fraley R, Papahadjopoulos D (1980) Science 210:539–540
83. Heath T, Macher B, Papahadjopoulos D (1981) Biochem Biophys Acta 640:66–81
84. Leserman L, Barbet J, Kourilsky F, Weinstein J (1980) Nature 288:602–604
85. Fraley R, Papahadjopoulos D (1981) Trends Biochem Sci 6 (3):77–80

Cauliflower Mosaic Virus on Its Way to Becoming a Useful Plant Vector

THOMAS HOHN*, KEN RICHARDS** AND GENEVIÈVE-LEBEURIER**

With an appendix compiled together with O. Salecker and V. Mysicka, CIBA-GEIGY Rechenzentrum, Basel, Switzerland

* Friedrich Miescher-Institut, P.O. Box 273, CH-4002 Basel, Switzerland
** Laboratoire de Virologie, Institut de Biologie Moléculaire et Cellulaire du C.N.R.S., 14, rue Descartes, F-67000 Strasbourg, France

1 Introduction

The improvement of crops with respect to yield, nutrient value, climatic tolerance and pest resistance together with the more ambitious task of rendering them independent of costly fertilizer is among the goals of genetic engineering with the greatest potential for beneficial impact on human life. So far, however, relatively little is known about gene organization and regulation in higher plants, and generalized techniques for manipulating plant genes and transferring them from one host to another are not yet available. Studies on plant DNA viruses should help to fill both these gaps as the genes of the virus are not only able to serve as a model for host gene function but also represent a convenient source of replicons, promoters, and other signals needed to successfully establish a given gene in a new environment.

Most plant viruses have an RNA genome; of the several hundred recognized groups only two, the geminiviruses and the caulimoviruses, are definitely known to employ DNA as genetic material. The geminiviruses contain a probably bipartite genome of single-stranded DNA with a total genome length of about 6000 bases (see e.g., *Haber* et al. 1981). The group name is derived from the fact that the virus particle is composed of two shells which are fused together, probably at pentameric corners (*Haber* et al. 1981). Caulimoviruses are icosahedral viruses containing double-stranded circular DNA of 8000 basepairs length. The caulimoviruses generally have moderately restricted host ranges. For example, the type virus, cauliflower mosaic virus, is mainly limited to plants of the *Cruciferae* family. Members of the geminivirus group, on the other hand, can often infect

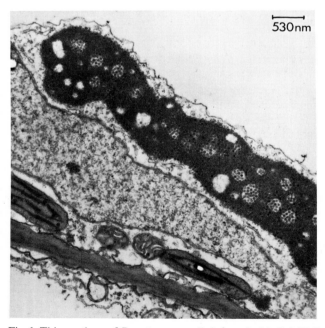

Fig. 1. Thin sections of *Brassica rapa* cells infected with CaMV. Inclusion bodies harboring virus particles, 21 days after inoculation. Electron micrographs were kindly provided by *L. Stoeckel*

plants across a broad spectrum of families including monocotyledoneous ones. For this reason geminiviruses may eventually become the more useful plant vectors. However, their handling is complicated by their very low yield and the fact that double-stranded replicative intermediate DNA must be isolated from infected tissue while double-stranded caulimovirus DNA is available from purified virus.

This review deals with cauliflower mosaic virus (CaMV), the best studied of the caulimoviruses, and its potential to become a useful vector. The development of a successful host-vector system for genetic engineering falls into three phases: 1. acquisition of as much information as possible about the viral infection cycle, particularly DNA replication and expression; 2. DNA cloning as well as transformation of plant cells and whole plants with exogenous DNA, a necessary step if genes are first to be manipulated in vitro and then introduced into a host; and 3. introduction of new beneficial genes into the vector genome and elimination of harmful ones. The three following sections of this article treat each of these steps in turn. In the first section the more recent information concerning the viral life cycle is found. There are excellent reviews dealing with the state of knowledge prior to 1979 (*Shepherd* 1976; 1979). For the third step little experimental evidence exists and thus the corresponding section will deal with prospects rather than facts. A number of as yet unpublished observations by the authors are included in this publication, such as observations on the instability of cloned CaMV DNA in bacteria and on the in vivo ligation and recombination of CaMV DNA introduced into the plant cell.

2 Virus Life Cycle

2.1 Physical and Phytopathological Properties of CaMV

Cauliflower mosaic virus infection is mainly limited to species of the family *Cruciferae* although some strains can infect *Solanaceae* as well (see *Shepherd* 1979). In the laboratory CaMV can be transmitted mechanically while in nature the virus can be propagated by aphids and perhaps by leaf-leaf contact. Once infection is established, virus can spread systemically through the tissues of the afflicted plant. Nonaphid transmissible variants of CaMV exist. It has been reported that these defective strains can be transmitted by aphids which have previously been allowed to feed on plants infected with a transmissible strain, suggesting that the transmissible strain produces a substance in the course of infection which can abet transmission of the defective strain (*Lung* and *Pirone* 1973; 1974).

Virus infection causes chlorotic lesions giving a mosaic appearance to older leaves and a mosaic-mottle appearance to younger ones. Growth of the newer leaves is also retarded. Light and electron microscopy reveal the existence of "inclusion bodies" associated with CaMV-infected cells. Inclusion bodies are compact electron-dense masses of irregular size and shape. Comparable structures are not elicited by any other type of plant virus and their appearance may be taken as diagnostic of caulimovirus infection. Two types of inclusion bodies have been described: 1. a vacuolated structure in which virus particles are frequently seen embedded within the matrix material and packed within vacuoles (Fig. 1) and 2. granular bodies, which are nonvacuolated and are devoid of virus particles (*Shepherd* et al. 1980). The relative amounts of the two bodies present vary from one strain to another (*Shalla* et al. 1980).

Inclusion bodies are rather stable structures but may be solubilized, with release of

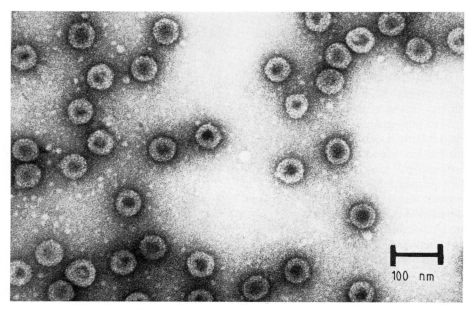

a

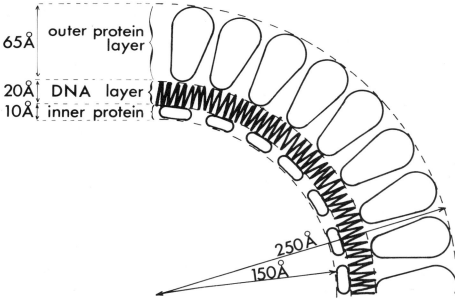

b

Fig. 2a, b. Cauliflower mosaic virus. a) Electron micrograph after treatment with 67% acetic acid and staining with 3% uranylacetate. b) Model according to the neutron diffraction data (*Chauvin* et al. 1980) showing radii of protein and DNA. Protein outside and inside of the DNA belong to the same series of molecules

occluded virions, by such denaturing agents as detergents, urea, and guanidinium hydrochloride (*Hull* et al. 1976; *Lebeurier* et al. 1978; *Al Ani* et al. 1979a; *Shepherd* et al. 1980). Purified solubilized inclusion bodies contain, in addition to virion protein(s), two major polypeptides. The larger one is described as having a molecular weight of 62 kdaltons (*Shockey* et al. 1980; *Al Ani* et al. 1980), or 66 kdaltons (*Odell* and *Howell* 1980). This protein is encoded by the virus (see Sect. 2.4). The smaller with a molecular weight of 58 kdaltons (*Shepherd* et al. 1980) is not encoded by the virus (*Daubert, Richins* and *Shepherd*, personal communication). Because they harbor virus particles it has been suggested by several authors that inclusion bodies may be virus assembly factories but at present there is no evidence either for or against such an idea.

The CaMV virion (Fig. 2) has a diameter of about 50 nm and a molecular weight of 22.8×10^6 daltons (*Hull* et al. 1976). Of this, 4.8×10^6 daltons can be attributed to the DNA (*Franck* et al. 1980) and, hence, 18×10^6 daltons remain for the capsid shell. As the major capsid protein is thought to have a molecular weight of 42 kdaltons (*Al Ani* et al. 1979) the best fit for the icosahedral triangulation number is $T = 7$, i.e., an outer shell composed of 420 subunits. There is no evidence for an inner shell. Indeed, neutron diffraction studies (*Chauvin* et al. 1979) have revealed that the DNA is layered against the inside surface of the capsid shell and that the center of the particle is free of DNA and protein.

2.2 CaMV DNA and Its Replication

Cauliflower mosaic virus DNA is double-stranded and circular (Fig. 3). The DNA of several strains has been mapped by restriction analysis (*Meagher* et al. 1977b; *Hull* and *Howell* 1978; *Hull* 1980; *Gardner* et al. 1980). The most detailed map is for strain S (*Hohn* et al. 1980, Fig. 4). A comparison of the most widely used strains is shown in Fig. 5.

The DNA of strain S has now been sequenced (*Franck* et al. 1980, Appendix C) and the restriction map deduced from the sequence shown to correspond closely to the experimentally determined map. The sequence of strain CM1841 will be published soon (*Gardner* et al. 1981). Comparison with strain S reveals 4.4% changes, mostly nucleotide substitutions with a few small insertions and deletions. Work on strain B-JI (*Hull*, personal communication) is also in progress. In addition, about $\frac{3}{4}$ of the Isolate D/H DNA sequence has been established (*Balazs* et al. to be published), revealing about a 5% base substitution frequency between it and strain S. Substitutions are most commonly of the type $A/T \lessgtr G/C$.

The DNA used to determine the CaMV-S sequence had not been cloned, but in their sequencing gels *Franck* et al. (1980) found no evidence of sequence heterogeneity either between or within the two DNA preparations studied. It is evident, however, that a low level of sequence drift at any one position could be easily overlooked on Maxam-Gilbert sequencing gels as the chemical reactions specific for the various bases sometimes give faint signals in the other lanes of the gel. Evidence suggesting that such sequence variation does in fact exist has come from restriction analysis of CaMV-S DNA cloned in bacteria (see p. 208 Table 2).

Cauliflower mosaic virus DNA is rich in A-T base pairs (Table 1) and accordingly, restriction enzymes recognizing A-T rich sequences cut CaMV DNA more frequently than those recognizing G-C-rich sequences (Appendix A, B). The base composition of the complementary strand is asymmetrical with the β-strand being much richer in A than the α-strand (Table 1).

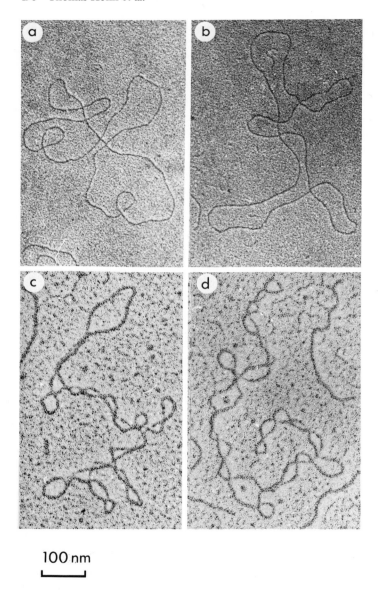

100 nm

Fig. 3a–d. Cauliflower mosaic virus DNA. a, b) Circular relaxed molecules extracted from virions. c, d) Supercoiled molecules prepared from infected plant cells. Electron micrographs were prepared according to the method of *Davis* et al. (1972)

DNA is isolated from virus particles in both linear and circular forms in proportions varying from one preparation to another (*Hull* and *Shepherd* 1977). It is likely that circularity is the natural state of the encapsidated DNA and that the linear molecules derive from the circular from by accidental physical or enzymatic cleavage.

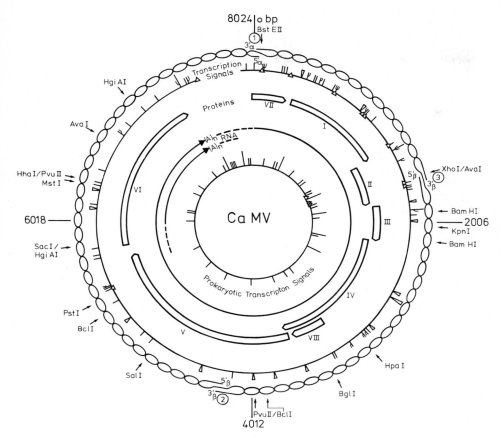

Fig. 4. Physical map of cauliflower mosaic virus DNA. *Outermost portion*: The double-stranded DNA with the three overlaps (*single-strand interruptions*) 1, 3, and 2. The α- and β-strand and the 3'- and 5' free ends are shown. Rare restriction sites are indicated. *Second ring:* Possible transcription signals. $\mathbb{V}$, "CAT boxes" (sequence UUCCAATCU (U is A or G)); $\mathbb{V}$ as $\mathbb{V}$ but with one sequence deviation; |, "TATA boxes" (TATARAR (R is A or I)); I, as | but with one sequence deviation; ⊻, indicates two closely located sites; ▽ "Polyadenylation signals" (AATAAA). For the meaning of these signal sequences in mammalian cells see *Corden* et al. (1980). *Frames*: Open reading frames in the three phases. *Arrows* indicate their direction. Each frame starts with an ATG and ends with one of the three stop codons. *Bent arrows*: Approximate location of the RNA species according to the findings of *Odell* et al. (1981), *Covey* et al. (1981) and other authors as described in the text. *Innermost circle:* Signal sequence that might act as promoters and terminators in prokaryotes. |, "Pribnow box" (sequence: TATATTG) with 0.5 deviation points; I, as | but with 1 deviation point. (Deviation in position 3, 4, 5, 7: 0.5 deviation points; in position 1, 2, 6: 1 deviation point); ▽, termination signal (sequence TAAAAA). See *Rosenberg* and *Court* (1977) for the meaning of these signal sequences. For more details and exact locations see Appendix

When analyzed by sucrose gradient centrifugation (*Civerolo* and *Lawson* 1978), circular CaMV DNA molecules isolated from virus particles behave like supercoiled DNAs of SV40 or bacterial plasmids. However, no band corresponding to supercoiled CaMV DNA is observed in CsCl-ethidium bromide density gradients or upon electrophoresis

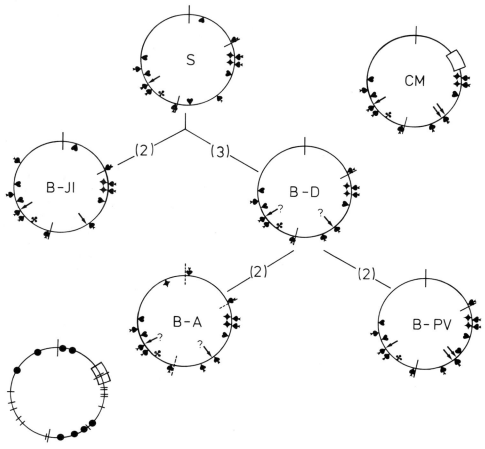

Fig. 5. Comparison of restriction maps of various CaMV strains. Strains *S* (*Hohn* et al. 1980); *B-JI*, *BD* (*Hull* 1980); *BA* (*Meagher* et al. 1977); *B-PV* (BPV147, *Volovitch* et al. 1979), and *CM* (CM4-184, *Gardner* et al. 1980) are shown. Restriction sites shown are EcoRI (♥), HindIII (♠), BamHI (♦), SalI (♣) and PstI (↓); presence or not of some sites are unknown [?] position of overlaps reported (|) or assumed by analogy (¦); □, deletion. *Connection lines* symbolize possible relationships based on minimized numbers of changes; the *number in brackets* represents the numbers of differences observed. The *circle* in the lower left summarizes the "variant" (●) and "constant" sites (ı). Redrawn and suplemented after *Hohn* et al. (1980)

Table 1. Base composition of CaMV DNA (β-strand)[a]

Adenine	2946	36.69%
Thymine	1871	23.31%
Guanine	1558	19.38%
Cytosine	1649	20.50%
Others	0	0%
	8024	

[a] Computed from the sequence published by *Franck* et al. (1980)

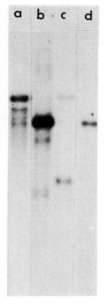

Fig. 6. CaMV DNA as seen in 0.6% agarase electrophoresis gels. *a* DNA extracted from virus particles; *b*, as *a* but digested with BstEII; *c*, DNA extracted from infected cells and purified by EtBr-CsCl density gradient centrifugation; *d*, as *c* but digested with BstEII. *Arrows* indicate from top to bottom positions of relaxed circular, linear and supercoiled molecules (*Lebeurier*, unpublished result)

through ethidium bromide-prestained agarose gels (Fig. 6a). This suggests that the twisted strands relax upon intercalation of ethidium bromide molecules because the double-stranded circular molecules might have one or more single-stranded interruptions, an idea now supported by several lines of experimental evidence (*Volovitch* et al. 1978, 1979; *Hull* and *Howell* 1978; *Hull* 1980; *Hohn* et al. 1980). Melting the double-stranded DNA molecule by heat or alkali treatment gives rise to three linear single-stranded species, one species of full-length molecule, and two species of $\frac{1}{3}$- and $\frac{2}{3}$-unit length. The two shorter chains are complementary to the full-length molecule. These observations can be explained if there are discrete discontinuities in the DNA chains of the double strand, one in one strand and two in the other. Nuclease Sl digestion produces three double-stranded fragments of 0.2-, 0.32-, and 0.48-unit length. The sites of nuclease attack have been mapped with respect to restriction sites and shown to coincide with the discontinuities detected by the aforesaid denaturation experiments (Fig. 4). Sequence analysis (*Franck* et al. 1980) has established that the discontinuities are not nicks or gaps as was first suggested but are instead sites of sequence overlap (Fig. 7). The degree of overlap varies within the population of DNA molecules but is predominantly eight residues in the α-strand for interruption 1 and predominantly 18 and 15 residues in the β-strand for interruptions 2 and 3, respectively.

To explain the supercoiled appearance in the electron microscope we would like to suggest that the three strands in the overlap region can entangle in some type of hydrogen-bonded superstructure permitting the CaMV DNA molecule to assume or retain a twisted conformation under mild conditions, whereas more drastic conditions, such as treatment with ethidium bromide, would disrupt the superstructure and cause the molecule to relax.

In that context it is interesting to note that, in the overlapping region 1, the noninterrupted strand consists of a stretch of six A's and hence the sequence present in two

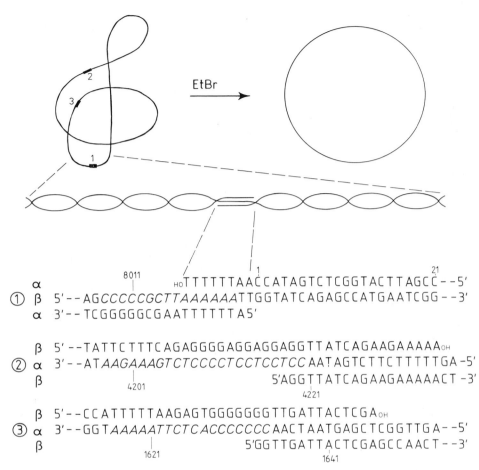

Fig. 7. Physical forms of CaMV DNA and the overlaps. Supercoiled DNA (*upper left*) is relaxed by the action of ethidium bromide. Responsible for this behavior might be the overlapping regions (*1, 2, 3*) which perhaps can form some sort of stabilizing super structure (triple helix?) as indicated. The sequences in these regions (*Franck* et al. 1980 and unpublished results) are drawn in the lower part. While the positions of the 5′ OH ends in the overlaps are at constant sites, the positions of the 3′ ends can extend further by one and two bases for ①, and by 24 bases or more for ③

copies on the discontinuous complementary strand consists of stretches of six T's each (Fig. 7), since triple helices consisting of a poly dA (or poly rA) chain and two hydrogen-bonded poly dT (or poly rU) chains have been shown by a number of workers to be stable under certain conditions (Fig. 8) (*Riley* et al. 1966; *Blake* et al. 1967; *Marsoulié* 1968a, b; *Arnott* and *Bond* 1973; *Arnott* abd *Selsing* 1974). The presence of these putative triple-helical regions or other superstructures in CaMV DNA as well as the length hetero-geneity of the overlaps could also account for the unusual migration in agarose gels of small restriction fragments containing discontinuities (*Hohn* et al. 1980).

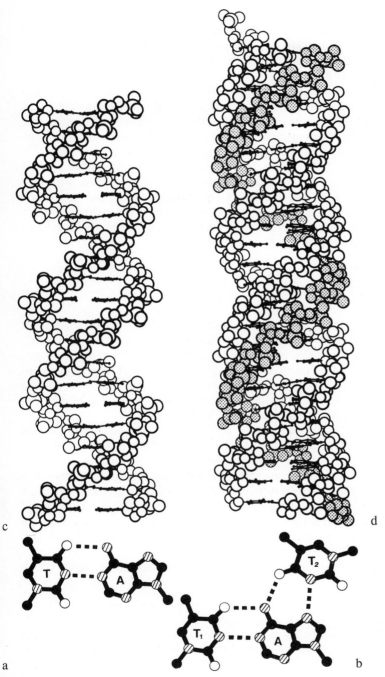

Fig. 8. Polynucleotide triple helices. The hydrogen-bonding schemes for a) poly(dA)· poly(dT)
and b) poly(dT)· poly(dA)· poly(dT). *Solid circles* represent carbon atoms, *hatched circles,* nitrogen
atoms, and *open circles,* oxygen atoms. Hydrogen bonds are shown as *broken lines.* The two

Additional forms of CaMV DNA have been seen in preparations of naked DNA isolated from infected leaves (*Menissier* et al. 1981). One of these species (Fig. 6c) has the properties of true covalently closed supercoils unlike the "pseudo-supercoils" of encapsidated CaMV DNA. This species may represent a precursor of the encapsidated interrupted form of the DNA.

Originally it was assumed that inclusion bodies are the site for CaMV DNA replication since they accumulate tritiated thymidine (*Kamei* et al. 1969; *Faveli* et al. 1973). More recently, *Guilfoyle* (1980) reported that CaMV DNA is found in isolated nuclei, too, but the author cautions that the preparations of nuclei could be contaminated with inclusion bodies. *Ansa, Bawyer* and *Shepherd* (personal communication) have shown that [32]P dATP is incorporated into CaMV DNA in isolated nuclei. The radioactive product was characterized by restriction nuclease mapping. No incorporation was seen in either isolated inclusion bodies or isolated chloroplasts. Thus DNA replication takes place most likely in the nucleus and accumulates then in the inclusion bodies. Which mechanism of transport from nucleus to the bodies is used and which enzymes are used for the DNA replication is not yet known.

2.3 Transcription and CaMV RNA

Based upon the nucleotide sequence it can be predicted that CaMV DNA codes for at least six if not eight polypeptides (see below) but the question remains open as to whether these putative products are specified by 1. a single polycistronic mRNA as in procariotes, 2. a family of spliced mRNA's sharing one or two promoters as in SV40, or 3. monocistronic mRNA's each equipped with its own promoter. An additional alternative, a single mRNA encoding a polyprotein, which is then processed as in poliovirus, can be ruled out since coding regions alternate in reading phase (Fig. 4).

RNA containing CaMV-specified mRNA activity has been isolated from infected leaves (*Hull* et al. 1979; *Al Ani* et al. 1980; *Odell* and *Howell* 1980), infected protoplasts (*Howell* and *Hull* 1978), or crude preparations of nuclei isolated from infected plants (*Guilfoyle* 1980). There is general agreement that the major fraction if not all of the virus-specific RNA hybridizes to the DNA α-strand, the strand containing one interruption. Thus transcription proceeds in a clockwise direction on the conventional map (Fig. 4). All authors find that the major portions of the viral mRNA is polyadenylated while *Guilfoyle* (1980) has shown that the fraction associated with his nuclei preparations is also capped.

There is still little evidence concerning the number and size of the viral RNA(s) with

---

poly(dT) chains are distinguished by the use of subscripts.
Projections perpendicular to the helix axis depict two helical repeats of c) poly(dA)·-poly(dT) and d) poly(dT)· poly(dA)· poly(dT). The atoms of the sugar-phosphate backbone are represented by larger circles to highlight the helical nature of the molecules; phosphorus atoms have been drawn largest, and the carbon atoms smallest. The base atoms have been represented as *small, solid circles* to illustrate more clearly the base plane orientations. In the triple-stranded complex, the T_2 chain has, for clarity, been stippled. It occupies what is normally the deeper groove of DNA double-helices. The two projections are drawn to the same scale. *Arnott* and *Selsing* (1974)

messenger activity. Several authors have observed full-length or near full-length viral RNA molecules (*Howell* and *Hull* 1978; *Hull* et al. 1979) as well as a 2.3 kb (19S) RNA species (*Corey* and *Hull* 1981; *Odell* et al. 1981), which codes for a 66-kdalton translation product, an inclusion body protein, which spans about the last one-quarter of the DNA molecule. This mRNA is apparently not spliced and its polyadenylated end maps near position 0 of the conventional map (Fig. 4). The polyadenylation site of the "full-length" viral RNA has been located between position 6000 and 8000 by hybridization of alkali partially cleaved RNA reselected for polyadenylated (3') fragments to DNA restriction fragments (*Hull* et al. 1979). *Hull* et al. (1979) suggest that, assuming the interruption of the α-strand is never closed and acts as a barrier, transcription must begin at or shortly after this interruption (i.e., position 0 of our map, Fig. 4). Other transcripts might however exist from the uninterrupted form of CaMV DNA as mentioned which in fact might be regulated by introducing the interruption.

The intracellular sites of CaMV DNA transcription, capping and polyadenylation have not been established. One possible site is the inclusion body which, as noted above, is a repository of virus particles. Another possibility is the nucleus, as *Guilfoyle* (1980) was able to detect virus-specific RNA transcription in vitro with nuclei prepared from infected plants. Finally, it is possible that nucleus and inclusion bodies are linked together during at least a portion of the viral development cycle. Electron micrographs of thin sections from virus-infected leaves occasionally reveal what could be nucleoplasmid bridges joining the two structures. Such a hypothesis is attractive in that it would allow easy access to nuclear replicases and polymerases and other factors to synthetic centers in the inclusion bodies.

Wheat germ RNA polymerases I, II, and III are all capable of transcribing *both* strands of CaMV DNA although there is a preference for the α-strand (*Teissère* et al. 1979). The above-mentioned nuclei preparations of *Guilfoyle* (1980), on the other hand, appear to be more specific, as only the α-strand is transcribed and the reaction is blocked by low concentrations of α-amanitin, suggesting that RNA polymerase II is resposible.

Volovitch et al. (1980) have shown that *Escherichia coli* RNA polymerase can transcribe CaMV DNA in vitro and *Meagher* et al. (1977b) have shown that viral RNA transcripts in *E. coli* transformed with hybrid plasmids containing CaMV DNA fragments can direct polypeptide synthesis, although the translation products do not appear to be related to known viral polypeptides (*Meagher* et al. 1977b). Promoters that might be responsible for this or other transcription activity have been isolated by *McKnight* and *Meagher* (1981) as certain CaMV DNA EcoRI* restriction fragments. *Daubert* et al. (pers. comm.) have shown that capsid protein is produced in *E. coli* from hybrid CaMV plasmids. Production was independent of the vector orientation and not effected by the cloning site (*Xho*I or *Sal*I) suggesting that a promoterlike sequence on the CaMV genome was used. As expected in view of the numerous prokaryotic transcription "stop" signals (*Rosenberg* and *Court* 1979) in the antisense direction (Appendix; *Honigman* and *Th. Hohn,* to be published), transcription in the *E. coli* system occurs primarily in the sense direction.

Whether transcriptional activity in bacteria will have any relevance to plant systems remains to be seen. Likewise, the significance of the various "TATA"- and "CAT"-boxes (*Corden* et al. 1980) in the CaMV DNA sequence will remain problematic until it is determined whether promoters and transcription termination signals in plant DNA resemble their counterparts in other eukaryotes. Nevertheless, both the prokaryotic-type and

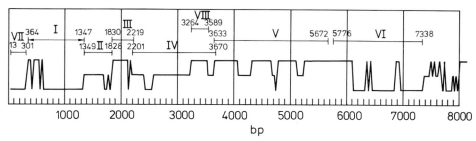

Fig. 9. The purine-pyrimidine sequence of CaMV DNA analyzed by J.C.W. Shepherd according to his method (*JCW Shepherd* 1981) for determining the likely reading frames of "paleomessages" UNYUNYUNY . . ., where U is a purine, Y a pyrimidine, and N any base. *Line,* shifts of reading phase of "paleomessage"; *low position,* corresponding to that of reading frames I, VI, and VII; *middle position,* corresponding to that of reading frames II and IV and *upper position,* corresponding to that of reading frames III, V, and VIII (see Fig. 4). ⊢ open reading frames according to *Franck* et al. (1980), each starting with ATG and ending with a stop codon

eukaryotic-type transcription signals present in the CaMV-S DNA sequence are indicated in Fig. 4, and Appendix D and E as they may prove useful in interpreting future results.

2.4 Translation and CaMV Proteins

In light of the asymmetrical pattern of RNA transcription discussed above it is not surprising that the protein coding capacity of the CaMV nucleotide sequence is found exclusively on the β-strand, that is, the strand complementary to the transcribed α-strand. Analysis of the CaMV-S DNA sequence (*Franck* et al. 1981) shows that only short stretches of sequence in the α-strand are free of translation stop codons whereas in the β-strand, long regions are devoid of termination signals in one or another reading phase. Indeed, with the exception of a 700 base-pair region encompassing the map origin (Fig. 4), virtually the entire β-strand sequence can code. If we assume that a viral cistron begins with the first in-phase ATG triplet in each open reading frame and stops at the first in-phase TAA, TGA, or TAG stop codon encountered, then six large (I–VI) and two small (VII and VIII) "genes" can be discerned. These putative genes generally alternate in reading frame and overlap one another little if at all. One exception is the short "gene" VIII, which is located wholly within open region IV but in another reading phase. *J. CW. Shepherd* (1981) has found a periodic rhythm in the purine-pyrimidine frequency for a large number of coding nucleic acid sequences and has shown that the rhythm may be correlated with the protein reading frame (*J.CW.Shepherd*1981). His prediction of reading frames for the CaMV-S DNA sequence is in good accord with the location of the various "genes" deduced in the straightforward manner described above; even the small internal coding region VIII shows up in such an analysis (Fig. 9). DNA sequences of other CaMV strains (CM1841, *Gardner* et al. 1981; D/H, *Balazs* et al., to be published) reveal at least the same six major open reading frames and probably the two minor ones also.

Is it possible to assign authentic virus-coded polypeptides to any of the aforesaid open reading frames? There is a natural variant of CaMV, termed CM4-184, which is partially

deleted in open reading frame II (*Hull* and *Howell* 1978, *Gardner* et al. 1980; *Howarth* et al., 1981). This modification does not affect virus multiplication but does impair transmission of the virus by aphids. "Gene" II, then, may code for an aphid acquisition factor. However, full length isolates of CaMV exist that are not aphid transmissible either.

Early investigation of CaMV structure suggested an extremely complicated protein composition for the virion, with various authors reporting the existence of from two to seven different polypeptides with molecular weights ranging from 30–85 kdaltons (*Tezuka* and *Taniguchi* 1972; *Kelly* et al. 1974; *Brunt* et al. 1975; *Hull* et al. 1976), some species of which are phosphorylated (*Hahn* and *Shepherd* 1980). *Al Ani* et al. (1979b), however, have presented evidence that only one major structural polypeptide exists, a species of 42 kdaltons from which the smaller proteins reported by others are derived by degradation and the larger by artifactual dimerization. *Brunt* et al. (1975) have shown that CaMV coat protein is very rich in lysine. Examination of the CaMV-S DNA sequence reveals that only coding region IV, however, is capable of encoding a polypeptide of 42 kdaltons with such a high lysine content (Appendix F). Translation of the entire coding region IV would produce a protein of 55 kdaltons in which the 42-kdalton lysine-rich central region thought to correspond to mature coat protein would be flanked by sequences rich in acidic amino acids. *Franck* et al. (1980) have suggested that the 55-kdalton polypeptide could be a precursor of viral coat protein, in which the portion of the chain rich in acidic amino acid residues would help neutralize the central core of basic amino acids. Elimination of the acidic region by proteolytic processing would engender the mature 42-kdalton chain in which the lysine-rich core is free to react with viral DNA. Final evidence that the coat protein is encoded by "gene IV" comes from *Daubert* et al. (personal communication), who show that plasmids containing this region in a subclone produce coat protein antigen in the bacterial cell.

Coding region VI is the best candidate to code for the 60–66 kdalton polypeptide which is the major structural protein of the viral inclusion body. Note that the molecular weight of about 61 kdaltons of the longest possible continuous translation product deriving from this region (Fig. 4) corresponds closely to the size estimated for inclusion body protein by certain authors (e.g., *Shockey* et al. 1980). A protein of 62–66 kdaltons is the major virus-specific translation product directed by poly-A-containing RNA isolated from CaMV-infected plants (*Odell* and *Howell* 1980; *Al Ani* et al. 1980). Hybrid-arrested translation techniques have been used to show that cloned restriction fragments covering region VI inhibit in vitro translation of this protein (*Odell* and *Howell* 1980; *Odell* et al. 1981; *Covey* et al. 1981). There is still no formal proof, however, that the 62–66 kdalton translation product and the inclusion body protein are the same.

Nothing is known about the other presumptive gene products, whether they exist at all and, if so, what their function is. The "gene" III region, in any event, appears to be essential as CaMV DNA from which it has been eliminated by restriction does not replicate in inoculated leaves (*Lebeurier* et al. 1980). Possible functions for this and other CaMV genes could be: to encode proteins involved in DNA replication and transcription; perhaps a second inclusion body protein (*Shepherd* et al. 1980); factors causing cross protection against other strains (*Tomlinson* and *Shepherd* 1978); or factors involved in the spread of infection from one cell to another. Since several groups are working on the production of CaMV mutants in vitro, there will certainly accumulate more knowledge on that matter in the near future.

Table 2. Abnormalities found in cloned CaMV DNA

Cloning site	"Normal"	Restriction site						Deletion in		IS-element	Total
		Missing in stretch of				Gained	Change of any other[a]	Mono-mer	Dimer		
		CaMV			pBR322						
		SacI	HindIII	PstI	BamHI	PstI					
PstI	4	1	0	0	0	1	0	3	1	0	10
SalI	6	0	1	0	0	0	0	0	0	1(IS27)	8
BamHI	5	0	0	1	0	0	0	0	0	5 (IS1)	11
KpnI	5	0	0	0	0	0	0	0	0	0	5
EcoRI	8	0	0	0	0	0	0	0	0	0	8
HindIII	10	0	0	0	1	0	0	0	0	0	11
BstEII	10	0	0	0	0	0	0	0	0	0	10
Total	48	1	1	1	1	1	0	3	1	6	63

[a] EcoRI, HincII, KpnI, SalI, BglII, XhoI

3 DNA Cloning

3.1 Cloning of CaMV DNA and Its Fate in Bacteria

The various unique restriction sites of CaMV DNA (Fig. 4) offer many possibilities for cloning the complete virus genome into standard bacterial vectors; other sites can, of course, also be used to clone CaMV fragments and study their specific properties. The hybrids can then be amplified in *E. coli* and the viral DNA part released by restriction at the cloning site. This approach has been employed using a wide combination of vectors (plasmid pGM706, *Szeto* et al. 1977; pBR322 *Hohn* et al. 1980; *Howell* et al. 1980; *Meagher* et al. 1977; *Shepherd* et al., personal communication; cosmid pHC79, *B.* and *Th. Hohn*, unpublished; single strand bacteriophage M13mp7, *Howarth* et al. 1981; bacteriophage λ, *Hohn* et al. 1980) and restriction sites (Table 2, Fig. 4). In the bacterial host the overlaps which exist in virion DNA are trimmed and the interruptions closed by ligation. Consequently, the cloned hybrids are true supercoils.

The majority of the clones represent faithful copies of the original DNA as far as can be tested by detailed comparison of the restriction maps. A minority of the clones, however, differ in sequence from original DNA (Table 2, *Th.* and *B. Hohn,* unpublished results). Some of the variants in which a single restriction site has been altered may reflect heterogeneity of the original virus population. One such variant, for example, contains a new PstI site around position 3000, close to the region where PstI sites are found in other CaMV strains (Fig. 5).

Events leading to a different class of clone variation probably occurred in the bacterium. Members of this group contain major insertions and deletions. Most of the insertions could be identified as the bacterial IS1 element (*Bukhari* et al. 1977) by length

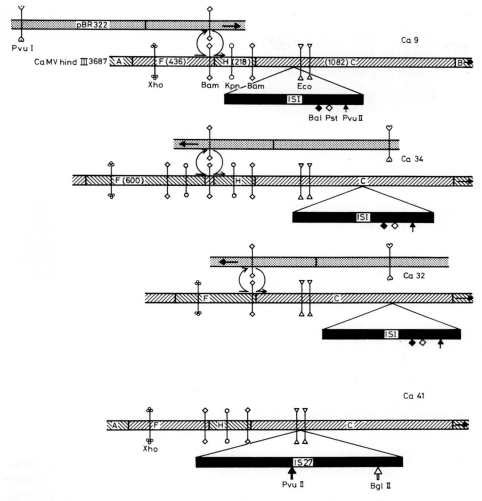

Fig. 10. Localization of bacterial IS elements 1 and 27 in various clones of CaMV DNA in plasmid pBR322. Only the relevant portions of the CaMV and pBR322 genomes are shown. *Frames* indicate HindIII restriction fragments. *Xho*–XhoI; *Bam*–BamHI; *Kpn*–KpnI; *Eco*–EcoRI; IS27 is 1100 base pairs long and lacks HindIII, EcoRI, PstI, and BamHI restriction sites. (Unpublished result by *Th. Hohn* and *B. Hohn*)

measurement, restriction analysis, and hybridization to the appropriate probes. All IS1 insertions analyzed occurred in the pRB322-recombinants cloned in the BamHI site regardless of the relative orientation of pBR322 and CaMV and both in the presence and absence of the small BamHI fragment of CaMV DNA. Although integration sites vary, they had been always located in "gene IV" (coat protein) and, in the three cases analyzed in detail, in the same orientation relative to CaMV DNA (Fig. 10). A different IS element (IS 27; Fig. 10; *Th. Hohn* and *Reif*, unpublished result) was detected in "gene IV" of a SalI clone.

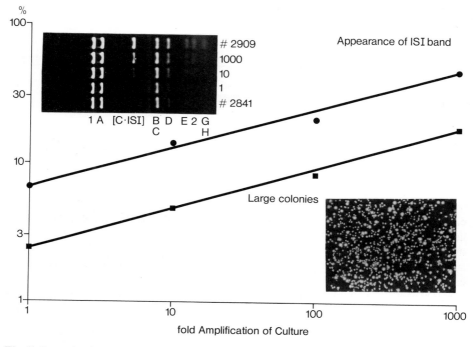

Fig. 11. In a mixed population of bacteria harboring CaMV hybrid plasmids those with a bacterial insertion element in the CaMV DNA outgrow the others. Bacteria containing CaMV hybrid plasmids with and without ISl inserted into the CaMV portion were mixed and repeatedly grown to stationary phase and diluted tenfold. From each dilution step an aliquot was used to determine the proportion of bacteria forming large and small colonies (*lower right insert* and *lower curve* (■)) showing percentage of colonies exhibiting large size). Each of ten sample tests showed that large colony formers contain hybrid plasmids with the IS element and small colony formers contain hybrid plasmids without. Other aliquots of the cultures were used to isolate plasmid, which was analyzed by gel electrophoresis for appearance of restriction fragment containing the IS element (*upper left insert* and *upper curve* (●)) showing percentage of ISl containing restriction fragment. CaMV HindIII restriction fragments A, B+C, D, E, G+H; pBR322-CaMV conjugal fragments 1 and 2; and the fragment containing the IS element C·ISl are indicated in the *insert*. 2841 and 2909 are purified clones with and without IS element). The density of electrophoreses bands were measured by optical scanning of photographs. Proportion of large colony formers and of HindIII restriction fragment C containing IS element are plotted against cycles of replication. (Unpublished results by *Th. Hohn* and *B. Hohn*)

Deletions, usually of about 2000 bp length and again affecting the "gene IV", were found in PstI clones (Table 2). The prevalence of deviations affecting the same region of the CaMV DNA sequence could arise if the region in question directs the synthesis of a protein which is somewhat toxic to the bacterial host. Yields of soluble CaMV capsid protein antigen from extracts of *E. coli* synthesizing it are quite sensitive to DNAse treatment of the lysate (*Daubert,* personal communication). This implies that the coat protein interacts with DNA, possibly constituting a selective disadvantage to the bacteria.

Additional evidence for the toxicity of the CaMV genome for *E. coli* is provided by our failure, despite repeated efforts, to clone a complete CaMV dimer in any plasmid, or

Table 3. Infectivity of encapsidated and naked CaMV DNA[a]

| | | Plants infected/10 plants inoculated Observed at | | |
		Day 13	Day 21	Day 25
Encapsidated	(1.6 µg/ml)	10	10	10
Encapsidated	(0.16 µg/ml)	2	8	8
Naked	(1.0 µg/ml)	4	10	10
Naked	(0.1 µg/ml)	2	7	7

[a] Data from *Lebeurier* et al. (1980)

to clone a complete monomer in the sense direction in the BamHI site of a plasmid enhancing expression of integrated DNA (*Wystup, Oppenheim, B.* and *Th. Hohn*, unpublished results). Furthermore, in our experience, bacteria harboring CaMV hybrid plasmids grow slower than those harboring the vector plasmid only. Colonies on agar are small and survival on plates or in stabs is poor. Although exact data are not available, we estimate that 99% of the bacteria with CaMV-hybrid plasmids die within 2 months when stored at room temperature in stabs. Survival, however, is good if such bacteria are stored at –70 °C and 50% glycerol.

Survivors from old stabs are frequently found to be cured from the hybrid plasmid if the antibiotic present decomposes; or to contain altered plasmids if the antibiotic is stable. In both cases, large colonies are regained by streaking on new agar. Altered plasmids contain insertions and deletions in the coat protein region similar to the ones described above and will outgrow the unaltered hybrid upon continued cultivation (Fig. 11).

Table 4. Infectivity of CaMV DNA

Type of CaMV DNA	Infectivity, % of complete virus
1 Encapsidated	100%
2 Naked	100%
3a Restricted with SalI, PstI, XhoI, or BstEII (long cohesive ends)	90%
3b Restricted with HhaI (short cohesive ends)	40%
4 Cloned in the SalI, PstI, or BstEII site of bact. plasmids, intact	0%
5 Cloned as above and released by restriction	90%
6 Cloned in the BamHI site of pBR322 and released by restriction	5%
7 As 6, but *large* BamHI fragment only	0%
8 As 6, but *short* BamHI fragment only	0%
9 Mixture of the two samples above	10%
10 Mixture of clones in the SalI site and BamHI site of pBR322, intact	25%
11 One complete and one incomplete unit of CaMV cloned in tandem in the PstI site of pBR322, intact	30%

Figures are compiled from several experiments and relate to the CaMV DNA molarities. The exact data, based on at least ten test plants for each case can be found in *Lebeurier* et al. (1980), *Lebeurier* et al. (to be published), and *Hohn* et al. (to be published). Data presented in lines 6, 9, 10, and 11 are based on at least three infected plants obtained in at least two independent experiments

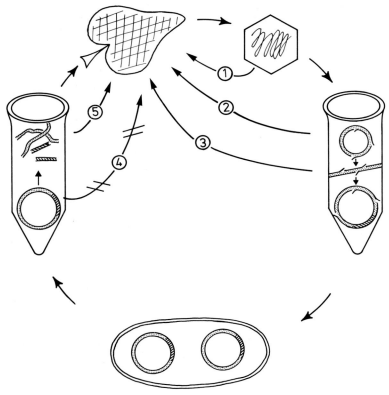

Fig. 12. Artificial life cycle of CaMV. Schematic representation of manipulations of CaMV DNA in vitro and its passage through bacteria. *Numbered arrows* refer to experiments shown in Table 4. See text for details. ▨, CaMV DNA stretches; ▨, bacterial plasmid DNA stretches

3.2 Reintroduction of Original and Cloned CaMV DNA into Plants

Any use of CaMV DNA as a future plant vector depends on our ability to reintroduce the DNA into the plant after passage through other organisms and treatments in vitro. Principally, two possibilities can be envisioned: transformation of plant protoplasts (next section) or infection of total plants.

Unlike many plant RNA viruses the molar infectivity of naked and encapsidated CaMV DNA is similar when leaves are infected mechanically by gently rubbing with an abrasive, like celite, and the inoculum (*Hull* and *Shepherd* 1977; *Szeto* et al. 1977; *Civerolo* and *Lawson* 1978; Table 3, Table 4, lines 1 and 2; Fig. 12). Generally, symptoms such as mosaic formation and wrinkling of the leaves are first discernible 12–15 days after inoculation and become fully developed at day 21–25 (Table 3; *Lebeurier* et al. 1980). Time of appearance and severity of the symptoms are somewhat dependent on the concentration of inoculum but 1 µg of CaMV DNA is generally sufficient to infect ten plants (Table 3). While only circular DNA was originally recognized as being infectious, it has been shown recently (*Lebeurier* et al. 1980; *Howell* et al. 1980) that linearization by the action of

restriction enzymes does not significantly reduce infectivity (Table 4, line 3). Damage to the ends of linear DNA may well have been responsible for the original negative results. The infectivity is independent of the restriction site used for linearization (Table 4, line 3), i.e., of the presumptive gene hit. However, digestion with restriction enzymes which create long (4 or 5 nucleotide) cohesive single-stranded ends, whether 5'- or 3'-overlapping, give linear DNA which is more infectious than DNA obtained by digestion with enzymes creating short (2-nucleotide) cohesive ends (Table 4, lines 3 and 4). Progeny virus isolated from plants infected with linearized DNA contains circular DNA again and thus recircularization must occur via ligation at some early stage of infection.

Cloned hybrid DNA, restricted in the cloning site, is presumably identical in sequence to linearized virion CaMV DNA except that the typical single-strand interruptions are missing. These interruptions do not seem to be necessary for early phases of infection since cloned DNA, after liberation from its plasmid by restriction, is fully infectious (Table 4, line 6). Progeny viral DNA, however, regains the interruptions (*Lebeurier* et al. 1980; *Howell* et al. 1980).

Can the intercellular ligase activity in plant cells also be used to join two separate fragments? *Lebeurier* et al. (to be published) showed that this is indeed the case. The two BamHI fragments of CaMV DNA were cloned in bacteria separately, released from the hybrids by restriction and offered to the plant. None of the single fragments was infectious, whereas the mixture showed a low but significant infectivity (compare lines 7–10 in Table 4).

Unrestricted CaMV DNA cloned in different sites of plasmid pBR322 or bacteriophage λ is not infectious (*Lebeurier* et al. 1980; *Howell* et al. 1980; Table 4, line 5). Several possible reasons for this lack of infectivity can be envisaged: 1. The increased size of the DNA does not allow packaging into the capsid, horizontal transmission, or other preconditions for the appearance of symptoms; 2. an essential gene or site on the CaMV DNA is interrupted by the inserted plasmid in all the cases tested; 3. messenger RNA production is perturbed; and 4. a "toxic sequence" within the plasmid inhibits its amplification as was observed for hybrid plasmids in animal cells (*Schaffner* 1979, *Luskey* and *Botcham* 1981).

Inoculation of plants with pairs of hybrids obtained in different cloning sites as well as infection with an unrestricted hybrid containing redundant CaMV DNA stretches leads to systemic symptoms and production of viruses containing CaMV DNA but not plasmid stretches (*Hohn* et al., to be published; Table 4, lines 10 and 11). In this case it is evident that some type of recombination has led to elimination of the plasmid insertions, as suggested in Fig. 13.

4 Prospects for the Domestication of CaMV

So far we have considered the complete CaMV genome, transfection of total plants, and assay of pathogenic systemic symptoms. But if CaMV is to become useful, pathogenic functions must be eliminated and space created for a payload. Finally, selected properties of the virus must be tested and assigned to regions of the genome, work which will no doubt be facilitated when a plant protoplast system which can be readily infected and tight mutants become available.

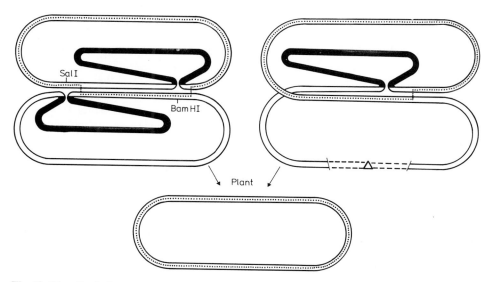

Fig. 13. Hypothetical recombination events explaining the findings that infection with a CaMV DNA hybrid clone containing redundant CaMV information, or with a mixture of two different CaMV hybrid clones, leads to the appearance and amplification of plasmid-free CaMV DNA. ⎓, CaMV DNA stretches; ■■■, bacterial plasmid DNA stretches, ==△==, deletion

4.1 In Vitro Mutagenesis

It is likely that only a few functions of CaMV will be needed to make a plant vector. These are a replication origin, possibly a replicase and/or replication cofactors; perhaps the inclusion body protein may prove necessary for CaMV DNA replication and transcription. The vector will also need one or more strong promoters, which must certainly exist in the viral DNA in view of the large amounts of virus coat protein and inclusion body protein produced upon infection. To situate the above features on the physical map recourse can be had to the techniques of in vitro mutagenesis, following the strategies worked out in bacterial and animal genomic systems (for reviews see *Peden* et al. 1980; *Timmis* 1979).

In the first attempts mutants should preferably be designed that conserve DNA length approximately, that each affect only one gene, and that rarely revert. Whether mutants are pleiotropic or not and whether they are located in different cistrons can be provisionally tested by in vitro complementation, although the efficient recombination systems described above may complicate this approach.

By cloning DNA from a population of molecules one would expect naturally occurring variants to be picked up and amplified as pure species. Other variants might originate in the bacterial cell. By screening a number of our clones such variants have indeed been detected (Table 2). One of these with a single base change in the SacI restriction site causes only retarded and weak symptoms and a very low yield of virus particles in plants (*Franck, Lebeurier* and *Th. Hohn*, unpublished result). Upon serial passage, however, normal symptoms are restored and the mutant reverts). This experiment suggests that more drastic changes may be required in order to make a variant stable.

Table 5. Presumptive genes on the CaMV map and handles to mutate them in vitro

Gene	Phase	Coordinates	Size	Protein[b] MW (KD)	Unique restriction sites	Deletions possible of	Mutants available[a]	Possible function
I	3	364–1347	983	38				
II	1	1349–1828	479	18	XhoI		CM4–184	Aphid transmission?
III	2	1830–2219	389	15	KpnI	BamHI fragment	"Bamone"	An essential one
IV	1	2201–3670		57	BglI, HpaI			Coat protein
V	2	3633–5672		79	SalI, PstI	BclI fragment	"Bcl Km"	
VI	3	5776–7338		61	HhaI, SacI	HgiAI fragment		Inclusion body?
VII	3	13– 300		11	BstEII			
VIII	2	3264–3587		12	BglI			
0		7339– 12						

[a] CM4–184 variant of CaMV (see *Gardner* et al. 1980); "Bamone", large BamHI fragment cloned in pBR322 (*Lebeurier* et al. 1980); "Bcl Km", replacement in cloned CaMV of the small BclI fragment by a BamHI fragment containing Km (Tn5) resistance (unpublished)
[b] Only presumptive proteins larger than 10 KD are considered

The mutants which would be easiest to obtain and test are those containing deletions of small restriction fragments. Some examples affecting "gene" III, V, and VI are listed in Table 5 along with the natural deletion CM4-184 of "gene" II (*Gardner* et al. 1980; *Howarth* et al. 1981). So far only the "gene" III deletion has been tested biologically (*Lebeurier* et al. 1980). This mutant does not cause pathogenic symptoms nor can its replication be detected in inoculated plants. Thus it is possible that, in this case, a major replication function has been hit.

Other approaches involve linearization of the DNA, either at a unique site by digestion with a single-site restriction endonuclease, or alternatively quasi randomly by partially cleaving with a restriction endonuclease which cuts frequently (e.g., SauIIIA). *Shortle* and *Nathans* (1978) report that in the presence of ethidium bromide supercoiled DNA is much more easily digested than linear or relaxed circular forms. Hence partial digestion is easily achieved since only the first cut on the original supercoiled DNA is introduced efficiently. The ends of linearized DNA can then be manipulated before ligation, e.g., either filled up by DNA polymerase, digested with Sl endonuclease or Bal31 3'/5' exonuclease (*Talmadge* and *Gilbert* 1980), or extended by introducing linkers. The success of such manipulation can be tested by screening for loss (or gain) of a restriction site or, alternatively, colonies arising from modified DNA may be obtained by plate selection if a selective marker is introduced at the site in question.

The nature of a defect of any CaMV mutant will be much more easily characterized if single cells can be transformed and assayed. Production of various virus-coded proteins could then be screened by immunofluorescence (*Furusawa* et al. 1980) or by a radioimmune assay (*Ghabrial* and *Shepherd* 1980; *Melcher* et al. 1980); effects on DNA replication may be detected by rapid DNA isolation procedures (*Gardner* and *Sherpherd* 1980) followed by Southern hybridization, while effects on transcription can be studied by "Sl

mapping" of isolated transcripts (*Berk* and *Sharp* 1977; *Grosschedl* and *Birnstiel* 1980). Finally, expression of selective markers incorporated into the CaMV genome could be used to select cells that have been transformed successfully (see below).

4.2 Protoplast Transformation

There are many reports on physical uptake of radioactively labeled DNA or RNA by protoplasts (review *Lurquin* and *Kado* 1979), and on transfection of plant protoplasts with plant RNA viruses (review, *Takebe* 1977) and ti plasmid (*Krens* et al., to be published). Methods of promoting uptake involve the use of polycations (e.g., poly L ornithine), polyethylene glycol, and certain cations such as Ca^{++}, Zn^{++}, and Cu^{++}. Recently, fusion of plant protoplasts with liposomes containing the DNA has been used successfully (*Fraley* et al. 1980).

With regard to CaMV, infection of protoplasts with total virus has been reported and measured in terms of transcriptional activity (*Howell* and *Hull* 1978), production of inclusion bodies and capsids (*Furusawa* et al. 1980), and of total virus. Improvement of the transformation systems will certainly allow the successful introduction of naked CaMV DNA and its mutants in the near future.

4.3 Employment of Selective Markers

A number of selective markers have recently been used in transformation of mammalian and yeast protoplasts. Thus for instance the Kanamycin resistance markers of bacterial transposons Tn5 and Tn903 were introduced and expressed in yeast and mammalian cells (*Jimenez* and *Davies* 1980, and personal communication), conferring resistance to the toxic substance G418, a derivative of Gentamycin (*Daniels* et al. 1973); the prokaryotic dihydrofolate reductase gene coupled to SV40 promoters and polyadenylation sites and mammalian splicing signals renders mammalian cells resistant to methotrexate (*O'Hare* et al. 1981); a hybrid with similar signals but containing the prokaryotic xanthine-guanine phosphoribosyl transferase gene is expressed in mammalian cells (*Mulligan* and *Berg* 1981); Herpes thymidine kinase genes complement the corresponding defect in mammalian cell mutants (*Pellicer* et al. 1980); and yeast metabolic genes complement yeast anxotrophic mutants (*Hinnen* and *Meyhak*, this volume). Work is in progress to introduce these or other selective markers as hitchhikers onto the cauliflower-mosaic virus genome employing promoters of prokaryotic or heterologous eukaryotic origin or to fish for CaMV promoters by insertion of promoterless fragments at the appropriate positions. An especially useful marker might be the T stretch of the ti plasmid (*Montague* and *Schell*, this volume), which renders cells harboring it independent of plant growth hormones. Finally host cells with appropriate auxotrophic markers would provide another selective system.

4.4 Autonomous Virus or Integrative Plasmid?

At present CaMV DNA and ti plasmid each have advocates as the vector of choice for genetic manipulation of plants. CaMV DNA is smaller and therefore easier to handle

than the ti plasmid or even only the T stretch of it. As a result of the larger size, e.g., no unique restriction site is present on the T stretch. On the other hand, total size of a CaMV-based vector might be limited by the amount packageable into the virus particle while no principal limitation of size exists for the ti plasmid. We envisage the ti plasmid system as the most suitable method for introducing a previously isolated gene into the plant genome in a stable fashion. CaMV derived plasmids, on the other hand, will ultimately prove to be more useful for introducing a variety of DNA pieces from a pool or library into plants, maintaining them in high copy number, and allowing a high rate of expression. Thus, both systems or a combination of them will find a place in the tool chest of plant genetic engineers.

Acknowledgments. We are extremely grateful for the constant help of Drs. B. Hohn and L. Hirth. As yet unpublished experiments of the authors were performed with the expert technical assistance of H. Grob. We thank Drs. H. Saedler and P. Starlinger for providing cloned IS elements and the many colleagues for contributing the preliminary information mentioned in the text.

Concerning the appendix the advice and software of G. Osterburg & R. Sommer (Institut für Dokumentation, Information und Statistik, Deutsches Krebsforschungsinstitut, Heidelberg, BRD) and R. Staden (1977) is highly appreciated.

References

Al Ani R, Pfeiffer P, Lebeurier G, Hirth L (1979a) The structure of cauliflower mosaic virus. Part 1: pH induced structural changes. Virology 93: 175–187

Al Ani R, Pfeiffer P, Lebeurier G (1979b) The structure of cauliflower mosaic virus. Part 2: identity and location of the viral polypeptides. Virology 93: 188–197

Al Ani R, Pfeiffer P, Whitechurch O, Lesot A, Lebeurier G, Hirth L (1980) A virus specified protein produced upon infection by cauliflower mosaic virus. Ann Virol (Inst. Pasteur) 131 E: 33–53

Arnott S, Bond PJ (1973) Structures for poly(U) · poly (A) · poly(U) triple stranded polynucleotides. Nature 244: 99–101

Arnott S, Selsing E (1974) Structures for the polynucleotide complexes poly(dA) · poly(dT) and poly(dT) · poly(dA) · poly(dT) · J Mol Biol 88: 509–521

Berk AJ, Sharp PA (1977) Sizing and mapping of early adenovirus mRNAs by gel electrophoresis of Sl endonuclease-digested hybrids. Cell 12:721–733

Blake RD, Massoulié J, Fresco JR (1967) Polynucleotides VIII. A spectral approach to the equilibria between polyriboadenylate and polyribouridylate and their complexes. J Mol Biol 30: 291–308

Brunt AA, Barton RJ, Tremaine JH, Stace-Smith R (1975) The composition of cauliflower mosaic virus protein. J Gen Virol 27: 101–106

Chauvin C, Jacrot B, Lebeurier G, Hirth L (1979) The structure of cauliflower mosaic virus. A neutron diffraction study. Virology 96: 640–641

Civerolo EL, Lawson RH (1978) Topological forms of cauliflower mosaic virus nucleic acid. Phytopathology 68: 101–109

Corden J, Wasylyk B, Buchwalder A, Sassone-Corsi P, Kedinger C, Chambon P (1980) Promoter Sequences of Eukaryotic Protein-Coding Genes. Science 209

Covey SN, Hull R (1981) Transcription of cauliflower mosaic virus DNA. Detection of transcripts, properties and location of the gene encoding the virus inclusion body protein. Virology 111:463–474

Daniels PJL, Yehaskel AS, Morton JB (1973) The Structure of antibiotic G-418 abstr 137 13th Interscience conf antimicrobiol agents & chemotherapy, Washington

Davis RW, Soim M, Davidson ND (1972) Methods Enzymol 21: 413–428

Favali MA, Bassi M, Conti GG (1973) A quantitative autoradiographic study of intracellular sites for replication of cauliflower mosaic virus. Virology 53: 115–119

Franck A, Jonard G, Richards K, Hirth L, Guiley H (1980) Nucleotide sequence of cauliflower mosaic virus DNA. Cell 21: 285–294

Fraley R, Papahadijopoulos D (1981) New generation liposomes: the engineering of an efficient vehicle for intracellular delivery of nucleic acids. TIBS 6:77–81

Furusawa I, Yamaoka N, Okuno T, Yamamoto M, Kohno M, Kunoh H (1980) Infection of turnip brassica-rapa cultivar perviribis protoplasts with cauliflower mosaic virus. J Gen Virol 48: 431–436

Gardner CO Jr, Melcher U, Shockey MW, Essenberg RC (1980) Restriction enzyme cleavage maps of the DNA of 2 cauliflower mosaic virus isolates. Virology 103:250–254

Gardner RC, Howarth AJ, Hahn P, Brown-Luedi M, Shepherd RJ, Messing JC (1981) The complete nucleotide sequence of an infectious clone of cauliflower mosaic virus by M 13mp7 shotgun sequencing. Nucleic Acid Res 9:287

Gardner RC, Shepherd RJ (1980) A procedure for rapid isolation and analysis of cauliflower mosaic virus DNA. Virology 106:159–161

Gebhardt C, Schnebli V, King PJ (1981) Isolation of biochemical mutants using haploid mesophyll protoplasts of Hyoscyamus muticus. II. Auxotrophic and temperature sensitive clones. Planta (in press)

Ghabrial SH, Shepherd RJ (1980) A sensitive radio immuno sorbent assay for the detection of plant viruses. J Gen Virol 48:311–318

Grosschedel R, Birnstiel ML (1980) Identification of regulatory sequences in the prelude sequences of an H2A histone gene by the study of specific deletion mutants in vivo. Proc Natl Acad Sci USA 77:1432–1436

Guilfoyle TJ (1980) Transcription of the cauliflower mosaic virus genome in isolated nuclei from turnip brassica-rapa cultivar just-right leaves. Virology 107:71–80

Haber S, Ikegami M, Bajet NB, Goodman RM (1981) Evidence for a devided genome in bean golden mosaic virus, a geminivirus. Nature 289:324–326

Hahn P, Shepherd RJ (1980) Phosphorylated proteins in cauliflower mosaic virus. Virology 107: 295–297

Hohn Th, Hohn B, Lesot A, Lebeurier G (1980) Restriction map of native and cloned cauliflower mosaic virus DNA. Gene 11:21–31

Hohn Th, Hohn B, Hirth L, Lebeurier G (1981) Recombination of cauliflower mosaic virus DNA in plant cells (to be published)

Howarth AJ, Gardner RC, Messing J, Shepherd RJ (1981) Nucleotide sequence of naturally occurring deletion mutants of cauliflower mosaic virus. Virology 112:678–685

Howell SA, Hull R (1978) Replication of cauliflower mosaic virus and transcription of its genome in turnip leaf protoplasts. Virology 86:468–481

Howell SH, Walker LL, Dudley RK (1980) Cloned cauliflower mosaic virus DNA infects turnips brassica-rapa. Science 208:1265–1267

Hull R (1980) Structure of the cauliflower mosaic virus genome. Part 3: Restriction endonuclease mapping of 33 isolates. Virology 100:76–90

Hull R, Covey SN, Stanley J, Davies JW (1979) The polarity of the cauliflower mosaic virus genome. Nucleic Acid Res 7:669–677

Hull R, Howell SH (1978) Structure of the cauliflower mosaic virus genome. Part 2: Variation in DNA structure and sequence between isolates. Virology 86:482–493

Hull R, Shepherd RJ (1977) The structure of cauliflower mosaic virus genome. Virology 79:216–230

Hull R, Shepherd RJ, Harvey RD (1976) Cauliflower mosaic virus: an improved purification procedure and some properties of the virus particles. J Gen Virol 31:93–100

Jimenez A, Davies J (1980) Expression of a transposable antibiotic resistance element in Saccharomyces. Nature 287:869–871

Kamei T, Rubio-Huertos M, Matsui C (1969) Thymidine-^{3}H uptake by X-bodies associated with cauliflower mosaic virus infection. Virology 37:506–508

Kelley DC, Cooper V, Walkey DGA (1974) Cauliflower mosaic virus structural proteins. Microbios 10:239–245

Krens FA, Wullems GJ, Molendijk L, Schilperoort RA (1981) Transformation of tobaccoprotoplasts with ti plasmid DNA (to be published)

Lebeurier G, Hirth L, Hohn Th, Hohn B (1980) Infectivities of native and cloned DNA of cauliflower mosaic virus. Gene 12:139–146

Lebeurier G, Hirth L, Hohn B, Hohn Th (1981) Ligation of cauliflower mosaic virus DNA fragments in plant cells (to be published)

Lung MCJ, Prione TP (1973) Studies on the reason for differential transmissibility of cauliflower mosaic virus isolates by aphids. Phytophatology 63:910–914

Lung MCY, Pirone TP (1974) Acquisition factor required for aphid transmission of purified cauliflower mosaic virus. Virology 60:260–264

Lusky M, Botcham M (1981) Inhibitory effect of specific pBR 322 DNA sequences upon SV40 replication in Simian cells. Nature 293:79–85

Lurquin PF, Kado CI (1979) Recent advances in the insertion of DNA into higher plant cells. Plant, Cell and Environment 2:199–203

Massoulié J (1968) Associations de poly A et poly U en milieu acid. Eur J Biochem 3:439–447

McKnight TD, Meagher RB (1981) Isolation and mapping of small cauliflower mosaic virus DNA fragments active as promoters in E. coli. J Virol 37:673–682

Meagher RB, Shepherd RJ, Boyer HW (1977) The structure of cauliflower mosaic virus. Part 1: restriction endonuclease map of cauliflower mosaic virus DNA. Virology 80:362–375

Meagher RB, Tait RC, Betlach M, Boyer HW (1976) Protein expression in Escherichia-coli mini cells by recombinant plasmids. Cell 10:521–536

Melcher U, Hein RA, Gardner CO Jr, Shockey MW, Essenberg RC (1980) An indirect radio immuno assay of cauliflower mosaic virus. Phytopathology 70:954–957

Menissier U, Lebeurier G, Hirth L (1981) Supercoiled CaMV DNA in infected cells. Virology (in press)

Mulligan RC, Berg P (1981) Selection for animal cells that express the Escherichia coli gene coding for xanthineguanine phosphoribosyltransferase. Proc Natl Acad Sci USA 78:2072–2076

Odell JT, Dudley K, Howell SH (1981) Structure of the 19S RNA Transcript encoded by the cauliflower mosaic virus genome. Virology 111:377–385

Odell JT, Howell SH (1980) The identification mapping and characterization of messenger RNA for p-66 a cauliflower mosaic virus coded protein. Virology 102:349–359

O'Hare K, Benoist C, Breathnach B (1981) Transformation of mouse fibroblasts to methotrexate resistance by a recombinant plasmid expressing a prokaryotic dihydrofolate reductase. Proc Natl Acad Sci USA 78:1527–1531

Pellicer A, Robins D, Wold B, Sweet R, Jackson J, Lowy I, Roberts JM, Sim GK, Silverstein S, Axel R (1980) Altering genotype and phenotype by DNA-mediated gene transfer. Science 209: 1414–1421

Peden KWC, Pipas JM, Pearson-White S, Nathans D (1980) Isolation of mutants of an animal virus in bacteria. Science 209:1392–1395

Riley M, Maling B (1966) Physical and chemical characterization of two- and three-stranded adenine-thymine and adenine-uracil homopolymer complexes. J Mol Biol 20:359–389

Rosenberg M, Court D (1979) Regulatory sequences involved in the promotion and termination of RNA transcription. Annu Rev Genet 13:319–353

Schaffner W (1980) Direct transfer of cloned genes from bacteria to mammalian cells. Proc Natl Acad Sci USA 77:2163–2167

Shalla TA, Shepherd RJ, Petersen LJ (1980) Comparative cytology of 9 isolates of cauliflower mosaic virus. Virology 102:381–388

Shepherd JCW (1981) Method to determine the reading frame of a protein from the purine/pyrimidine genome sequence and its possible evolutionary justification. Proc Natl Acad Sci USA 78:1596–1600

Shepherd RJ (1976) DNA viruses of higher plants. Adv Virus Research 20:305–339

Shepherd RJ (1979) DNA plant viruses. Annu Rev Plant Physiol 30:405–423

Shepherd RJ, Richins R, Shalla TA (1980) Isolation and properties of the inclusion bodies of cauliflower mosaic virus. Virology 102:389–400

Shockey MW, Gardner CO Jr, Melcher U, Essenberg RC (1980) Polypeptides associated with inclusion bodies from leaves of turnip infected with cauliflower mosaic virus. Virology 105: 575–581

Shortle D, Nathans D (1978) Local mutagenesis: A method for generating viral mutants with base substitutions in preselected regions of the viral genome. Proc Natl Acad Sci USA 75:2170–2174

Staden R (1977) Sequence data harboring by computer. Nucleic Acid Res 4:4037–4051

Szeto WW, Hamer DH, Carlson PS, Thomas CA Jr (1977) Cloning of cauliflower mosaic virus

DNA in Escherichia-coli. Science 196:210–212

Szybalski W (1977) IS elements in E. coli, plasmids and bacteriophages. In: Bukhari AI, Shapiro JA, Adhya SL (eds) DNA insertion elements, plasmids and episomes. Cold Spring Harbor, pp 583–590

Takebe I (1975) The use of protoplasts in plant virology. Annu Rev of Phytopathology 13:105–125

Talmadge K, Gilbert W (1980) Construction of plasmid vectors with unique PstI cloning sites in a signal sequence coding region. Gene 12:235–241

Timmis KN, Glover SW, Hopwood DA (1981) Gene manipulation in vitro. In: Glover SW, Hopwood DA (eds) Genetics as a tool in microbiology. Society for General Microbiology Symposium 31. Cambridge University Press, Cambridge, pp 49–109

Teissère M, Durand R, Ricard J, Cooke R, Penon P (1979) Transcription in vitro of cauliflower mosaic virus DNA by RNA polymerase I, II and III purified from wheat embryos. Biochem Biophys Res Commun 89:526–533

Tomlinson JA, Stepherd RJ (1978) Studies on mutagenesis and cross protection of cauliflower mosaic virus. Ann Appl Biol 90:223–232

Volovitch M, Chouikh Y, Kondo H, Yot P (1980) Asymmetric transcription of cauliflower mosaic virus genome by the Escherichia-coli RNA polymerase in-vitro. FEBS Letters 116:257–260

Volovitch M, Drugeon G, Yot P (1978) Studies on the single-stranded discontinuities of the cauliflower mosaic virus genome. Nucleic Acid Res 5:2913–2925

Volovitch M, Drugeon G, Dumas JP, Haenni AL, Yot P (1979) A restriction map of cauliflower mosaic virus DNA strain PV-147. Mapping of the cleavage sites of HhaI, SacI, AvaI, PvuII, PstI, XbaI, EcoRI, BglII, HincII, HpaII and HindII plusIII. Eur J Biochem 100:245–256

Appendix A–G
Computer Printouts Concerning CaMV DNA

Appendix A. Restriction sites (recognition sites) of CaMV DNA

ENZYME	SITE	POS. (LEN.)	POS. (LEN.)	POS. (LEN.)	POS. (LEN.)	POS. (LEN.)
ACCI	GTAGAC	3764(8024)				
	GTCTAC	7044(8024)				
ALUI	AGCT	706(1002)	774(68)	1206(432)	1284(78)	1514(230)
		1678(164)	1773(95)	1907(134)	1950(43)	1993(43)
		2168(175)	2189(21)	2955(766)	2961(6)	3115(154)
		3250(135)	3494(244)	3864(370)	3991(127)	4069(78)
		4337(268)	4348(11)	4367(19)	4553(186)	4565(12)
		4766(201)	5119(353)	5123(4)	5152(29)	5266(114)
		5318(52)	5376(58)	5384(8)	5825(441)	5851(26)
		5862(11)	5970(108)	6321(351)	6381(60)	6400(19)
		7146(746)	7728(582)			
ASUI	GGNCC	1170(1322)	2180(1010)	2321(141)	2544(223)	2585(41)
		4443(1858)	4746(303)	6069(1323)	6212(143)	6934(722)
		7256(322)	7268(12)	7872(604)		
ASUII	TTCGAA	784(3266)	2492(1708)	5542(3050)		
AVAI	CYCGUG	1643(2976)	6691(5048)			
AVAII	GGRCC	2321(3077)	2544(223)	2585(41)	4443(1858)	4746(303)
		6212(1466)	6934(722)	7256(322)	7268(12)	
AVAIII	ATGCAT	1037(3777)	3782(2745)	5284(1502)		
BAMHI	GGATCC	1927(7802)	2149(222)			
BBV1	GCRGC	862(2222)	1965(1103)	2696(731)	5385(2689)	6253(868)
		6316(63)	6319(3)	6664(345)		
BCII	TGATCA	3905(6659)	5270(1365)			
BGLI	GCCNNNNNGGC	3417(8024)				
BGIII	AGATCT	92(452)	221(129)	1306(1085)	4634(3328)	6495(1861)
		7644(1149)	7664(20)			
BSTEII	GGTNACC	127(8024)				
ClAI	ATCGAT	822(864)	1785(963)	2857(1072)	3962(1105)	7982(4020)
ECORI	GAATTC	408(2325)	2417(2009)	2477(60)	3928(1451)	5649(1721)
		6045(396)	6107(62)			
ECORII	CCRGG	2176(2495)	2618(442)	3041(423)	5992(2951)	6705(713)
		7705(1000)				
FNUDII	CGCG	3150(4845)	4409(1259)	6329(1920)		
HAEI	RGGCCR	2093(2891)	2937(844)	3463(526)	6413(2950)	6908(495)
		7226(318)				
HAEIII	GGCC	1172(1323)	2093(921)	2181(88)	2205(24)	2937(732)
		3463(526)	6070(2607)	6413(343)	6908(495)	7226(318)
		7755(529)	7873(118)			
HGAI	GACGC	6335(6633)	7376(1041)	7726(350)		
	GCGTC	598(8024)				
HGIAI	GRGCRC	5829(6821)	7032(1203)			
HGICI	GGSCC	1169(1322)	2179(1010)	6068(3889)	7871(1803)	
HHAI	GCGC	6283(8024)				
HINDII	GTYUAC	3097(3327)	4500(1403)	4545(45)	4839(294)	7018(2179)
		7794(776)				
HINDIII	AAGCTT	1514(3687)	1950(436)	2168(218)	3250(1082)	4337(1087)
		5152(815)	5376(224)	5851(475)		
HINFI	GANTC	16(245)	191(175)	238(47)	456(218)	598(142)
		831(233)	1219(388)	1572(353)	1782(210)	2208(426)
		2307(99)	2394(87)	2412(18)	2460(48)	2519(59)
		3654(1135)	3660(6)	3719(59)	4111(392)	4664(553)
		4690(26)	5078(388)	5539(461)	5691(152)	6002(311)
		6164(162)	6753(589)	6829(76)	6873(44)	6899(26)
		6914(15)	6922(8)	6979(57)	6990(11)	7134(144)
		7795(661)				

Appendix A (continued)

ENZYME	SITE	POS. (LEN.)	POS. (LEN.)	POS. (LEN.)	POS. (LEN.)	POS. (LEN.)
HPAI	GTTAAC	3097(8024)				
HPAII	CCGG	125(1030)	2580(2455)	2588(8)	2833(245)	3914(1081)
		5912(1998)	6222(310)	6273(51)	7119(846)	
HPHI	GGTGA	139(477)	1013(874)	2550(1537)	2864(314)	3687(823)
		5223(1536)	6584(1361)	7686(1102)		
	TCACC	45(630)	3678(3633)	3864(186)	6580(2716)	7439(859)
KPNI	GGTACC	2045(8024)				
MSTI	TGCGCA	6279(8024)				
PSTI	CTGCAG	5391(8024)				
PVUII	CAGCTG	3993(5694)	6323(2330)			
RSA1	GTAC	210(325)	285(75)	319(34)	1142(823)	2041(899)
		2590(549)	2866(276)	3058(192)	5338(2280)	6245(907)
		6809(564)	7631(822)	7736(105)	7909(173)	
SACI	GAGCTC	5829(8024)				
SA1I	GTCGAC	4837(8024)				
SAUIIIA	GATC	92(452)	102(10)	221(119)	231(10)	466(235)
		474(8)	484(10)	1306(822)	1534(228)	1795(261)
		1847(52)	1927(80)	2149(222)	2282(133)	2333(51)
		2498(165)	2744(246)	2854(110)	2941(87)	3360(419)
		3445(85)	3636(191)	3701(65)	3905(204)	4308(403)
		4404(96)	4634(230)	4683(49)	4872(189)	4895(23)
		5270(375)	5830(560)	6478(648)	6495(17)	6513(18)
		6544(31)	6684(140)	6711(27)	6801(90)	7644(843)
		7664(20)				
SFANI	GATGC	643(899)	2858(2215)	3267(409)	4578(1311)	5283(705)
		6739(1456)	7237(498)	7768(531)		
	GCATC	3129(3867)	3240(111)	3789(549)	3958(169)	5286(1328)
		5367(81)	6767(1400)	7286(519)		
TAQI	TCGA	169(210)	190(21)	615(425)	785(170)	823(38)
		1645(822)	1785(141)	1795(9)	2282(487)	2483(201)
		2493(10)	2502(9)	2673(171)	2687(14)	2744(57)
		2858(114)	3489(631)	3701(212)	3963(262)	4269(306)
		4624(355)	4650(26)	4839(189)	5277(438)	5543(266)
		5830(287)	6562(732)	7983(1421)		
XBAI	TCTAGA	838(3883)	1702(864)	2549(847)	3586(1037)	4979(1393)
XHOI	CTCGAG	1643(8024)				

Appendix B. Restriction sites not found in CaMV

NAME	ENZYM
AVRII	CCTAGG
BA1I	TGGCCA
ECOB	TGANNNNNNNNTGCT
ECOB	AGCANNNNNNNTCA
HAEII	UGCGCY
PVUI	CGATCG
SACII	CCGCGG
SPHI	GCATGC
XMAI	CCCGGG
XMAIII	CGGCCG

N = A,G,T,C
U = A,G
Y = T,C
R = A,T
S = G,C

Appendix C. Sequence of CaMV DNA including restriction sites

```
              10        20        30        40        50        60        70        80        90        100
   1 | GGTATCAGAG CCATGAATCG GTTTAAGACC AAAACTCAAG AGGGTAAAAC CTCACCAAAA TACGAAAGAG TTCTTAACTC TAAAAATAAA AGATCTTTCA |
     |             HINFI                                       HPHI                                           BGIII
     |                                                                                                        SAUIIIA
     |
 101 | AGATCAAACA TAGTTCCCTC ACACCGGTGA CCGACAGGAT TACCACCGTA AGGTTTCAGA ACAACATCGA AAGCGTTTAC GCCAACTTCG ACTCTCAACT |
     | SAUIIIA               BSTEII                                                  TAQI                   HINFI
     |                       HPAII                                                                          TAQI
     |                       HPHI
     |
 201 | CAAGTCGTCG TACGATGGTA GATCTAAAAA GATCAAGACT CTAAGCCTTA AAAATCTTAG ATGTTACGAA GCCTTCCTCA GGAAGTACCT TCTGGAACAA |
     |      RSA1            BGIII                HINFI                                                  RSA1
     |                      SAUIIIA     SAUIIIA
     |
 301 | TAAATCTCTC TGAGAATAGT ACTCTATTGA GTATCCACAG GAAAAATAAC CTTCTGTGTT GAGATGGATT TGTATCCAGA AGAAAATACC CAAAGCGAGC |
     |                RSA1
     |
 401 | AATCGCAGAA TTCTGAAAAT AATATGCAAA TATTTAAATC AGAAAATTCG GATGGATTCT CCTCCGATCT AATGATCTCA AACGATCAAT TAAAAAATAT |
     |      ECORI                                            HINFI                 SAUIIIA    SAUIIIA    SAUIIIA
     |
 501 | CTCTAAAACC CAATTAACCT TGGAGAAAGA AAAGATATTT AAAATGCCTA ACGTTTTATC TCAAGTTATG AAAAAAGCGT TTAGCAGGAA AAACGAGATT |
     |                                                                                                            HINFI
     |
 601 | CTCTACTGCG TCTCGACAAA AGAATTATCA GTGGACATTC ACGATGCCAC AGGTAAGGTA TATCTTCCCT TAATCACTAA GGAAGAGATA AATAAAAGAC |
     |      HGAI    TAQI                            SFANI
     |
 701 | TTTCCAGCTT AAAACCTGAA GTCAGAAAGA CCATGTCCAT GGTTCATCTT GGAGCGGTCA AAATATTGCT TAAAGCTCAA TTTCGAAATG GGATTGATAC |
     |      ALUI                                                                    ALUI       ASUII
     |                                                                                         TAQI
     |
 801 | CCCAATCAAA ATTGCTTTAA TCGATGATAG AATCAATTCT AGAAGAGATT GTCTTCTTGG TGCAGCCAAA GGTAATCTAG CATACGGTAA GTTTATGTTT |
     |                      C1AI            HINFI      XBAI                          BBV1
     |                      TAQI
     |
 901 | ACTGTATACC CTAAGTTTGG AATAAGCCTT AACACCCAAA GACTTAACCA AACCCTAAGC CTTATTCATG ATTTTGAAAA TAAAAATCTT ATGAATAAAG |
     |                                                                                                            HP
     |
1001 | GTGATAAAGT TATGACCATA ACCTATGTCG TAGGATATGC ATTAACTAAT AGTCATCATA GCATAGATTA TCAATCAAAT GCTACAATTG AACTAGAAGA |
     | HI                               AVAIII
     |
1101 | CGTATTTCAA GAAATTGGAA ATGTCCAGCA ATCTGAGTTC TGTACAATAC AGAATGATGA ATGCAATTGG GCCATTGATA TAGCCCAAAA CAAAGCCTTA |
     |                                                        RSA1                  ASUI
     |                                                                              HAEIII
     |                                                                              HGICI
     |
1201 | TTAGGAGCTA AAACCAAGAC TCAAATTGGT AATAACCTTC AAATAGGTAA CAGTGCTTCA TCCTCTAATA CTGAAAATGA ATTAGCTAGG GTAAGCCAGA |
     |      ALUI            HINFI                                                              ALUI
     |
1301 | ACATAGATCT TTTAAAGAAT AAATTAAAAG AAATCTGTGG AGAATAATAT GAGCCATTACG GGACAACCGC ATGTTTATAA AAAAGATACT ATTATTAGAC |
     | BGIII
     | SAUIIIA
     |
1401 | TAAAACCATT GTCTCTTAAT AGTAATAATA GAAGTTATGT TTTTAGTTCC TCAAAAGGGA ACATTCAAAA TATAATTAAT CATCTTAACA ACCTCAATGA |
     |
1501 | GATTGTAGGA AGAAGCTTAC TCGGAATATG GAAGATCAAC TCATACTTCG GATTAAGCAA AGACCCTTCG GAGTCCAAAT CAAAAAACCC GTCAGTTTTT |
     |            ALUI                  SAUIIIA                                     HINFI
     |            HINDIII
     |
1601 | AATACTGCAA AAACCATTTT TAAGAGTGGG GGGGTTGATT ACTCGAGCCA ACTAAAGGAA ATAAAATCCC TTTTAGAAGC TCAAAACACT AGAATAAAAA |
     |                                            AVAI                             ALUI
     |                                            TAQI
     |                                            XHOI
     |
1701 | GTCTAGAAAA AGCAATTCAA TCCTTAGAAA ATAAGATTGA ACCAGAGCCC TTAACTAAAG AGGAAGTTAA AGAGCTAAAA GAATCGATTA ACTCGATCAA |
     | XBAI                                                                         ALUI       C1AI       SAUIIIA
     |                                                                                         HINFI      TAQI
     |                                                                                         TAQI
     |
1801 | AGAAGGATTA AAGAATATTA TTGGCTAAAA TGGCTAATCT TAATCAGATC CAAAAAGAAG TCTCTGAAAT CCTCAGTGAC CAAAAATCCA TGAAAGCGGA |
     |                                            SAUIIIA
     |
1901 | TATAAAAGCT ATCTTAGAAT TATTAGGATC CCAAAATCCT ATTAAAGAAA GCTTAGAAAC CGTTGCAGCA AAAATCGTTA ATGACTTAAC CAAGCTCATC |
     | ALUI                  BAMHI                           ALUI       BBV1                              ALUI
     |                       SAUIIIA                          HINDIII
```

```
              10        20        30        40        50        60        70        80        90       100
2001 | AATGATTGTC CTTGTAACAA AGAGATATTA GAAGCCTTAG GTACCCAACC TAAAGAGCAA CTAATAGAAC AACCTAAAGA AAAAGGTAAA GGCCTTAACT |
     |                                          KPNI                                                   HAEI         |
     |                                          RSA1                                                   HAEIII       |

2101 | TAGGAAAATA CTCTTACCCC AATTACGGAG TAGGAAATGA AGAATTAGGA TCCTCTGGAA ACCCTAAAGC TTTAACCTGG CCCTTCAAAG CTCCAGCAGG |
     |                                          BAMHI                   ALUI       ASUI       ALUI                 |
     |                                          SAUIIIA                 HINDIII    ECORII                          |
     |                                                                             HAEIII                          |
     |                                                                             HGICI                           |

2201 | ATGGCCGAAT CAATTTTAGA CAGAACCATT AATAGGTTTT GGTATAATCT GGGAGAAGAT TGTCTCTCAG AAAGTCAATT CGATCTTATG ATAAGATTGA |
     | HAEIII                                                                                 SAUIIIA              |
     | HINFI                                                                                  TAQI                 |

2301 | TGGAAGAGTC CCTTGACGGG GACCAAATTA TTGATCTAAC CTCTCTACCT AGTGATAATT TGCAGGTTGA ACAGGTTATG ACAACTACCG AAGACTCAAT |
     | HINFI          ASUI         SAUIIIA                                                              HINFI      |
     |                AVAII                                                                                        |

2401 | CTCGGAAGAA GAATCAGAAT TCCTTCTAGC AATAGGAGAA ACATCTGAAG AAGAAAGCGA TTCAGGAGAA GAACCTGAAT TCGAGCAAGT TCGAATGGAT |
     |                ECORI                                         HINFI                   ECORI TAQI    ASUII    SAUI |
     |           HINFI                                                                              TAQI      TA  |

2501 | CGAACAGGAG GAACGGAGAT TCCAAAAGAA GAAGATGGTG AAGGACCATC TAGATACAAT GAGAGAAAGA GAAAGACCCC GGAGGACCGG TACTTTCCAA |
     | IIA             HINFI                     HPHI       ASUI XBAI                         HPAII ASUI RSA1       |
     | QI                                                   AVAII                                   AVAII          |
     |                                                                                             HPAII          |

2601 | CTCAACCAAA GACCATTCCA GGACAAAAGC AAACGTCTAT GGGAATGCTC AACATTGACT GCCAAACCAA TCGAAGAACT CTAATCGACG ACTGGGCAGC |
     |                ECORII                                                                 TAQI       TAQI     BBV1 |

2701 | AGAAATCGGA TTGATAGTCA AGACCAATAG AGAAGACTAT CTCGATCCAG AAACAATTCT ACTCTTGATG GAACACAAAA CATCAGGAAT AGCCAAGGAG |
     |                                          SAUIIIA                                                            |
     |                                          TAQI                                                               |

2801 | TTAATCCGAA ATACAAGATG GAACCGCACT ACCGGAGACA TCATAGAACA GGTGATCGAT GCGATGTACA CCATGTTCTT AGGACTAAAC TACTCCGACA |
     |                          HPAII                      C1AI       RSA1                                       |
     |                                                     HPHI  SFANI                                             |
     |                                                     SAUIIIA                                                 |
     |                                                     TAQI                                                    |

2901 | ACAAAGTTGC TGAGAAGATT GACGAGCAAG AGAAGGCCAA GATCAGAATG ACCAAGCTCC AGCTCTGCGA CATCTGCTAC CTTGAGGAAT TTACATGTGA |
     |                          HAEI       SAUIIIA    ALUI       ALUI                                              |
     |                          HAEIII                                                                            |

3001 | TTATGAAAAG AACATGTATA AGACAGAACT GGCGGATTTC CCAGGATATA TCAACCAGTA CCTGTCAAAA ATCCCCATCA TTGGAGAAAA AGCGTTAACA |
     |                                          ECORII      RSA1                                      HINDII       |
     |                                                                                               HPAI         |

3101 | CGCTTTAGGC ATGAAGCTAA CGGAACCAGC ATCTACAGTT TAGGTTTCGC GGCAAAGATA GTCAAAGAAG AACTATCTAA AATCTGCGAC TTATCCAAGA |
     |                ALUI             SFANI                 FNUDII                                                |

3201 | AGCAGAAGAA GTTGAAGAAA TTCAACAAGA AGTGTTGTAG CATCGGAGAA GCTTCAACAG AATATGGATG CAAGAAGACA TCCACAAAGA AGTATCACAA |
     |                          SFANI             ALUI                 SFANI                                      |
     |                                           HINDIII                                                          |

3301 | GAAGCGATAC AAGAAAAAAT ATAAGGCTTA CAAACCTTAT AAGAAGAAAA AGAAGTTCCG ATCAGGAAAA TACTTCAAGC CCAAAGAAAA GAAGGGCTCA |
     |                                                                 SAUIIIA                                    |

3401 | AAGCAAAAGT ATTGCCCAAA AGGCAAGAAA GATTGCAGAT GTTGGATCTG CAACATTGAA GGCCATTACG CCAACGAATG TCCTAATCGA CAAAGCTCGG |
     |                BGLI                            SAUIIIA         HAEI                  TAQI       ALUI       |
     |                                                              HAEIII                                        |

3501 | AGAAGGCTCA CATCCTTCAA CAAGCAGAAA AATTGGGTCT CCAGCCCATT GAAGAACCCT ATGAAGGAGT TCAAGAAGTA TTCATTCTAG AATACAAAGA |
     |                                                                                                XBAI        |

3601 | AGAGGAAGAA GAAACCTCTA CAGAAGAAAG TGATGGATCA TCTACTTCTG AAGACTCAGA CTCAGACTGA GCAGGTGATG AACGTCACCA ATCCCAATTC |
     |                                          SAUIIIA            HINFI HINFI            HPHI        HPHI        |
     |                                                                                                        TAQ |

3701 | GATCTACATC AAGGGAAGAC TCTACTTCAA GGGATACAAG AAGATAGAAC TTCACTGTTT CGTAGACACG GGAGCAAGCC TATGCATAGC ATCCAAGTTC |
     | SAUIIIA              HINFI                                       ACCI                 AVAII SFANI          |
     | I                                                                                                          |

3801 | GTCATACCAG AAGAACATTG GGTCAATGCA GAAAGACCAA TTATGGTCAA AATAGCAGAT GGAAGCTCAA TCACCATCAG CAAAGTCTGC AAAGACATAG |
     |                                                                             ALUI       HPHI               |

3901 | ACTTGATCAT AGCCGGCGAG ATATTCAGAA TTCCCACCGT CTATCAGCAA GAAAGTGGCA TCGATTTCAT TATCGGCAAC AACTTCTGTC AGCTGTATGA |
     | BC1I       HPAII            ECORI                                C1AI                           ALUI      |
     | SAUIIIA                                                          SFANI                          PVUII     |
     |                                                                 TAQI                                      |
```

Appendix C (continued)

```
              10        20        30        40        50        60        70        80        90        100
4001 | ACCATTCATA CAGTTTACGG ATAGAGTTAT CTTCACAAAG AACAAGTCTT ATCCTGTTCA TATTGCGAAG CTAACCAGAG CAGTGCGAGT AGGCACCGAA |
     |                                                                      ALUI                                        |
     |                                                                                                                  |
4101 | GGATTTCTTG AATCAATGAA GAAACGTTCA AAAACTCAAC AACCAGAGCC AGTGAACATT TCTACAAACA AGATAGAAAA TCCACTAGAA GAAATTGCTA |
     |         HINFI                                                                                                     |
     |                                                                                                                  |
4201 | TTCTTTCAGA GGGGAGGAGG TTATCAGAAG AAAAACTCTT TATCACTCAA CAAAGAATGC AAAAAATCGA AGAACTACTT GAGAAAGTAT GTTCAGAAAA |
     |                                                           TAQI                                                    |
     |                                                                                                                  |
4301 | TCCATTAGAT CCTAACAAGA CTAAGCAATG GATGAAAGCT TCTATCAAGC TCAGCGACCC AAGCAAAGCT ATCAAGGTTA AACCCATGAA GTATAGCCCA |
     |      SAUIIIA                      ALUI        ALUI                ALUI                                             |
     |                                  HINDIII                                                                          |
     |                                                                                                                  |
4401 | ATGGATCGCG AAGAATTTGA CAAGCAAATC AAAGAATTAC TGGACCTAAA AGTCATCAAG CCCAGTAAAA GCCCTCACAT GGCACCAGCC TTCTTGGTCA |
     |      FNUDII                                    ASUI                                                       HINDI   |
     |      SAUIIIA                                   AVAII                                                              |
     |                                                                                                                  |
4501 | ACAATGAAGC CGAGAAGCGA AGAGGAAAGA AACGTATGGT AGTCAACTAC AAAGCTATGA ACAAAGCTAC TGTAGGAGAT GCCTACAATC TTCCCAACAA |
     | I                                                HINDII        ALUI        ALUI        SFANI                      |
     |                                                                                                                  |
4601 | AGACGAGTTA CTTACACTCA TTCGAGGAAA GAAGATCTTC TCTTCCTTCG ACTGTAAGTC AGGATTCTGG CAAGTTCTGC TAGATCAAGA ATCAAGACCT |
     |                      TAQI        BG1II        TAQI                HINFI                 HINFI                     |
     |                                  SAUIIIA                                               SAUIIIA                    |
     |                                                                                                                  |
4701 | CTAACGGCAT TCACATGTCC ACAAGGTCAC TACGAATGGA ATGTGGTCCC TTTCGGCTTA AAGCAAGCTC CATCCATATT CCAAAGACAC ATGGACGAAG |
     |                                              ASUI                  ALUI                                           |
     |                                              AVAII                                                                |
     |                                                                                                                  |
4801 | CATTTCGTGT GTTCAGAAAG TTCTGTTGCG TTTATGTCGA CGACATTCTC GTATTCAGTA ACAACGAAGA AGATCATCTA CTTCACGTAG CAATGATCTT |
     |                                  HINDII                                     SAUIIIA                 SAUIIIA      |
     |                                  SA1I                                                                             |
     |                                  TAQI                                                                             |
     |                                                                                                                  |
4901 | ACAAAAGTGT AATCAACATG GAATTATCCT TTCCAAGAAG AAAGCACAAC TCTTCAAGAA GAAGATAAAC TTCCTTGGTC TAGAAATAGA TGAAGGAACA |
     |                                                                                        XBAI                      |
     |                                                                                                                  |
5001 | CATAAGCCTC AAGGACATAT CTTGGAACAC ATCAACAAGT TCCCCGATAC CCTTGAAGAC AAGAAGCAAC TTCAGAGATT CTTAGGCATA CTAACATATG |
     |                                                                            HINFI                                 |
     |                                                                                                                  |
5101 | CCTCGGATTA CATCCCGAAG CTAGCTCAAA TCAGAAAGCC TCTGCAAGCC AAGCTTAAAG AAAACGTTCC ATGGAGATGG ACAAAAGAGG ATACCCTCTA |
     |            ALUI  ALUI                                  ALUI                                                       |
     |                                                       HINDIII                                                     |
     |                                                                                                                  |
5201 | CATGCAAAAG GTGAAGAAAA ATCTGCAAGG ATTTCCTCCA CTACATCATC CCTTACCAGA GGAGAAGCTG ATCATCGAGA CCGATGCATC AGACGACTAC |
     |          HPHI                                                        ALUI        TAQI        AVAIII              |
     |                                                                     BC1I                    SFANI                |
     |                                                                     SAUIIIA                 SFANI               |
     |                                                                                                                  |
5301 | TGGGGAGGTA TGTTAAAAGC TATCAAAATT AACGAAGGTA CTAATACTGA GTTAATTTGC AGATACGCAT CTGGAAGCTT TAAAGCTGCA GAAAAGAATT |
     |            ALUI                            RSA1                             SFANI        ALUI        ALUI         |
     |                                                                                         HINDIII     BBV1        |
     |                                                                                                     PSTI        |
     |                                                                                                                  |
5401 | ACCACAGCAA TGACAAAGAG ACATTGGCGG TAATAAATAC TATAAAGAAA TTTAGTATTT ATCTAACTCC TGTTCATTTT CTGATTAGGA CAGATAATAC |
     |                                                                                                                  |
5501 | TCATTTCAAG AGTTTCGTTA ATCTCAATTA CAAAGGAGAT TCGAAACTTG GAAGAAACAT CAGATGGCAA GCATGGCTTA GCCACTATTC ATTTGATGTT |
     |                                            ASUII                                                                 |
     |                                            HINFI                                                                 |
     |                                            TAQI                                                                  |
     |                                                                                                                  |
5601 | GAACACATTA AAGGAACCGA CAACCACTTT GCGGACTTCC TTTCAAGAGA ATTCAATAAG GTTAATTCCT AATTGAAATC CGAAGATAAG ATTCCCACAC |
     |                                             ECORI                                                   HINFI        |
     |                                                                                                                  |
5701 | ACTTGTGGCT GATATCAAAA GGCTACTGCC TATTTAAACA CATCTCTGGA GACTGAGAAA ATCAGACCTC CAAGCATGGA GAACATAGAA AAACTCCTCA |
     |                                                                                                                  |
5801 | TGCAAGAGAA AATACTAATG CTAGAGCTCG ATCTAGTAAG AGCAAAAATA AGCTTAGCAA GAGCTAACGG CTCTTCGCAA CAAGGAGACC TCTCTCTCCA |
     |                      ALUI SAUIIIA                     ALUI        ALUI                                           |
     |                      HGIAI                            HINDIII                                                    |
     |                      SACI                                                                                        |
     |                      TAQI                                                                                        |
     |                                                                                                                  |
5901 | CCGTGAAACA CCGGAAAAAG AAGAAGCAGT TCATTCTGCA CTGGCTACTT TTACGCCATC TCAAGTAAAA GCTATTCCAG AGCAAACGGC TCCTGGTAAA |
     |          HPAII                                                              ALUI                    ECORII       |
```

Appendix C (continued)

```
            10        20        30        40        50        60        70        80        90        100
6001 | GAATCAACAA ATCCGTTGAT GGCTAATATC TTGCCAAAAG ATATGAATTC AGTTCAGACT GAAATTAGGC CCGTAAAGCC ATCGGACTTC TTACGTCCAC |
     | HINFI                                      ECORI                 ASUI                                        |
     |                                                                  HAEIII                                      |
     |                                                                  HGICI                                       |

6101 | ATCAGGGAAT TCCAATCCCA CCAAAACCTG AACCTAGCAG TTCAGTTGCT CCTCTCAGAG ACGAATCGGG TATTCAACAC CCTCATACCA ACTACTACGT |
     |     ECORI                                                        HINFI                                       |

6201 | CGTGTATAAC GGACCTCATG CCGGTATATA CGATGACTGG GGTTGTACAA AGGCAGCAAC AAACGGTGTT CCCGGAGTTG CGCATAAGAA GTTTGCCACT |
     |           ASUI       HPAII                              RSA1      BBV1             HPAII    HHAI              |
     |           AVAII                                                                   MSTI                       |

6301 | ATTACAGAGG CAAGAGCAGC AGCTGACGCG TATACAACAA GTCAGCAAAC AGATAGGTTG AACTTCATCC CCAAAGGAGA AGCTCAACTC AAGCCCAAGA |
     |                BBV1   ALUI FNUDII                                                       ALUI             AL |
     |                BBV1   HGAI                                                                                   |
     |                PVUII                                                                                         |

6401 | GCTTTGCGAA GGCCTTAACA AGCCCACCAA AGCAAAAAGC CCACTGGCTC ATGCTAGGAA CTAAAAAGCC CAGCAGTGAT CCAGCCCCAA AAGAGATCTC |
     | UI        HAEI                                                                         SAUIIIA            BGIII |
     |           HAEIII                                                                                        SAUIIIA |

6501 | CTTTGCCCCA GAGATCACAA TGGACGACTT CCTCTATCTC TACGATCTAG TCAGGAAGTT CGACGGAGAA GGTGACGATA CCATGTTCAC CACTGATAAT |
     |           SAUIIIA                              SAUIIIA           TAQI      HPHI               HPHI            |

6601 | GAGAAGATTA GCCTTTTCAA TTTCAGAAAG AATGCTAACC CACAGATGGT TAGAGAGGCT TACGCAGCAG GTCTCATCAA GACGATCTAC CCGAGCAATA |
     |                                                                 BBV1               AVAI                     |
     |                                                                           SAUIIIA                           |

6701 | ATCTCCAGGA GATCAAATAC CTTCCCAAGA AGGTTAAAGA TGCAGTCAAA AGATTCAGGA CTAACTGCAT CAAGAACACA GAGAAAGATA TATTTCTCAA |
     | ECORII    SAUIIIA                              SFANI                HINFI      SFANI                        |

6801 | GATCAGAAGT ACTATTCCAG TATGGACGAT TCAAGGCTTG CTTCACAAAC CAAGGCAAGT AATAGAGATT GGAGTCTCTA AAAAGGTAGT TCCCACTGAA |
     | SAUIIIA RSA1           HINFI                                                 HINFI                      HINF |

6901 | TCAAAGGCCA TGGAGTCAAA GATTCAAATA GAGGACCTAA CAGAACTCGC CGTAAAGACT GGCGAACAGT TCATACAGAG TCTCTTACGA CTCAATGACA |
     | I     HAEI      HINFI      HINFI      ASUI                                            HINFI      HINFI       |
     |       HAEIII                          AVAII                                                                 |

7001 | AGAAGAAAAT CTTCGTCAAC ATGGTGGAGC ACGACACGCT TGTCTACTCC AAAAATATCA AAGATACAGT CTCAGAAGAC CAAAGGGCAA TTGAGACTTT |
     |           HINDII      HGIAI                ACCI                                                             |

7101 | TCAACAAAGG GTAATATCCG GAAACCTCCT CGGATTCCAT TGCCCAGCTA TCTGTCACTT TATTGTGAAG ATAGTGGAAA AGGAAGGTGG CTCCTACAAA |
     |                     HPAII      HINFI        ALUI                                                            |

7201 | TGCCATCATT GCGATAAAGG AAAGGCCATC GTTGAAGATG CCTCTGCCGA CAGTGGTCCC AAAGATGGAC CCCCACCCAC GAGGAGCATC GTGGAAAAAG |
     |                     HAEI        SFANI                  ASUI       ASUI                  SFANI               |
     |                     HAEIII                             AVAII      AVAII                                     |

7301 | AAGACGTTCC AACCACGTCT TCAAAGCAAG TGGATTGATG TGATATCTCC ACTGACGTAA GGGATGACGC ACAATCCCAC TATCCTTCGC AAGACCCTTC |
     |                                                                 HGAI                                        |

7401 | CTCTATATAA GGAAGTTCAT TTCATTTGGA GAGGACACGC TGAAATCACC AGTCTCTCTC TACAAATCTA TCTCTCTCTA TAATAATGTG TGAGTAGTTC |
     |                                           HPHI                                                              |

7501 | CCAGATAAGG GAATTAGGGT TCTTATAGGG TTTCGCTCAT GTGTTGAGCA TATAAGAAAC CCTTAGTATG TATTTGTATT TGTAAAATAC TTCTATCAAT |
     |                                                                                                            |

7601 | AAAATTTCTA ATTCCTAAAA CCAAAATCCA GTACTAAAAT CCAGATCCTC TAAAGTCCCT ATAGATCTTT GTGGTGAATA TAAACCAGAC ACGAGACGAC |
     |                               RSA1          BGIII                          BGIII      HPHI                 |
     |                                             SAUIIIA                         SAUIIIA                         |

7701 | TAAACCTGGA GCCCAGACGC CGTTTGAAGC TAGAAGTACC GCTTAGGCAG GAGGCCGTTA GGGAAAAGAT GCTAAGGCAG GGTTGGTTAC GTTGACTCCC |
     |      ECORII    HGAI          ALUI          RSA1          HAEIII           SFANI                       HINDII |
     |                                                                                                      HINFI  |

7801 | CCGTAGGTTT GGTTTAAATA TCATGAAGTG GACGGAAGGA AGGAGGAAGA CAAGGAAGGA TAAGGTTGCA GGCCCTGTGC AAGGTAAGAC GATGGAAATT |
     |                                                                           ASUI                             |
     |                                                                           HAEIII                           |
     |                                                                           HGICI                            |

7901 | TGATAGAGGT ACGTTACTAT ACTTATACTA TACGCTAAGG GAATGCTTGT ATTTACCCTA TATACCCTAA TGACCCCTTA TCGATTTAAA GAAATAATCC |
     |     RSA1                                                                               C1AI                 |
     |                                                                                        TAQI                 |

8001 | GCATAAGCCC CCGCTTAAAA AATT                                                                                   |
```

Appendix D. Eukaryotic transcription signals

a) Matches of TATARAR in CaMV DNA (R = T or A)

Pos.	Sequ.	Dev.	Pos.	Sequ.	Dev.	Pos.	Sequ.	Dev.
79	TCTAAAA	(1)	1660[+]	AATAAAA	(1)	6042	TATGAAT	(1)
81	TAAAAAT	(1)	1693	AATAAAA	(1)	6203	TGTATAA	(1)
85	AATAAAA	(1)	1704	TAGAAAA	(1)	6225	TATATAC	(1)
223	TCTAAAA	(1)	1725	TAGAAAA	(1)	6331	TATACAA	(1)
249	TAAAAAT	(1)	1899	GATATAA	(1)	6787	GATATAT	(1)
299	AATAAAT	(1)	1901	TATAAAA	(0)	6789	TATATTT	(1)
431	TATTTAA	(1)	1940	TATTAAA	(1)	6877	TCTAAAA	(1)
433	TTTAAAT	(1)	3002	TATGAAA	(1)	7056	TATCAAA	(1)
491	TAAAAAA	(1)	3015	TGTATAA	(1)	7195	TACAAAT	(1)
502	TCTAAAA	(1)	3017	TATAAGA	(1)	7402	TCTATAT	(1)
536	TATTTAA	(1)	3045	GATATAT	(1)	7404	TATATAA	(0)
538	TTTAAAA	(1)	3047	TATATCA	(1)	7461	TACAAAT	(1)
567	TATGAAA	(1)	3114	TATCTAA	(1)	7477	TCTATAA	(1)
659	TATATCT	(1)	3116	TCTAAAA	(1)	7479	TATAATA	(1)
687	GATAAAT	(1)	3318	AATATAA	(1)	7549	CATATAA	(1)
689	TAAATAA	(1)	3338	TATAAGA	(1)	7551	TATAAGA	(1)
691	AATAAAA	(1)	4174	TAGAAAA	(1)	7567	TATGTAT	(1)
979	AATAAAA	(1)	4900	TACAAAA	(1)	7581	TGTAAAA	(1)
981	TAAAAAT	(1)	4981	TAGAAAT	(1)	7594	TATCAAT	(1)
990	TATGAAT	(1)	5321	TATCAAA	(1)	7598	AATAAAA	(1)
1069	TATCAAT	(1)	5432	AATAAAT	(1)	7660	TATAGAT	(1)
1318	AATAAAT	(1)	5441	TATAAAG	(1)	7677	AATATAA	(1)
1374	TTTATAA	(1)	5456	TATTTAT	(1)	7679	TATAAAC	(1)
1376	TATAAAA	(0)	5460	TATCTAA	(1)	7813	TTTAAAT	(1)
1378	TAAAAAA	(1)	5713	TATCAAA	(1)	7815	TAAATAT	(1)
1469	AATATAA	(1)	5731	TATTTAA	(1)	7959	TATATAC	(1)
1471	TATAATT	(1)	5786	TAGAAAA	(1)	8016	TAAAAAA	(1)

b) Matches of UUCCAATC (U = A or G)

Pos.	Sequ.	Dev.	Pos.	Sequ.	Dev.	Pos.	Sequ.	Dev.
397	GAGCAATC	(1)	2665	AACCAATC	(0)	4000	AACCATTC	(1)
946	AACCAAAC	(1)	2722	GACCAATA	(1)	4423	AGCAAATC	(1)
1832	GGCTAATC	(1)	2950	GACCAAGC	(1)	5147	AGCCAAGC	(1)
1988	AACCAAGC	(1)	3686	CACCAATC	(1)	6006	AACAAATC	(1)
2611	GACCATTC	(1)	3835	GACCAATT	(1)	6160	GACGAATC	(1)

c) Polyadenylation Sequences AATAAA

85   299   691   979   1318   1660   1693   5432   7598

Appendix E Prokaryotic transcription signals

a) matches for TATAAATG in CaMV DNA

Pos.	Sequence	Dev.*	Pos.	Sequence	Dev.*	Pos.	Sequence	Dev.*
211	TACGATG	(1)	1915	TAGAATT	(1)	6579	TACCATG	(1)
420	TAATATG	(1)	2243	TATAATC	(0.5)	6595	GATAATG	(1)
468	TCTAATG	(1)	2287	TATGATA	(1)	6789	TATATTT	(1)
540	TAAAATG	(0.5)	2555	TACAATG	(0.5)	7161	TATTGTG	(1)
828	TAGAATC	(1)	3178	TAAAATC	(1)	7406	TATAAGG	(1)
1083	TACAATG	(1)	3320	TATAAGG	(1)	7479	TATAATA	(0.5)
1143	TACAATA	(1)	3588	TAGAATA	(1)	7482	AATAATG	(1)
1345	TAATATG	(1)	4050	TATCCTG	(1)	7564	TAGTATG	(1)
1390	TATTATT	(1)	4584	TACAATC	(1)	7583	TAAAATA	(1)
1471	TATAATT	(0.5)	5018	TATCTTG	(1)	7600	TAAAATT	(1)
1662	TAAAATC	(1)	5441	TATAAAG	(1)	7635	TAAAATC	(1)
1690	TAGAATA	(1)	6027	TATCTTG	(1)	7819	TATCATG	(0.5)
1816	TATTATT	(1)	6205	TATAACG	(1)	7918	TATACTT	(1)
1826	TAAAATG	(0.5)	6229	TACGATG	(1)	7924	TATACTA	(1)

*Deviations in Position 3, 4, 5, 7: 0.5 points in 1, 2, 6: 1 point

b) Matches for TTTTTA in CaMV DNA   (Stop signals sense)

Pos.	Sequence	Dev.	Pos.	Sequence	Dev.	Pos.	Sequence	Dev.
71	TTCTTA	(1)	1392	TTATTA	(1)	5079	TTCTTA	(1)
369	TTTGTA	(1)	1440	TTTTTA	(0)	5456	TATTTA	(1)
431	TATTTA	(1)	1596	TTTTTA	(0)	5731	TATTTA	(1)
536	TATTTA	(1)	1597	TTTTAA	(1)	5948	CTTTTA	(1)
553	GTTTTA	(1)	1617	TTTTTA	(0)	6088	TTCTTA	(1)
896	TGTTTA	(1)	1618	TTTTAA	(1)	6614	TTTTCA	(1)
972	TTTTGA	(1)	1670	CTTTTA	(1)	7098	TTTTCA	(1)
1198	TTATTA	(1)	1920	TTATTA	(1)	7520	TTCTTA	(1)
1309	CTTTTA	(1)	2213	ATTTTA	(1)	7573	TTTGTA	(1)
1310	TTTTAA	(1)	2876	TTCTTA	(1)	7579	TTTGTA	(1)
1372	TGTTTA	(1)	4159	TTTCTA	(1)	7605	TTTCTA	(1)
1374	TTTATA	(1)	4237	TCTTTA	(1)	7950	TATTTT	(1)

c) Matches for TAAAAA in CaMV DNA (Stop signals antisense)

81   225   249   491   981   1378   1695   6462   6879   8016

126 additional sites found if one deviation is allowed.

Appendix F. Protein sequences derived from CaMV DNA (one letter code)

```
         G I R A M N R F K T K T Q E G K T S P K Y E R V L N S K N K R S F K I K H S S L
         V S E P * I G L R P K L K R R V K P H Q N T K E F L T L K I K D L S R S N I V P S
         Y Q S H E S V * D Q N S R G * N L T K I R K S S * L * K K * K I F Q D Q T * F P H
GGTATCAGAGCCATGAATCGGTTTAAGACCAAAACTCAAGAGGGTAAAACCTCACCAAAATACGAAAGAGTTCTTAACTCTAAAAATAAAAGATCTTTCAAGATCAAACATAGTTCCCTC
         10        20        30        40        50        60        70        80        90       100       110       120

         I P V T D R I T T V R F Q N N I E S V Y A N F D S Q L K S S Y D G R S K K I K I
         H R * P T G L P P * G F R T T S K A F T P T S L N S S R R T M V D L K R S R L
         T G D R Q D Y H R K V S E Q H R K R L R Q L R L S T Q V V R W * I * K D Q D S
ACACCGGTGACCGACAGGATTACCACCGTAAGGTTTCAGAACAACATCGAAAGCGTTTACGCCAACTTCGACTCTCAACTCAAGTCGTCGTACGATGGTAGATCTAAAAAGATCAAGACT
        130       140       150       160       170       180       190       200       210       220       230       240

         I S L K N L R C Y E A F L R K Y L L E Q * I S L R I V L Y * V S I G K I T F C V
         * A L K I L D V T K P S S G S T F W N N K S L * E * S Y S I E Y P Q E K * P S V L
         K P * K S * M L R S L Q P G E V P S G T I N L S E N S T L L S I H R K N N L L C *
CTAAGCCTTAAAAATCTTAGATGTTACGAAGCCTTCCTCAGGAAGTACCTTCTGGAACAATAAATCTCTCTGAGAATAGTACTCTATTGAGTATCCACAGGAAAAATAACCTTCTGTGTT
        250       260       270       280       290       300       310       320       330       340       350       360

         E M D L Y P E E N T Q S E Q S Q N S E N N M Q I F K S E N S D G F S S D L M I S
         R W I C I Q K K I P K A S N R R I L K I I C K Y L N Q K I R M D S P P I * * S Q
         D G F V S R R K Y P K R A I A E F * K * Y A N I * I R K F G W I L L R S N D L K
GAGATGGATTTGTATCCAGAGGAAAATACCCAAAGCGAGCAATCGCAGAATTCTGAAAATAATATGCAAATATTTAAATCAGAAAATTCGGATGGATTCTCCTCCGATCTAATGATCTCA
        370       380       390       400       410       420       430       440       450       460       470       480

         N D G L K N I S K T Q L T L E K E K I F K M P N V L S Q V M K K A F S R K N E I
         T I N * K K I S L K P N L * P W R K K R D Y L K C L T F Y L K L * K K R L A G K T R F
         R S I K K Y L * N P I N L G E R K D I * N A * R F I S S Y E K S V * Q E K R D S
AACGATCAATTAAAAAATATCTCTAAAACCCAATTAACCTTGGAGAAAGAAGATATTTAAAATGCCTAACGTTTTATCTCAAGTTATGAAAAAAGCGTTTAGCAGGAAAAACGAGAGATT
        490       500       510       520       530       540       550       560       570       580       590       600

         L Y C V S T K E L S V D I H D A T G K V V Y L P L I T K E E I N K R L S S L K P E
         S T A S R Q K N Y Q W T F T M P Q V R Y I F P * S L R K R * I K D F P A * N L K
         L L R L D K R I I S G H S R C H R * G I S S L N H * G R D K * K T F Q L K T * S
CTCTACTGCGTCTCGACAAAAGAATTATCAGTGGACATTCACGATGCCACAGGTAAGGTATATCTTCCCTTAATCACTAAGGAAGAGATAAATAAAAGACTTTCCAGCTTAAAACCTGAA
        610       620       630       640       650       660       670       680       690       700       710       720

         V R K T M S M V H L G A V K I L L K A Q F R N G I D T P I K I A L I D D R I N S
         S E R P C P W F I L E R S K Y C L K L N F E M G L I P Q S K L L * S M I E S I L
         Q K D H V S S S W S G Q N I A * S S I S K W D * Y P N Q N C F N R * * N Q F *
GTCAGAAAGACCATGTCCATGGTTCATCTTGGAGCGGTCAAAATATTGCTTAAAGCTCAATTTCGAAATGGGGATTGATACCCCAATCAAAATTGCTTTAATCGATGATGAATCAATTCT
        730       740       750       760       770       780       790       800       810       820       830       840

         R R D C L L G A A K G N L A Y G K F M F T V Y P K F G I S L N T Q R L N Q T L S
         E E I V L S S W C S Q R * * H T V S L C L L Y T L S L E * A L T P K D L T K P * A
         K R L S S W C S Q R * S S I R * V Y V C Y C I P * V W N K P * H P K T * P N P K P
AGAAGAGATTGTCTTCTTGGTGCAGCCAAAGGTAATCTAGCATACGGTAAGTTTATGTTTACTGTATACCCTAAGTTTGGAATAAGCCTTAACACCCAAAGACTTAACCAAACCCTAAGC
        850       860       870       880       890       900       910       920       930       940       950       960

         L I H D F E N K N L M N K G D K V M T I T Y V V G Y A L T N S H H S I D Y Q S N
         L F M I L K I K I L * I K V I K L * P * P M S * D M H * L I V I I A * I I N Q H
         Y S * F * K K * S Y E * R * S Y D H N L C R R I C I N * S S * H R L S I K C
CTTATTCATGATTTTGAAAATAAAAATCTTATGAATAAAGGTGATAAAGTTGACCATAACCTATGTCGTGGATATGCATTAACTAATAGTCATCATAGCATAGATTATCAATCAAAT
        970       980       990      1000      1010      1020      1030      1040      1050      1060      1070      1080

         A T I E L E D V F Q E I G N V Q Q S E F C T I Q N D E C N W A I D I A Q N K L
         L Q L N * K D V F R K L E M S S N L S S V Q Y R M D N M Q L G P L I * P K T K P Y
         Y N * T R R R I S R N W K C P A I * V L Y N T E * M Q L G H * Y S P K Q S L I
GCTACAATTGAACTAGAAGACGTATTTCAAGAAATTGGAAATGTCCAGCAATCTGAGTTCTGTACAATACAAGATGATGAATGCAATTGGGCCATTGATATAGCCCAAAACAAAGCCTTA
       1090      1100      1110      1120      1130      1140      1150      1160      1170      1180      1190      1200

         L G A K T K T Q I G N N L Q I G N S A S S S N T E N E L A R V S Q N I D L L K N
         * E L R P R L K L V I T F K * V I V H L P L I L K M N * L G * A R T * I F * R I
         R S * N Q D S N W * * P S N R * Q C F I L * Y * K * I S * G K P E H R S F K E *
TTAGGAGCTAAAACCAAGACTCAAATTGGTAATAACCTTCAAATAGGTAACAGTGCTTCATCCTCTAATACTGAAAATGAATTAGCTAGGGTAAGCCAACATAGATCTTTTAAAGAAT
       1210      1220      1230      1240      1250      1260      1270      1280      1290      1300      1310      1320

         K L K E I C G E * Y E H Y G T T A C L * K R Y Y Y * T K T I V S * * * * * * K L C
         N * K K S V E N N M S I T Q G Q P H V Y K K D T I I R L K P L S L N S N N R S Y V
         I K R N L W R I I * A L R D N R M F I K K I L L L D * N H C L L I V I I E V M F
AAATTAAAAGAAATCTGTGGAGAATAATATGAGCATTACTCAGGGACAACCGCATGTTTATAAAAAAGATACTATTATTAGACTAAAACCATTGTCTCTTAATAGTAATAATAGAAGTTATGT
       1330      1340      1350      1360      1370      1380      1390      1400      1410      1420      1430      1440

         F * F L K R E H S K Y N * S S * Q P Q * D C R K K L T R N M E D Q L I L R I K Q
         F S S S K G N I Q N I H L N N L N E I V G R S L L G I W K I N S Y F G L S K
         L V P Q K G T F K I * I L I L T T S M R L * E E A Y S E Y G R S T H T S D * A K
TTTTAGTTCCTCAAAAGGGAACATTCAAAATATAATTAATCATCTTAACAACCTCAATGAGATTGTAGGAAGAAGCTTACTCGGAATATGGAAGATCAACTCATACTTCGGATTAAGCAA
       1450      1460      1470      1480      1490      1500      1510      1520      1530      1540      1550      1560

         R P F G V Q I K K P V S F * Y C K N H F * E W G G * L L E P T K G N K I P F R S
         D P S E S K S K N P S V F N T A K T I F K S G G V D Y S S Q L K E I K S L L E A
         T L R S P N Q K T R Q F L I L Q K P F L R V G G L I T R A N * R K * N P F * K L
AGACCCTTCGGAGTCCAAATCAAAAAACCCGTCAGTTTTTAATACTGCAAAAACCATTTTTAAGAGTGGGGGGTTGATTACTCGAGCCAACTAAAGGAAATAAAATCCCTTTTAGAAGC
       1570      1580      1590      1600      1610      1620      1630      1640      1650      1660      1670      1680

         S K H * N K K S R K S N S I L R K * D * T R A L N * R G S * R A K R I D * L D Q
         Q N T R I K S L E K A I Q S L E N K I E P E P L T K E E V K E L K E S I N S I K
         K T L E * K V * K K Q F N P * K I R I N Q S * L K R K L K S * K N R L T R S K
TCAAAACACTAGAATAAAAAGTCTAGAAAATCAACTCAATCCTTAGAAAATAAGATTGAACCAGAGCCCTTAACTAAAGAGGAAGTTAAAGAGCTAAAGAAATCCGATTAACTCGATCAA
       1690      1700      1710      1720      1730      1740      1750      1760      1770      1780      1790      1800

         R R I K E Y Y W L K W L I L I R S K K K S L K S S V T K N P * K R I * K L S * N
         E G L K N I I G * N G * S * S D P K R S L * N P Q * P K I H E S G Y K S Y L R I
         K D * R I L L A K M A N L N Q I Q K E V S E I L S D Q K S M K A D I K A I L E L
AGAAGGATTAAAGAATATTATTGGCTAAAATGGCTAATCTTAATCAGATCAAAAAGAAGTCTCTGAAATCCTCAGTGACCAAAAATCCATGAAAGCGGATATAAAAGCTATCTTAGAAT
       1810      1820      1830      1840      1850      1860      1870      1880      1890      1900      1910      1920

         Y * D P K I L L K K A * K P L Q Q K S L M T * P S S S M I V L V T K R Y * K P *
         I R I P S N Y * R K L R N R C S K N R * L N Q A H Q * L S L * Q R D I R S L R
         G S Q N P I K E S E T V A A K I V N D L T K L I N D C P C N K E I L E A L G
TATTAGGATCCCAAAATCCTATTAAAGAAAGCTTAGAAACCGTTGCAGCAAAAATCGTTAATGACTTAACCAAGCTCATCAATGATTGTCCTTGTAACAAAGAGATATTAGAAGCCTTAG
       1930      1940      1950      1960      1970      1980      1990      2000      2010      2020      2030      2040
```

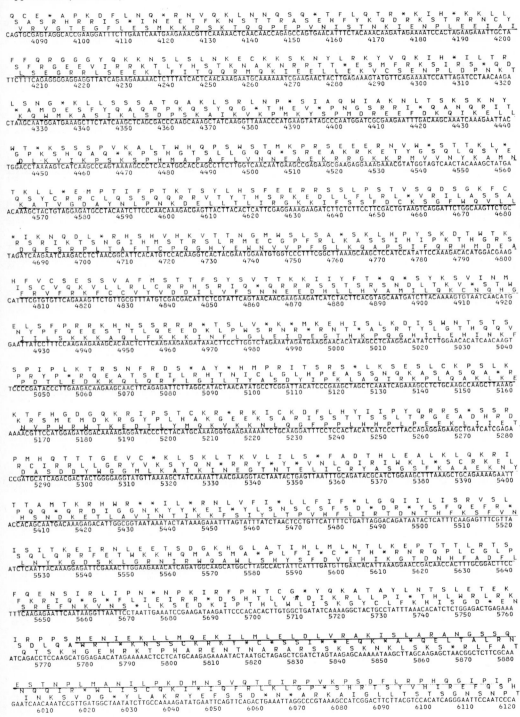

```
   P K P E P S S S V A P L R D E S G I Q H P H T N Y Y V V Y N G P H A G I Y D D W
  Q N L N L A V Q L L L S E T N R G V F N T L P I T T S C I I D L M G T K K P S S D
   K T * T * Q F S C S S Q R R I G Y S T P S Y Q L L R R V * R T S C R Y I R * L G
CCAAAACCTGAACCTAGCAGTTCAGTTGCTCCTCTCAGAGACGAATCGGGTATTCAACACCCTCATACCAACTACTACGTCGTGTATAACGGACCTCATGCCGGTATATACGATGACTGG
   6130      6140      6150      6160      6170      6180      6190      6200      6210      6220      6230      6240

   G C T K A A T N G V P G V A H K K F A T I T E A R A A A D A Y T T S Q Q T D R L
  V V Q R Q A Q T V F P E L R I R S L P L L Q R Q E Q Q L T R I Q V S K Q I G *
   L Y K G S N K R C S R S C A * E V C H Y Y R G K S S S * R V Y N K S A N R * V E
GGTTGTACAAAGGCAGCAACAAACGGTGTTCCCGGAGTTGCGCATAAGAAGTTTGCCACTATTACAGAGGCAAGAGCAGCAGCTGACGCGTATACAACAAGTCAGCAAACAGATAGGTTG
   6250      6260      6270      6280      6290      6300      6310      6320      6330      6340      6350      6360

   N F I P K G E A Q L K P K S F A K A L T S P P K Q K A H W L M L G T K K P S S D
  T S S P K E K L N S S P R A L R R P * Q A H Q S K K P Y G S C * E L K S P A V I
   L H P Q R R S T Q A Q E L C E G L N K P T K A K S P L A H A R N * K A Q Q * S
AACTTCATCCCCAAAGGAGAAGCTCAACTCAAGCCCAAGAGCTTTGCGAAGGCCTTAACAAGCCCACCAAAGCAAAAAGCCCACTGGCTCATGCTAGGAACTAAAAAGCCCAGCAGTGAT
   6370      6380      6390      6400      6410      6420      6430      6440      6450      6460      6470      6480

   Q G D L S L H R E T P E K E E A V H S A L A T F T P S Q V K A I P E Q T A P G K
  K E T S L S * V K H R K R K K Q F I L H L L L R H L K * K L F Q S K R L L V K
   R R P L S P P * N T G K R R S S F C T G Y F Y A I S S K S Y S R A N G S W * R
CAAGGAGACCTCTCTCTCCACCGTGAAACACCGGAAAAAGAAGAAGCAGTTCATTCTGCACTGGCTACTTTTACGCCATCTCAAGTAAAAGCTATTCCAGAGCAAACGGCTCCTGGTAAA
   5890      5900      5910      5920      5930      5940      5950      5960      5970      5980      5990      6000

   P A P K E I S F A P E I T M D D F L Y L Y D L V R K F D G E G D D T M F T T D N
  Q P Q R S P L P Q R S Q W T T S S I S T I * S G S S T E K V T I P C S P L I M
   S P K R D L L C P R D H N G R L P L S L R S S Q E V R R R * R Y H V H H * * *
CCAGCCCCAAAAGAGATCTCTTTGCCCCAGAGATCACAATGGACGACTTCCTCTATCTCTACGATCTAGTCAGGAAGTTCGACGGAGAAGGTGACGATACCATGTTCACCACTGATAAT
   6490      6500      6510      6520      6530      6540      6550      6560      6570      6580      6590      6600

   E K I S L F N F R K N A N P Q M V R E A Y A A G L I K T I Y P S N N L Q E I K Y
  R R L A F S I S E R M L T R H W L E R L T Q Q V S S R R S T R A I I S R R S N T
   E D * P F Q F Q K E C * P T D G * R G L R S R S H Q D L P E Q * S P G D Q I P
GAGAAGATTAGCCTTTTCAATTTCAGAAAGAATGCTAACCCACAGATGGTTAGAGAGGCTTACGCAGCAGGTCTCATCAAGACGATCTACCCGAGCAATAATCTCCAGGAGATCAAATAC
   6610      6620      6630      6640      6650      6660      6670      6680      6690      6700      6710      6720

   L P K K V K D A V K R F R T N C I K N T E K D I F L K I R S T I P V W T I Q G L
  F P R R R L K M Q S Q K I Q D S G L T A S R T Q R K I Y R S E V L F Q Y G F K A L
   S Q E G * R C S Q K I Q D * L H Q E H R E R Y I S Q D K Y Y S S M D D S R L A
CTTCCCAAGAAGGTTAAAGATGCAGTCAAAAGATTCAGGACTAACTGCATCAAGAACACAGAGAAAGATATATTTCTCAAGATCAGAAGTACTATTCCAGTATGGACGATTCAAGGCTTG
   6730      6740      6750      6760      6770      6780      6790      6800      6810      6820      6830      6840

   I H K P R Q V I E I G V S K K V V P T E S K A M E S K I Q I E D L T E L A V K T
  F T N Q G K * * * R L E S L R * P * L P L N Q R P W S Q R F K * R I * Q N S P * R L
   S Q T K A S N R D W S L * K G S S H * I K G H G V K D S N R G P N R T R R K D W
CTTCACAAACCAAGGCAAGTAATAGAGATTGGAGTCTCTAAAAAGGTAGTTCCCACTGAATCAAAGGCCATGGAGTCAAAGATTCAAATAGAGGACCTAACAGAACTCGCCGTAAAGACT
   6850      6860      6870      6880      6890      6900      6910      6920      6930      6940      6950      6960

   G E Q F I Q S L L R L N D K K K I F V N M V E H D T L V Y S K N I K D T V S E D
  A N S S Y R V S Y D S M T R R K S S S T H A W S I L T L P K L S K I Q S K Q X
   R T V H T E S L T T Q * Q E E N L R Q H G G A R H A C L L Q K Y Q R Y S L R R P
GGCGAACAGTTCATACAGAGTCTCTTACGACTCAATGACAAGAAGAAAATCTTCGTCAACATGGTGGACGACACGCTTGTCTACTCCAAAAATATCAAAGATACAGTCTCCAGAAGAC
   6970      6980      6990      7000      7010      7020      7030      7040      7050      7060      7070      7080

   Q R A I E T F Q Q R V I S G N L L G F H C P A I C H F I V K I V E K E G G S Y K
  K G Q L R L F N K G * Y P E T S S I A Q L S V T L L Y * R * W K R K V A P T N
   K G N * D F S T K G N I R K P P R I P L P S Y L S L Y C E D S G K G R W L L Q M
CAAAGGGCAATTGAGACTTTTCAACAAAGGGTAATATCCGGAAACCTCCTCGGATTCCATTGCCCAGCTATCTCTGCACTTTATTGTGAAGATAGTGGAAAAGGAAGGTGGCTCCTACAA
   7090      7100      7110      7120      7130      7140      7150      7160      7170      7180      7190      7200

   C H H C D K G K A I V E D A S A D S G P K D G P P P T R S I V E K E D V P T T S
  A I I A I K E R P S L K H P L P T V V P K M D P H P R G A S W K K K T F Q P R L
   P S L R * R K G H R * R C L C R Q W S Q R W T P T H E E H R G K R R S N H V F
TGCCATCATTGCGATAAAGGAAAGGCCATCGTTGAAGATGCCTCTGCCGACAGTGGTCCCAAAGATGGACCCCCACCCACGAGGAGCATCGTGGAAAAAGAAGCAGTTCCAACCACGTCT
   7210      7220      7230      7240      7250      7260      7270      7280      7290      7300      7310      7320

   S K Q V D * C D I S T D V R D D A Q S H Y P S Q D P S S I * G S S F H L E R T R
  Q S K W I D V I S P L T * G M T H N P T I L R A K T L P L Y K E V H F I W R G H A
   K A S G L M * Y L H * R K G * R T I P L S F A R P F L Y I R K F I S F G E D T L
TCAAAGCAAGTGGATTGATGTGATATCTCCACTGACGTAAGGGATGACGCACAATCCCACTATCCTTCGCAAGACCCTTCCTCTATATAAGGAAGTTCATTTCATTTGGAGAGGACACGC
   7330      7340      7350      7360      7370      7380      7390      7400      7410      7420      7430      7440

   * N H Q S L S T N L S L S I I M C E * F P D K G I R V L I G F R S C V E H I R N
  E I T S L S L S L Q I Y L S L Y N * * C V S S S Q I R E L G F L * G F A H V L S I * E T
   K S P V S L Y K S I S L Y N N V * V V P R * G N * G S Y R V S L M C * A Y K K P
TGAAATCACCAGTCTCTCTCTACAAATCTATCTCTCTCTATAATAATGTGTGAGTAGTCCCCAGATAAGGGAATTAGGGTTCTTCGGCTCATGTGTTGAGCATATAAGAAAC
   7450      7460      7470      7480      7490      7500      7510      7520      7530      7540      7550      7560

   P * Y V F V F V K Y F Y Q * N F * F L K P K S S T K I Q I S * S P Y R S L W * I
  L S M Y L Y L * N T S I N K I S N S * N Q N P V L K S R S P K V P I D L C G E Y
   L V C I C I C K I L L S * F L I P K T K I Q Y * N P D L L K S L * I F V V N I
CCTTAGTATGTATTTGTATTTGTAAAATACTTCTATCAATAAAATTTCTAATTCCTAAAACCAAAATCCAGTACTAAAATCCAGATCTCCTATAGATCTTTGTGGTGAATA
   7570      7580      7590      7600      7610      7620      7630      7640      7650      7660      7670      7680

   * T R H E T T K P G A Q T P F E A R S T A * A G G R * G K D A K A G L V T L T P
  K P D T R R L N L E P R R R L K L E V P L R Q E A V R E K M L R Q G W L R * L P
   N Q T R D D * T W S P D A V * S * K Y R L G R R R P L G K R C * G R V G Y V D S P
TAAACCAGACACGAGACGACTAAACCTGGAGCCCAGACGCCGTTTGAAGCTAGAAGTACCGCTTAGGCAGGAGGCCGTTAGGGAAAAGATGCTAAGGCAGGGTTGGTTACGTTGACTCCC
   7690      7700      7710      7720      7730      7740      7750      7760      7770      7780      7790      7800

   P * V R F G F K Y H E V D G R K E E D K E G * G C R P C A R * D D G N L I E V R Y Y
  R R F G L N I M K W T E G R R K T R K D K V A G P V Q G K T M E I * R Y V T I
   V G L V * I S * S G R K E G G R Q G R I R L Q A L C K V R R W K F D R G T L L Y
CCGTAGGTTTGGTTTAAATATCATGAAGTGGACGGAAGGAAGGAGGGAAGACAAGGAAGGATAAGGTTGCAGGCCCTGTGCAAGGTAAGACGATGGAAATTTGATGAGGTACGTTACTAT
   7810      7820      7830      7840      7850      7860      7870      7880      7890      7900      7910      7920

   T Y T I R * G N A C I Y P I Y P N D P L S I * R N N P H K P P L K K
  L I L Y A K G M L V F T L Y T L M T P Y R F K E I I R I S P G Y F
   Y Y T L R E C L Y L V F Y I P * * P L I D L K K * S A * A P A * K I
ACTATACTATACGCTAAGGGAATGCTTGTATTTACCCTATATACCCTAATGACCCCTTATCGATTTAAAGAAATAATCCGCATAAGCCCCCGCTTAAAAAATT
   7930      7940      7950      7960      7970      7980      7990      8000      8010      8020
```

Appendix G. Hairpins and inverted repeats
a) Pairing ≧ 8; loop size ≦ 6; deviations ≦ 2

Pairing	Loop Size	Position	Deviations
8	4	91	2
8	1	102	2
8	2	124	2
8	5	163	2
8	1	198	2
8	4	242	2
8	4	246	2
8	0	250	2
8	1	272	2
8	0	287	2
8	3	358	2
8	4	426	2
8	4	438	2
8	2	451	2
8	5	457	2
8	0	541	2
8	4	543	2
9	4	549	2
8	4	558	2
10	6	565	2
8	2	566	2
8	5	586	2
8	2	596	2
8	3	604	2
9	6	640	2
9	1	660	2
12	4	674	2
9	0	675	2
9	2	681	2
8	1	711	2
8	1	744	2
8	6	761	2
8	2	787	2
8	5	796	2
8	3	799	2
10	1	813	2
9	4	834	2
9	1	840	2
10	0	849	2
8	2	859	2
11	4	894	2
8	4	934	2
8	4	951	1
10	1	979	2
8	5	998	2
10	1	1008	2
9	6	1011	2
8	1	1015	1
11	0	1028	2
8	5	1060	2
8	3	1099	1
8	1	1109	2
8	0	1134	2
8	1	1142	2
8	2	1177	2
9	4	1221	2
9	4	1269	2
9	3	1319	2
11	4	1379	2
8	3	1386	2
8	4	1387	1
8	5	1405	2
8	4	1422	2
8	1	1434	2
9	0	1477	1
9	1	1497	1
9	4	1536	2
8	4	1590	2
8	5	1591	2
8	0	1592	2
9	3	1611	1
8	2	1614	2
9	4	1663	2
8	4	1728	1
9	4	1760	2
8	1	1829	2
8	4	1855	2
8	2	1870	2
9	0	1902	2
9	4	1928	2
10	3	3868	2
8	5	3890	2
8	2	3926	2
9	3	3946	2
9	4	4003	2
8	5	4036	2
8	5	4047	2
8	3	4049	2
9	0	4112	2
8	0	4114	2
9	3	4187	2
8	2	4197	2
11	4	4197	2
9	3	4227	2
9	4	4246	2
8	6	4282	2
8	0	4295	2
8	6	4303	2
9	1	4359	2
8	1	4380	2
8	1	4387	1
11	5	4432	2
8	6	4452	2
9	2	4497	2
9	1	4499	2
8	6	4503	2
8	5	4579	2
8	1	4634	2
9	1	4635	2
11	4	4636	2
9	2	4638	1
9	2	4744	2
9	4	4784	2
8	4	4799	2
8	4	4861	2
10	4	4875	2
10	0	4918	2
8	1	4933	2
10	3	4936	2
8	4	4946	2
8	0	4967	2
8	4	4969	2
12	2	4984	2
11	5	5033	2
8	6	5045	2
8	5	5064	2
8	5	5073	2
11	3	5093	2
9	1	5151	2
10	5	5226	2
8	6	5322	2
9	5	5336	2
12	4	5342	2
8	2	5366	2
8	4	5417	2
8	3	5444	2
12	3	5447	2
9	5	5454	2
8	1	5462	2
9	2	5479	2
8	1	5487	1
8	3	5506	2
9	6	5530	2
8	0	5599	2
9	2	5602	2
10	3	5664	2
8	0	5685	2
9	0	5725	2
8	3	5774	2
8	2	5788	2
9	1	5828	1
8	3	5926	2
9	6	5986	2
9	4	6034	2
8	0	6035	2
8	5	6050	2
10	0	6055	2
9	0	6058	2
9	1	6136	1
13	4	6216	2

Appendix G, a (continued)

PAIRING :	10	12	8	9	9	8	8	9	8	9
LOOP SIZE :	3	5	4	0	3	2	6	0	1	3
POSITION :	1932	1990	2002	2016	2022	2045	2050	2092	2092	2123
DEVIATIONS :	1	2	2	2	2	1	2	2	2	2

PAIRING :	8	9	8	9	8	10	10	8	8	8
LOOP SIZE :	4	1	6	3	1	0	0	4	3	5
POSITION :	2180	2186	2192	2259	2267	2269	2290	2367	2406	2417
DEVIATIONS :	1	2	2	1	2	2	1	2	2	2

PAIRING :	9	8	8	9	9	9	10	8	12	8
LOOP SIZE :	2	5	1	3	2	4	5	5	5	3
POSITION :	2430	2455	2487	2492	2497	2584	2757	2759	2775	2796
DEVIATIONS :	2	2	1	2	2	2	2	2	2	2

PAIRING :	9	8	8	11	10	10	8	8	8	12
LOOP SIZE :	1	1	6	2	6	2	4	0	1	1
POSITION :	2851	2889	2903	3005	3043	3194	3200	3220	3231	3246
DEVIATIONS :	2	2	2	2	2	2	2	2	2	2

PAIRING :	8	8	10	8	10	12	8	8	9	8
LOOP SIZE :	4	6	1	5	5	2	1	4	2	3
POSITION :	3274	3315	3330	3362	3388	3443	3447	3472	3476	3510
DEVIATIONS :	1	2	1	2	2	2	1	1	2	2

PAIRING :	8	9	8	8	9	8	8	9	8	8
LOOP SIZE :	2	3	4	3	1	0	0	3	5	1
POSITION :	3540	3546	3575	3581	3591	3660	3663	3806	3829	3846
DEVIATIONS :	2	2	2	1	2	2	1	2	2	2

PAIRING :	8	10	10	10	8	8	8	8	8	9
LOOP SIZE :	2	0	2	6	2	1	1	5	1	1
POSITION :	6264	6303	6334	6372	6406	6412	6496	6505	6528	6581
DEVIATIONS :	2	2	2	1	2	2	1	1	2	2

PAIRING :	8	8	8	9	10	9	8	8	8	9
LOOP SIZE :	2	5	5	3	5	0	5	0	1	1
POSITION :	6591	6595	6606	6609	6623	6644	6649	6707	6789	6801
DEVIATIONS :	2	2	2	2	2	2	2	1	1	2

PAIRING :	8	8	9	8	8	8	8	9	8	8
LOOP SIZE :	5	0	2	1	3	6	3	1	5	1
POSITION :	6802	6836	6845	6871	6907	6984	6985	7166	7196	7233
DEVIATIONS :	2	2	2	1	2	2	2	2	2	2

PAIRING :	8	8	8	8	8	8	9	8	8	8
LOOP SIZE :	5	1	6	3	4	1	2	1	2	3
POSITION :	7277	7285	7296	7311	7323	7329	7332	7417	7496	7505
DEVIATIONS :	2	1	2	0	2	2	2	2	2	1

PAIRING :	9	8	8	8	8	8	8	8	8	8
LOOP SIZE :	6	2	2	4	4	1	4	3	4	4
POSITION :	7515	7542	7546	7580	7589	7597	7679	7874	7915	7948
DEVIATIONS :	2	2	2	1	2	2	2	2	2	2

b) Pairing $\geqq$ 11; loop size $\leqq$ 6; deviations $\leqq$ 4

PAIRING :	11	13	11	12	12	12	11	12	12	12
LOOP SIZE :	4	5	3	2	1	1	3	4	2	4
POSITION :	15	81	82	90	98	102	163	198	227	259
PA. ERROR :	4	3	4	4	4	4	4	4	4	4

PAIRING :	12	12	11	11	11	12	12	11	11	11
LOOP SIZE :	1	4	4	4	3	3	4	2	1	6
POSITION :	287	313	325	347	348	358	366	371	377	381
PA. ERROR :	4	4	3	4	4	4	4	4	4	4

PAIRING :	11	12	12	13	11	14	11	11	11	18
LOOP SIZE :	2	5	2	2	2	1	3	0	1	1
POSITION :	3362	3388	3443	3476	3504	3510	3563	3581	3591	3635
PA. ERROR :	4	3	2	4	4	4	4	3	3	4

PAIRING :	11	11	11	11	11	11	11	13	11	11
LOOP SIZE :	1	2	0	3	1	3	2	5	5	2
POSITION :	3637	3653	3666	3721	3762	3806	3829	3868	3901	3926
PA. ERROR :	3	3	4	4	4	3	4	4	4	4

PAIRING :	12	11	17	11	11	13	11	11	11	12
LOOP SIZE :	0	1	1	0	2	3	1	4	5	2
POSITION :	3946	3962	4003	4013	4036	4049	4066	4068	4113	4114
PA. ERROR :	4	4	3	4	4	4	4	4	4	4

Appendix G, b (continued)

PAIRING	11	11	15	11	11	11	11	11	12	15
LOOP SIZE	4	4	4	3	4	3	5	5	5	6
POSITION	416	426	427	434	438	459	514	543	544	565
PA. ERROR	4	4	4	4	3	4	4	4	4	4

PAIRING	11	13	15	11	11	13	11	11	13	12
LOOP SIZE	0	4	4	4	0	5	6	6	2	1
POSITION	565	586	674	675	677	681	682	694	714	736
PA. ERROR	3	4	4	4	4	4	4	4	4	3

PAIRING	11	11	14	12	15	11	11	11	15	13
LOOP SIZE	3	4	0	1	4	4	3	2	1	3
POSITION	761	765	772	784	796	796	799	813	824	828
PA. ERROR	4	4	4	4	4	4	3	4	4	4

PAIRING	11	15	13	12	11	11	11	11	11	11
LOOP SIZE	0	1	1	2	5	5	4	0	4	2
POSITION	832	834	840	843	884	891	894	903	903	912
PA. ERROR	4	4	4	4	3	4	2	4	4	4

PAIRING	14	11	11	16	11	11	11	12	11	12
LOOP SIZE	4	3	1	4	5	2	2	3	5	1
POSITION	934	962	975	980	983	984	986	998	1005	1008
PA. ERROR	4	4	4	4	4	4	4	4	4	3

PAIRING	11	11	11	11	13	12	11	11	13	14
LOOP SIZE	6	1	1	1	1	0	2	1	1	0
POSITION	1011	1028	1031	1045	1060	1069	1071	1083	1118	1134
PA. ERROR	3	2	4	4	4	4	4	4	4	4

PAIRING	13	13	11	13	13	14	12	11	11	15
LOOP SIZE	1	3	5	0	5	4	3	3	5	4
POSITION	1142	1146	1177	1184	1195	1221	1269	1290	1296	1319
PA. ERROR	4	4	4	4	4	3	4	4	4	3

PAIRING	12	11	12	11	11	11	11	16	16	12
LOOP SIZE	4	4	2	4	1	0	3	3	5	3
POSITION	1320	1360	1379	1386	1397	1399	1400	1405	1410	1414
PA. ERROR	4	4	4	4	4	4	4	3	4	4

PAIRING	13	12	11	12	12	11	11	12	11	11
LOOP SIZE	3	4	3	3	3	2	2	3	0	6
POSITION	4123	4141	4187	4189	4197	4227	4233	4251	4295	4316
PA. ERROR	4	4	3	4	3	3	4	4	3	3

PAIRING	12	13	11	11	11	11	14	12	12	11
LOOP SIZE	1	3	0	0	5	5	1	2	3	4
POSITION	4432	4452	4497	4499	4499	4621	4636	4638	4650	4738
PA. ERROR	4	4	4	4	3	4	4	3	4	4

PAIRING	11	11	12	12	12	11	11	13	11	11
LOOP SIZE	4	3	1	1	4	5	2	0	0	1
POSITION	4744	4756	4782	4784	4787	4811	4825	4840	4846	4861
PA. ERROR	4	4	4	4	4	4	4	3	4	4

PAIRING	11	11	13	12	12	11	11	12	11	13
LOOP SIZE	3	1	1	6	4	6	4	2	5	3
POSITION	4927	4933	4936	4944	4946	4967	4969	4984	5033	5093
PA. ERROR	3	4	3	4	4	4	4	2	2	3

PAIRING	12	12	11	11	14	12	11	11	11	15
LOOP SIZE	4	2	6	6	4	5	2	3	6	4
POSITION	5119	5129	5146	5225	5226	5248	5257	5263	5322	5342
PA. ERROR	4	4	4	3	4	4	4	4	4	2

PAIRING	11	11	11	11	13	11	13	11	13	11
LOOP SIZE	2	3	0	3	3	3	3	3	6	5
POSITION	5366	5370	5381	5381	5447	5452	5487	5506	5529	5530
PA. ERROR	3	4	4	4	1	4	4	4	4	4

PAIRING	11	11	15	11	13	11	11	12	12	12
LOOP SIZE	1	4	2	4	3	2	2	2	6	1
POSITION	5574	5599	5602	5644	5664	5685	5725	5774	5788	5811
PA. ERROR	4	4	4	4	2	4	3	4	4	4

PAIRING	12	13	11	11	11	11	11	12	11	12
LOOP SIZE	4	1	3	3	3	1	1	6	4	3
POSITION	5825	5828	5867	5879	5891	5935	5959	5986	5989	6011
PA. ERROR	4	4	3	4	4	3	4	4	4	4

Appendix G, b (continued)

PAIRING :	11	11	13	13	11	11	11	11	12	11
LOOP SIZE :	2	3	0	1	3	2	6	2	5	5
POSITION :	1418	1422	1434	1448	1454	1491	1496	1508	1536	1563
PA. ERROR :	4	4	4	4	4	4	4	4	4	4

PAIRING :	12	12	11	11	11	11	11	11	11	12
LOOP SIZE :	4	6	3	2	1	2	0	1	6	1
POSITION :	1591	1610	1611	1614	1664	1668	1783	1800	1822	1829
PA. ERROR :	4	4	3	4	3	4	3	4	4	4

PAIRING :	11	11	14	11	12	11	11	15	12	12
LOOP SIZE :	5	5	5	1	1	5	2	2	5	3
POSITION :	1837	1885	1902	1913	1928	1929	1983	1990	2101	2110
PA. ERROR :	4	4	4	4	4	4	4	4	4	4

PAIRING :	11	11	12	12	11	13	11	12	12	11
LOOP SIZE :	0	0	0	4	5	3	4	6	0	2
POSITION :	2111	2156	2180	2231	2234	2259	2260	2271	2279	2287
PA. ERROR :	4	4	4	4	4	4	4	4	4	3

PAIRING :	13	11	13	12	11	11	11	11	11	11
LOOP SIZE :	4	6	3	3	4	4	4	1	1	2
POSITION :	2367	2375	2406	2417	2438	2456	2481	2487	2492	2542
PA. ERROR :	4	4	4	4	4	4	4	3	3	3

PAIRING :	11	12	14	12	11	11	11	12	11	12
LOOP SIZE :	2	6	4	5	1	3	1	2	2	3
POSITION :	2584	2601	2624	2630	2647	2681	2716	2727	2734	2745
PA. ERROR :	3	4	4	4	4	3	4	4	4	4

PAIRING :	12	12	11	12	11	12	11	11	11	11
LOOP SIZE :	1	5	3	3	4	5	1	1	3	3
POSITION :	2759	2775	2796	2829	2848	2862	2872	2889	2903	2999
PA. ERROR :	4	2	4	4	4	4	4	4	4	4

PAIRING :	15	11	14	11	11	12	14	11	12	11
LOOP SIZE :	0	4	4	3	3	1	4	1	1	5
POSITION :	3005	3006	3043	3132	3135	3196	3200	3231	3246	3350
PA. ERROR :	4	4	4	4	4	4	4	3	2	3

PAIRING :	11	11	11	11	12	11	11	11	15	11
LOOP SIZE :	3	0	6	3	3	0	0	5	4	6
POSITION :	6019	6034	6077	6136	6143	6156	6175	6179	6216	6252
PA. ERROR :	4	4	4	3	4	4	4	4	3	4

PAIRING :	11	12	11	11	13	11	11	11	12	13
LOOP SIZE :	2	2	2	4	3	1	5	1	1	5
POSITION :	6268	6307	6347	6369	6372	6412	6417	6463	6496	6505
PA. ERROR :	4	4	4	4	3	4	4	4	4	4

PAIRING :	12	11	11	11	11	12	14	12	11	11
LOOP SIZE :	6	4	2	5	2	1	3	5	2	2
POSITION :	6511	6537	6541	6543	6591	6606	6609	6623	6663	6677
PA. ERROR :	4	4	4	3	4	4	3	3	4	4

PAIRING :	12	13	11	11	12	11	11	11	12	12
LOOP SIZE :	3	2	3	5	2	4	1	3	3	2
POSITION :	6699	6726	6789	6802	6845	6848	6871	6907	6911	6964
PA. ERROR :	3	4	4	4	4	4	3	4	4	4

PAIRING :	11	14	11	12	11	11	11	12	11	11
LOOP SIZE :	1	1	3	1	6	1	3	3	4	1
POSITION :	7022	7083	7087	7166	7172	7233	7263	7277	7302	7309
PA. ERROR :	4	4	4	4	4	4	4	4	4	3

PAIRING :	11	11	11	12	12	13	11	11	11	11
LOOP SIZE :	1	1	5	3	3	5	3	1	0	0
POSITION :	7323	7341	7355	7398	7406	7417	7432	7439	7482	7483
PA. ERROR :	4	3	4	4	4	4	4	4	4	4

PAIRING :	11	15	11	11	11	14	11	11	11	11
LOOP SIZE :	3	3	2	0	1	1	1	1	5	1
POSITION :	7509	7545	7546	7566	7583	7589	7597	7605	7666	7679
PA. ERROR :	4	4	4	4	3	4	4	4	4	4

PAIRING :	12	11	11	13
LOOP SIZE :	5	4	0	4
POSITION :	7789	7915	7928	8039
PA. ERROR :	4	4	4	4

The Ti Plasmids of Agrobacterium

M. Van Montagu* and J. Schell* **

1 Introduction

It is generally thought that exchanges and transfers of genetic material between different types of organisms do not occur in nature and can only be obtained through in vitro manipulations. However, in the past 10 years it has become evident that bacteria have developed various types of sexual mechanisms to promote gene transfer among taxonomically unrelated genera (*Holloway* 1979). Furthermore, there is now clear evidence that a group of bacteria can transfer genes into unrelated organisms, namely plants. Apparently some parasitic and symbiotic situations represent a domain where the strength of evolution might break through the biological barriers prohibiting DNA transfers when this provides a selective advantage to the parasite. The crown gall formation by *Agrobacterium tumefaciens* is the first example of such a situation. These bacteria have evolved a process in which a DNA segment (T-DNA) present in an *A. tumefaciens* plasmid (Ti plasmid) becomes inserted into the plant nuclear DNA (*Chilton* et al. 1977; *De Beuckeleer* et al. 1978, 1981; *Schell* et al. 1979; *Thomashow* et al. 1980a, 1980b; *Zambryski* et al. 1980; *Lemmers* et al. 1980).

* Laboratorium voor Genetica, Rijksuniversiteit Gent, K.L. Ledeganckstraat 35, B-9000 Gent (Belgium)
** Max-Planck-Institut für Züchtungsforschung, D-5000 Köln 30 (Vogelsang) (F.R.G.)

Crown gall is a neoplastic disease of plants. The causative agents of these plant tumors are a group of gram-negative soil bacteria, belonging to the genus *Agrobacterium*. Upon wounding and infection of plants, *Agrobacteria* can transform plant cells into autonomously growing tumor cells. The disease is very widespread in nature, affecting most dicotyledonous plants. Crown gall formation on monocotyledonous plants has never been convincingly documented (*De Cleene* and *De Ley* 1976). That bacteria are responsible for the crown gall disease was already firmly established in 1911 (*Smith* et al. 1911). However, the elucidation and understanding of the precise role of the bacteria in bringing about this phenomenon took a very long time.

There have been several major steps in the development of our understanding of crown gall. *Braun* and coworkers established that the tumorous character of crown gall tissues could be maintained indefinitely in vitro in the absence of the bacteria (*Braun* and *White* 1943), thus focusing attention on a hypothetical tumor-inducing principle produced by the bacterium, responsible for the induction, if not for the maintenance, of the tumorous condition of the crown gall plant cells (*Braun* 1943; *Braun* and *Mandle* 1948).

The observations of *Morel* and his collaborators at Versailles further indicated the importance of the bacteria in the crown gall phenomenon. These authors (*Petit* et al. 1970) demonstrated that there are at least two different forms of pathogenic *Agrobacteria*. The distinction was based on the metabolism of two arginine derivatives, octopine and nopaline, that had previously been detected in crown gall tissue. The type of arginine derivative synthesized in the tumor was found to be specified by the particular strain of the tumor-inducing bacteria and to be independent of the host plant on which the tumors were induced. Furthermore, these authors demonstrated that *Agrobacterium* strains which induce the synthesis of octopine in crown galls can selectively use this product, but not nopaline, as sole source of energy, carbon and/or nitrogen, whereas *Agrobacterium* strains that induce the synthesis of nopaline in crown galls can selectively use it, but not octopine, as energy, carbon, and/or nitrogen source. These properties have been exploited as taxonomic criteria.

More recently, other classes of opines have been identified, such as a condensation product between glutamine and mannose called agropine (*Firmin* and *Fenwick* 1978; *Tempé* et al. 1980), and phosphorylated sugar derivatives called agrocinopines (*Ellis* and *Murphy* 1981). Agropine was isolated from octopine tumors but is not detected in nopaline tumors. Agrocinopines were detected in nopaline tumors but are not found in octopine tumors. Some *Agrobacterium* strains incite tumors which do not contain octopine or nopaline, but which do contain agropine and a special class of agrocinopines, and have been called null type or agropine strains (*Guyon* et al. 1980; *Ellis* and *Murphy* 1981). They are considered to form a third class of *A. tumefaciens* strains.

The next critical step in the identification of the elusive tumor-inducing principle was the discovery in our laboratory of the Ti plasmids and the demonstration of their involvement in oncogenicity (*Zaenen* et al. 1974; *Van Larebeke* et al. 1974, 1975). This discovery was soon confirmed and extended by others (*Watson* et al. 1975; *Kerr* and *Roberts* 1976).

Prior to the discovery of Ti plasmids, *Kerr* (1969, 1971) had observed that oncogenicity could be transferred from one strain of *Agrobacterium* to another by inoculating both strains together onto the same plant. Transfer of oncogenic potential was readily observed if the nononcogenic acceptor was kept on the crown gall for several weeks. It

was therefore assumed that oncogenicity was somehow correlated with an infectious entity.

Once efficient transfer of Ti plasmids between *Agrobacterium* strains became possible, it was demonstrated that the Ti plasmids determine not only the capacity for tumor induction, but also the capacity specifically to induce opine synthesis in the transformed plant cells and the capacity of the *Agrobacterium* strains harboring the Ti plasmids to catabolize the opines (*Bomhoff* et al. 1976; *Kerr* and *Roberts* 1976; *Kerr* et al. 1977; *Van Larebeke* et al. 1977; *Genetello* et al. 1977; *Montoya* et al. 1977; *Schell* and *Van Montagu* 1977; *Hooykaas* et al. 1977; *Holsters* et al. 1978; *Klapwijk* et al. 1978). Progress in molecular biological techniques such as the development of methods for cloning bacterial and plant genomes and the Southern *(Southern* 1975) and Northern (*Alwine* et al. 1977) blotting techniques for identification of Ti-DNA and Ti-encoded mRNAs has allowed rapid accumulation of information on the structure and function of the Ti plasmids. As a result it is possible to identify some crucial steps in the tumor induction and to understand the advantages which these Ti plasmids confer on the bacteria. It is also possible to take advantage of this natural transfer of DNA from a bacterial into a plant genome and to develop methods for genetic engineering in plants.

2 Structure of Ti Plasmids

2.1 Physical Structure

All Ti plasmids studied thus far are large (around 200 kb), low copy plasmids. They are conjugative plasmids which in natural isolates are nearly always repressed for autotransfer (*Genetello* et al. 1977; *Kerr* et al. 1977). The opines which are catabolites for the plasmid carrying bacteria are inducers for Ti plasmid transfer functions (*Petit* et al. 1978; *J. Ellis*, personal communication). Efficient methods have been developed both for the preparative isolation of these large plasmids and for analytical rapid screening for the presence of supercoils of the size of Ti plasmids in different isolates (*Casse* et al. 1979; *Hirsch* et al. 1980; *Eckhardt* 1978).

Most restriction enzyme digest to Ti plasmids yielded too many fragments to order them in a physical map by conventional double digest techniques. A map could be constructed for some rare enzymes which formed a limited number of fragments (*Chilton* et al. 1978b) by filter hybridization of fragments isolated from gels and radioactively labelled in vitro. For the construction of more extensive maps the Ti plasmids were first cloned as a set of overlapping *Hin*dIII fragments obtained by partial digests. By assembling the mapping data obtained with the individual clones, physical maps of the whole Ti plasmid were readily produced (*Depicker* et al. 1980; *De Vos* et al. 1981). These maps provided the basis for the localization of insertion and deletion mutants in the Ti plasmid (see Sect. 2.2) and for determining the homology between octopine and nopaline Ti plasmids in detail.

It was indeed intriguing to determine how closely related these two classes of plasmids are. Solution hybridization data (Cot curves) suggested that the homology was not very extensive (*Currier* and *Nester* 1976). Reciprocal hybridization of restriction fragments derived from the T-region of octopine and nopaline plasmids indicated that an important part of the plasmid DNA transferred to plants (T-DNA) was highly conserved

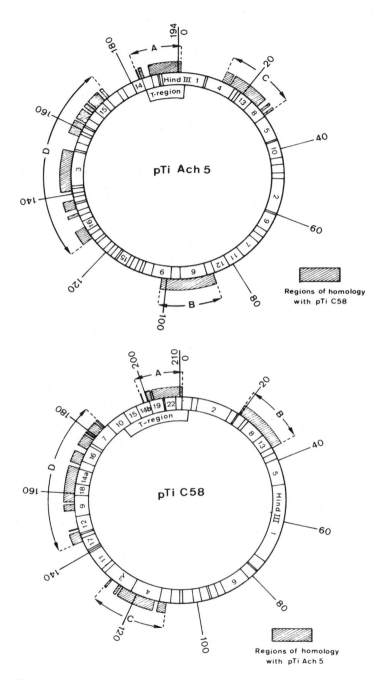

Fig. 1.a) Physical map of an octopine Ti plasmid, pTiAch5. b) Physical map of a nopaline Ti plasmid, pTiC58. The *shaded areas* on both maps indicate the regions of homology between these two plasmids. The four corresponding regions have been indicated by the letters *A, B, C* and *D*

(*Depicker* et al. 1978; *Chilton* et al. 1978a). In one area of the T-DNA, the analyzed restriction sites were conserved perfectly; a common *Sma*I site was chosen as zero point coordinate for the orientation of both the octopine and nopaline Ti plasmid maps (*Depicker* et al. 1980). More detailed information about regions of homology between octopine and nopaline Ti plasmids was obtained through electron microscope heteroduplex studies (*Engler* et al. 1981). This work demonstrated that roughly 30% of DNA contained in the two plasmids is common and is restricted to four major stretches of homology. Two of these homology regions are distributed in the same relative order as compared to a common reference point, and two are inverted.

A summary of these results is represented in Fig. 1. Several research groups are now determining the nucleotide sequence of large segments of the T-DNAs, so the exact degree of sequence conservation and the nature of the differences will soon be available. So far the sequence analysis of the regions encoding respectively for the nopaline synthase (*S. Stachel*, personal communication) and for the octopine synthase (*J. Seurinck* and *H. De Greve,* personal communication) have been completed. No homology in DNA sequence was observed.

2.2 Functional Maps

2.2.1 Transposon Insertion Mutagenesis

At the time of the discovery of the Ti plasmids, it was unrealistic to hope to identify the Ti-encoded genes promptly. Fortunately, both transposon insertion mutagenesis and DNA cloning techniques became available.

Insertion of an antibiotic resistance transposon is a very efficient way to tag a plasmid with a selectable marker. This is also a particularly efficient method for mutagenesis since no selection for a mutant phenotype has to be devised. It is sufficient to select directly for a plasmid carrying the transposon. Since the transposon is likely to be inserted into a functional DNA sequence, it follows that most plasmids harboring a transposon will be mutants. In principle, an approximate location of the insertion site can be determined by isolating the mutant plasmid on a preparative scale and analyzing the fragmentation pattern after digestion with different restriction endonucleases. A more efficient method is Southern blot hybridization with a filter containing DNA fragments obtained from a restriction enzyme digest of total *Agrobacterium* DNA (chromosomal plus plasmid DNA). This filter can be hybridized with different probes, such as the pure Ti plasmid, DNA from the drug transposon, and cloned Ti-DNA fragments. The latter probe was used to confirm the identification of the restriction fragment in which the insertion had occurred (*Dhaese* et al. 1979). A more precise localization may be obtained by analyzing in the electron microscope the heteroduplexes formed between mutant Ti plasmids and the corresponding cloned Ti-DNA segments. The exact location of the mutation can best be obtained by reisolation of the segment of the Ti plasmid which contains the insert. The drug resistance marker of the transposon can be used as a selectable marker to identify the cloned DNA fragment containing the transposon insert (*Holsters* et al., in preparation). These cloned fragments are the starting material for detailed mapping or DNA sequencing analysis.

The most important conclusion of this work was that the non-T-DNA part of the Ti

plasmid also contained extensive regions essential for tumor induction. These so-called Onc regions seem to be conserved among nopaline and octopine plasmids, since Onc⁻ insertion mutants were mapped in one of the four areas of homology mentioned above and represented in Fig. 1. Some of these regions possibly encode functions essential for the transfer of the T-DNA into the plant nucleus. Other functions may interfere with the balance of growth factors (plant hormones) of the infected tissue and therefore be essential for the stimulation of cell proliferation. It has been shown that *Agrobacterium tumefaciens* strains excrete transzeatin and that this production is Ti plasmid determined (*Kaiss-Chapman* and *Morris* 1977; *Claeys* et al. 1978; *McCloskey* et al. 1980). Ti plasmid deletion and insertion mutants which no longer mediate transzeatin synthesis confer an Onc⁻ phenotype (*R. Morris* and *E. Messens,* personal communication). With the same approach, transposon insertion mutants were isolated in most of the functions which were known to be Ti plasmid-determined (*Van Montagu* and *Schell* 1979; *Hooykaas* et al. 1979; *Holsters* et al. 1980; *Ooms* et al. 1980, 1981; *Garfinkel* and *Nester* 1980; *De Greve* et al. 1981). A large segment of Ti plasmid DNA is involved in the determination of catabolic functions. Ti plasmids can therefore be considered as a special type of catabolic plasmids. Some of the compounds catabolized by *Agrobacterium* under the control of Ti plasmid genes are not normally "found" in nature, but rather the *Agrobacterium* strains, via the Ti plasmid, oblige the transformed plants to synthesize the compounds which only they are able to utilize. Crown gall cells release these compounds into the rhizosphere, and bacteria able to utilize them have a selective advantage over other soil bacteria. This new type of parasitism can therefore be seen to be the result of a "genetic colonization" (*Schell* et al. 1979). While some of these catabolic functions can be mapped to single sites, other functions are more complex and many independent loci are involved.

2.2.2 Site-Specific Mutagenesis

Transposon insertion mutagenesis of Ti plasmids allowed a rudimentary localization of some relevant loci (*Holsters* et al. 1980; *Ooms* et al. 1980, 1981; *Garfinkel* and *Nester* 1980; *De Greve* et al. 1981). Due to a high site or regional specificity, transposons cannot be expected to integrate in all the genes of the Ti plasmid. By combining in vitro and in vivo recombinant DNA techniques, it is possible to efficiently mutate a predetermined DNA segment (*Heffron* et al. 1970; *Green* and *Tibbets* 1980; *Shortle* et al. 1980). Through the development of efficient cloning techniques, it is straightforward to isolate and order a complete clonal bank of any plasmid (*Collins* and *Brüning* 1978).

Site-specific mutagenesis of isolated fragments of DNA followed by exchange, through in vivo recombination, of the mutated DNA segment with the original plasmid or prokaryotic genome, allows the construction of mutants and hence identifies and delimits genes under study (*Ruvkun* and *Ausubel* 1981; *Leemans* et al. 1981a).

An extensive program of site-specific mutagenesis of cloned subfragments of the T-region of Ti plasmids is now under way. In an initial phase well-defined insertions and deletions were constructed in cloned T-DNA fragments using identified restriction sites as endpoints. To achieve the replacement of wild-type Ti plasmid sequence, by an altered DNA fragment, a cloning vector containing this altered fragment was transmitted from an *Escherichia coli* host to an *Agrobacterium* strain harboring a transfer constitutive derivative of the Ti plasmid from which the altered fragment was originally derived. Two consecutive conjugations readily allowed the isolation of the required Ti plasmid: the

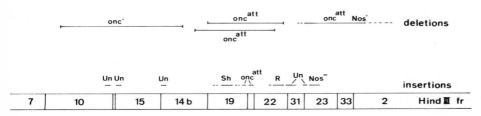

Fig. 2a, b. The following abnormal phenotypes were observed in tumors induced by mutant Ti plasmids with specific deletions or insertions. a) Octopine Ti plasmid mutants; b) nopaline Ti plasmids. *Un*, phenotype of tumors unaltered; *onc*, no tumor formation; *onc*^{att} attenuated tumor formation; *sh*, shoot induction; *R*, root induction; *Ocs*$^-$, no octopine synthase activity; *Nos*$^-$, no nopaline synthase activity.

first was followed by selection for markers of the cloning vector and resulted in a cointegrate of the two plasmids, and the second was followed by screening and selection for the loss of the markers of the cloning vector.

Alternatively, in vivo mutagenesis of cloned T-DNA fragments has been performed by using the insertion ability of movable elements. This type of insertion can be detected when a plasmid containing a cloned T-DNA fragment is mobilized from one bacterium to another by a autotransferable plasmid which is able to form IS-mediated fused replicons (*Leemans* et al. 1981b; *Hernalsteens* et al., in preparation). One practical advantage of the in vivo method is that many independent insertions can be isolated through a single replica conjugation. For this in vivo method the transposons Tn*1*, Tn*601* and the IS*8* were used extensively.

2.2.3 T-DNA Encoded Functions

The Ti plasmids containing mutants in the T-DNA obtained by these different approaches of site-specific mutagenesis were used for tumor induction on different test plants (tobacco, potato, kalanchoe). Preliminary results using these mutants are summarized in Fig. 2 and indicate the localization of the octopine and nopaline synthase genes and of several regions involved in the control of tumor morphology, host range, and organ morphogenesis (see Sect. 3). Some mutants did not show an altered pheno-

type, suggesting that some T-DNA functions are not essential for tumor induction and maintenance or for the pattern of cell proliferation; these mutants may intervene in the synthesis of a not yet identified opine. Only extensive deletions of the T-DNA were Onc⁻, so that no specific function involved in tumor maintenance has been identified. In fact the available evidence indicates strongly that several different functions are involved in tumor induction and maintenance.

The stable maintenance of the T-DNA in the nucleus of transformed plant cells is probably the direct consequence of its insertion into plant nuclear DNA (*Willmitzer* et al. 1980; *Chilton* et al. 1980; *Yadav* et al. 1980; *Zambryski* et al. 1980). It is not known whether the functions responsible for this integration are encoded by the T-DNA itself, by some other region of the Ti plasmid, or by bacterial or plant functions.

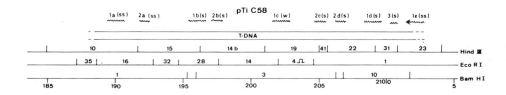

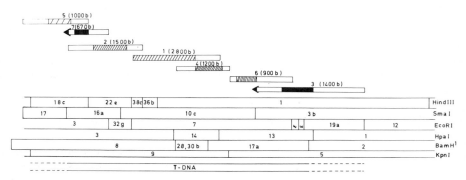

Fig. 3. a) Tentative location on a physical map of the T-region of pTiC58 of poly-A containing transcripts derived from the T-DNA of a nopaline tumor line of tobacco (BT 37). The location of the transcripts based on Northern analysis (*L. Willmitzer*) and cDNA mapping (*L. Willmitzer, P. Dhaese*, unpublished results) is indicated by ~ the relative abundancy of the different RNAs is indicated in parentheses (*ss* ≜ high; *s* ≜ middle; *w* ≜ low). The transcripts fall into mainly three different size classes (1, 0.6 kb; 2, 0.4 kb; 3, 0.3 kb). Due to this fact the location of the different transcripts is still very tentative and does not necessarily represent the final picture. b) Location on a physical map of the T-region of pTiAch5 of seven different poly-A containing transcripts derived from the T-DNA of the octopine tobacco tumor line A6-S1 based on Northern analysis (*L. Willmitzer*, in preparation). The size of the different transcripts in nucleotides is given in parenthese. The relative abundancy of the different transcripts is proportional to the *intensity of shading*. The *shaded area* indicates the most probable location of the different transcripts assuming the absence of splicing, the uncertainty still left for the exact location of the different transcripts is indicated by the *nonshaded area* of each box. The *black arrows* on transcripts 7 and 3 define the direction of the transcription

The determination of the tumorous mode of growth can best be explained by a model whereby T-DNA functions actively prevent normal differentiation. Indeed, various Ti plasmids with an altered T-region induce a pattern of transformed cell growth which is significantly different from that induced by wild-type plasmids. Thus, one has Ti plasmid mutants that produce tumors which proliferate either shoots (Fig. 2, sh) or roots (Fig. 2, R). Each of these mutant phenotypes is produced by mutations of a well-defined locus of the T-region of Ti plasmids. It is conceivable that the functions encoded by these loci directly or indirectly interfere with the level of plat growth factors. The present model is supported by the observation (*De Greve* et al., in preparation; *Otten* et al. 1981) that particular Ti plasmid mutants produce transformed cells, some of which can regenerate and form completely normal and fertile tobacco plants (see also Sect. 4.2). Other laboratories also observed that deletion of part of the T-DNA can result in the formation of fertile or partially fertile plants regenerating from tumors (*Wullems* et al. 1981; *Nester* et al. 1981). The observations definitively demonstrate that T-DNA transfer and tumorous growth can be separated and must therefore be controlled by different functions of the Ti plasmid (see also Sect. 3).

Another approach to the study of the functions encoded by T-DNA is to analyze its transcripts in crown gall tissue. Several such studies have been published (*Drummond* et al. 1977; *Gurley* et al. 1979; *Yang* et al. 1980; *Willmitzer* et al. 1981), and these results indicate that the T-DNA present in crown gall cells is transcribed by the α-amanitin sensitive RNA polymerase II over nearly its entire length (*Willmitzer* et al. 1981; *Willmitzer* et al., in preparation). Well-defined regions of the T-DNA are reproducibly transcribed actively, whereas other parts are only weakly transcribed. No extensive differences were found between nuclear transcripts and polysomal transcripts (*Willmitzer* et al. 1981). This might indicate that extensive processing steps do not occur in the formation of the polysomal RNA.

In most crown gall tissues analyzed thus far the level of nopaline synthase mRNA or of octopine synthase mRNA was about 50 times higher than that of the mRNAs transcribed from the other areas of the T-DNA. Northern blot analyses have been obtained both for octopine (*S. Galvin*, personal communication; *Willmitzer* et al. 1981) and for nopaline (*M. Bevan*, personal communication; *L. Willmitzer*, personal communication) crown gall cells. Some of these results are presented in Fig. 3.

Finally, the position of the nopaline synthase (Nos) and octopine synthase mRNA has been localized accurately by determining the DNA sequences from the corresponding area and by establishing the position of the mRNA by S1 mapping (*S. Stachel*, personal communication; *J. Seurinck* and *P. Dhaese*, personal communication). Both mRNAs are more eukaryotic than prokaryotic in their recognition signals. For example, they do not contain a *Shine* and *Dalgarno* sequence and an open reading frame starts at the first AUG codon after the 5′ terminus of the messenger. The open reading frames are not interrupted from initiation to termination codon, thus indicating that no introns interrupt the octopine and nopaline synthase structural genes. In vitro translation product of the octopine mRNA was shown to be immunoprecipitable with antiserum raised against the octopine synthase isolated from crown gall cells (*Schröder* et al. 1981).

3 Crown Gall Formation

3.1 The Transformed State

The finding that *Agrobacterium tumefaciens* strains have the capacity to insert stably one or several copies of DNA segments of 15–23 kb into a plant genome opened the crown gall field to researchers other than plant pathologists. Molecular and developmental biologists became interested in the mechanism of DNA transfer and integration and in the nature of the functions which stimulate neoplasmic proliferation or teratoma formation. Others became attracted by the potentials of this system for genetic engineering in plants.

One of the central questions is the relation of the T-DNA transfer and tumor induction, i.e., does each T-DNA transfer event lead to cell proliferation and does every proliferating cell contain T-DNA? The latter is not the case. It is well known that spontaneous tumors can arise in a variety of plants, and interspecific hybrids seem to be particularly prone to these neoplastic growths (*Smith* 1972). These tumors have been termed genetic tumors and are often difficult to distinguish from crown galls. With many plants it is possible to isolate habituated cells from cell cultures which have never been in contact with *Agrobacterium* or with crown gall tumors (*Binns* and *Meins* 1973). Habituated cells can grow without addition of growth factors. They differ from crown gall cells by the fact that they do not contain any *Agrobacterium* DNA sequences, with the result that they do not synthesize opines. One might therefore imagine that the T-DNA per se is not essential for tumor induction by *A. tumefaciens,* provided the bacteria activate a mechanism which creates the same type of growth factor imbalance as for the cases mentioned above.

The other question, whether or not it is possible to have T-DNA transfer without crown gall formation, has more practical implications, and the answer may be positive. Recent results (see Sect. 4) with plant regenerates demonstrate that plant cells may contain a large portion of the T-DNA, including the left and right borders, yet still grow and develop in a fully normal way. If the deletion found in the T-DNAs of these plant . regenerates could be introduced in the T-region of a Ti plasmid, it is not unlikely that this type of Ti plasmid would still be able to mediate the transfer of the mutated T-DNA. However, it may be difficult to identify and isolate the cells in which this T-DNA becomes integrated. Indeed, there is no isolated DNA yet which could be inserted into the T-DNA to provide a marker for direct selection of T-DNA containing cells without depending on the tumorous growth characteristics. Possibly conditional tumor growth could be exploited. Strains containing some pTi Onc^{att} mutants are no longer weakly oncogenic when growth hormones are added externally (unpublished results of this laboratory). Perhaps addition of growth hormones allows temporary development of a tumor which would revert to normal shoots upon hormone withdrawal. This situation prompts us to make a distinction in the uses of the expression "transformed plant cell". Oncologists use it to stress certain types of abnormalities in cell growth and development, whereas microbiologists use it to indicate that foreign DNA sequences have been newly inserted into a cell. Until now, crown gall formation was thought to include both types of transformation.

Thus we propose to reserve the term Onc⁻ for two classes of mutants that are not defective for integration functions but either cannot transfer Ti plasmid DNA to plant cells

or carry mutations in the T-region such that the T-DNA, although present in the "transformed" cells, does not stimulate tumorous growth. The terms weakly oncogenic and attenuated (Onc^{att}) or altered oncogenicity are used if small to barely visible tumors develop with the standard assay on tobacco seedlings (see Sect. 3.2) or if a limited number of small tumors are formed on the potato slice assay (*Anand* and *Heberlein* 1977).

These Onc^{att} mutants will (for reasons discussed in Sect. 3.2) often display a more restricted host range of plants susceptible to tumor induction. The term host range mutant (Onc^h), which has been used in the past (*Van Montagu* and *Schell* 1979), is interchangeable with Onc^{att}.

Mutants which allow transfer of the T-region to the plant cell but which no longer integrate this DNA might be called Int$^-$. It is likely that the Onc$^-$ and Onc^{att} (or Onc^h) phenotype may be caused by mutations either in the Ti plasmid or in the bacterial host genome. It is well known that many *Agrobacterium* auxotrophs are no longer able to establish themselves in a plant wound; such mutants would be characterized as chromosomal Onc$^-$.

The Int$^-$ mutants are most likely to be Ti plasmid mutants. Nevertheless, it is conceivable that a mutation in the bacterial chromosome might result in the transfer of Ti plasmid DNA into the plant cell in such a form that the T-region can no longer be integrated. Finally, it is conceivable that plant cell functions are also involved in the integration mechanism.

3.2 Host Range

The susceptibility of different plants to crown gall has been extensively analyzed (*De Cleene* and *De Ley* 1976). Some gymnosperms and the majority of the dicotyledon angiosperms tested are suceptible to the standard *Agrobacterium* strains. The monocotyledonous plants do not show tumor formation. As mentioned above, mutants can be obtained which are much more restricted in their ability to induce crown gall on some plants. There are natural isolates of *Agrobacterium* which also display such a limited host range and in several cases it was shown that this was a Ti plasmid-encoded property (*Loper* and *Kado* 1979; *Thomashow* et al. 1980c). Many pTi Onc^{att} *Agrobacteria* can induce tumors upon external addition of growth hormones, but wild-type. *A. tumefaciens* will no longer induce tumors upon external addition of the same growth hormones. These findings strongly indicate that there are optimal concentrations of growth hormones for tumor induction (unpublished results of this laboratory).

It is well known, that the concentration of growth hormones is different in the various tissues and parts of a plant. It is therefore not unexpected to find that tumors can sometimes only be induced on a particular part of the plant and that the consequences of an *Agrobacterium* inoculation can vary according to the age and growth conditions of the plant. Hence, comparisons of the relative tumor-inducing capacity of different *A. tumefaciens* strains should be done under strictly standardized conditions. Tobacco seedlings which have been germinated and grown under sterile conditions are the most frequently used test material (*Leemans* et al. 1981a).

It is not unlikely that by using either the right inoculating conditions or the appropriate *Agrobacterium* mutant, it will be possible to induce crown galls on nearly all dicotyledonous plants. Monocotyledonous plants may display additional barriers to

crown gall induction, such as the inability to make the correct plant cell wall contacts necessary for Ti plasmid transfer.

Because some monocotyledonous plants are known to be partially habituated for hormone-independent growth, it is conceivable that the hormone level in the plant is already a barrier to crown gall formation. Once appropriately marked T-DNA regions are available it will be possible to ascertain whether transfer of the T-DNA can occur in plants which do normally allow crown gall development.

4 Ti Plasmids as Vectors for Genetic Engineering in Plants

From the evidence described in the previous sections, it follows that *Agrobacteria* and their Ti plasmids constitute an efficient system for DNA transfer to plants and in fact constitute a natural example of genetic engineering to the benefit of *A. tumefaciens*. In order to adopt the Ti plasmids as experimental gene vectors for plants, two modifications are important: a) the replacement of T-DNA sequences by experimentally chosen genes, and b) the inactivation of the tumor-inducing properties of the T-DNA without affecting its DNA transfer properties. Both goals have been obtained in model experiments.

4.1 Substitution of the Opine Synthase Gene

In view of the observed involvement of the "ends" of the T-region in the integration of T-DNA (*Zambryski* et al. 1980), one could expect that any DNA segment inserted between these ends would be cotransferred, provided no function essential for T-DNA transfer and stable maintenance was inactivated by the insertion. The genetic analysis of the T-region by transposon insertion provided us with Ti plasmid mutants suited to test this hypothesis (see Sects. 2.2.1, 2.2.2). A Tn7 insertion in the nopaline synthase locus produced a Ti plasmid able to initiate T-DNA transfer and tumor formation. Analysis of the DNA extracted from these tumors showed that the T-region containing the Tn7 segment had been transferred as a single 38×10^3 base pair segment without any major rearrangements (*Hernalsteens* et al. 1980).

The next question was whether genes inserted via T-DNA into plant nuclei could be expressed. The expression of the Tn7 insertion in the nopaline synthase locus was assayed as a model system for other genes inserted in this area. Tn7 codes for a dihydrofolate reductase which is resistant to methotrexate (*Tennhammer-Ekman* and *Sköld* 1979). Suspension cultures of both untransformed and crown gall tobacco tissue were found to be completely inhibited by 2 μg/ml methotrexate. In contrast, a culture established with crown gall cells containing Tn7 grew well on media containing 2 μg/ml methotrexate. In addition, nuclei isolated from these methotrexate resistant tobacco lines were shown to synthesize Tn7 transcripts homologous to Tn7 and a poly A⁻ mRNA fraction derived from purified polysomes was also shown to contain Tn7 transcripts.

The size of Ti plasmids (about 200 kb) and the large number of Ti plasmid genes involved in the transformation mechanism, prevent the development of a "mini-Ti" cloning vector with unique cloning sites at appropriate locations within the T-region, and with

all functions essential for T-DNA transfer and stable maintenance. An alternative way was therefore developed to introduce genes at specified sites in the T-region of a functional Ti plasmid. The principle is to construct an "intermediate vector" consisting of a common *E. coli* cloning vehicle, for example the pBR322 plasmid, into which an appropriate fragment of the T-region of a Ti plasmid is inserted. Single restriction sites in this T-region fragment can then be used to insert chosen DNA sequences. This intermediate vector is subsequently introduced, via transformation or mobilization, into an *A. tumefaciens* strain carrying a Ti plasmid which has been made constitutive for transfer (*Leemans* et al. 1981a), and carrying antibiotic resistance markers (e.g., streptomycin, sulfonamide) cloned into the same T-region restriction site as that chosen for insertion of the DNA to be transferred. These resistance markers were introduced by a procedure essentially identical to the one described here. Recombination in vivo will transfer the DNA of interest into the appropriate site of the Ti plasmid (*Leemans* et al. 1981a).

It should be noted that the introduction of inserts and substitutions in all parts of the T-region by these methods is equivalent to site-specific mutagenesis and should allow us to probe further for the different T-DNA functions.

4.2 Can Normal Plants be Regenerated from Plant Cells Containing T-DNA?

The ultimate aim of many gene transfer attempts in plants is to produce fertile cultivars harboring and transmitting new genetic properties. It was, therefore, essential to determine whether T-DNA transfer could be dissociated from neoplastic transformation and whether normal fertile plants could be derived from plant cells transformed by T-DNA. In order to answer this question, a large set of insertion mutants were obtained with an octopine TiB6S3 plasmid (*De Greve* et al. 1981). One of the mutant plasmids (pGV2100) was clearly less oncogenic on tobacco and sunflower hypocotyls than was the wild-type plasmid. Tumors appeared only after prolonged incubation. Furthermore, shoots proliferated from the greenish tumors, in contrast to the undifferentiating white tumors induced by strains harboring the wild-type TiB6S3 plasmid. The Tn7 insertion in pGV2100 was mapped and found to be located in the left arm of the common DNA of the T-region (Fig. 2).

As a test for transformation, tobacco tumor tissue derived from infection with pGV2100 and the shoots regenerating from this tissue were assayed for the presence of octopine TiB6S3 plasmid (*De Greve* et al. 1981). One of the mutant plasmids (pGV2100) positive. Most of the shoots were negative but some of the proliferating shoots were positive. One such shoot was grown further on media free of growth hormone and was found to develop roots and later to grow into a fully normal flowering plant. Each part of this plant, leaves, stem, and roots, was found to contain lysopine dehydrogenase activity, and polysomal RNA was found to contain T-DNA transcripts homologous to the opine synthesis locus. No transcripts of the conserved segment of the T-region were observed.

These observations therefore demonstrate that normal plants can be obtained from plant cells transformed with Ti plasmids genetically altered in specific segments of the T-region. Furthermore, seeds obtained by self-fertilization of these plants produced new plants with active T-DNA linked genes; thus, genes introduced into plant nuclei via the Ti plasmid can be sexually inherited. The inheritance pattern was shown to be as expected for a single dominant Mendelian factor (*Schell* et al. 1981; *Otten* et al. 1981).

5 Conclusion

The phenomenon of crown gall formation on plants is of fundamental importance in the study of the molecular biology of plant development. The abnormal growth pattern of crown gall plant cells has been shown to be the direct consequence of the presence and expression of specific genes which have been transferred into the plant nuclei by gram-negative bacteria. These bacteria contain natural gene vectors in the form of large Ti plasmids. Because these plasmids carry genes which, when mutated directly, affect the growth and morphogenetic properties of the transformed plant cells, the crown gall system represents a model system uniquely suited to the study of the genes and gene products involved in the control of cellular growth and differentiation. The Ti plasmid, because of its gene vector properties, is also suited to the analysis, via reversed genetics, of the relation between structure and function of plant genes. Finally, the Ti plasmid can be used to introduce selected and isolated genes into plants and may therefore open new avenues for plant breeding.

Acknowledgments. The investigations reported here were supported by grants from the "Kankerfonds van de A.S.L.K", from the "Instituut tot aanmoediging van het Wetenschappelijk Onderzoek in Nijverheid and Landbouw" (I.W.O.N.L.) (248/A), from the "Fonds voor Wetenschappelijk Geneeskundig Onderzoek" (3.0052.78), and from the "Onderling Overledge Akties" (O.O.A. 12054179).

We collectively thank all the members of our laboratories for their help and contributions to this project.

References

Alwine JC, Kemp DJ, Stark GR (1977) Method for detection of specific RNA's in agarose gels by transfer to diazobenzyloxymethyl-paper and hybridization with DNA probes. Proc Natl Acad Sci USA 74:5350–5354

Anand VK, Heberlein GT (1977) Crown gall tumorigenesis in potato tuber tissue. Am J Bot 64: 153–158

Binns A, Meins F Jr (1973) Habituation of tobacco pith cells for factors promoting cell division is heritable and potentially reversible. Proc Natl Acad Sci USA 70:2660–2662

Bomhoff G, Klapwijk PM, Kester HCM, Schilperoort RA, Hernalsteens JP, Schell J (1976) Octopine and nopaline synthesis and breakdown genetically controlled by a plasmid of *Agrobacterium tumefaciens*. MGG 145:177–181

Braun AC (1943) Studies on tumor inception in crown gall disease. Am J Bot 30:674–677

Braun AC, Mandle RJ (1948) Studies on the inactivation of the tumor-inducing principle in crown-gall. Growth 12:255–269

Braun AC, White PR (1943) Bacteriological sterility of tissues derived from secondary crown gall tumors. Phytopathol 33:85–100

Casse F, Boucher C, Julliot JS, Michel M, Dénarié J (1979) Identification and characterization of large plasmids in *Rhizobium meliloti* using agarose gel electrophoresis. J Gen Microbiol 113: 229–242

Chilton M-D, Drummond HJ, Merlo DJ, Sciaky D, Montoya AL, Gordon MP, Nester EW (1977) Stable incorporation of plasmid DNA into higher plant cells: the molecular basis of crown gall tumorigenesis. Cell 11:263–271

Chilton M-D, Drummond MH, Merlo DJ, Sciaky D (1978a) Highly conserved DNA of Ti-plasmids overlaps T-DNA, maintained in plant tumors. Nature 275:147–149

Chilton M-D, Montoya AL, Merlo DJ, Drummond MH, Nutter R, Gordon MP, Nester EW (1978b) Restriction endonuclease mapping of a plasmid that confers oncogenicity upon *Agrobacterium tumefaciens* strain B6-806. Plasmid 1:254–269

Chilton M-D, Saiki RK, Yadav N, Gordon MP, Quetier F (1980) T-DNA from *Agrobacterium* Ti plasmid is in the nuclear DNA fraction of crown gall tumor cells. Proc Natl Acad Sci USA 77:4060–4064

Claeys M, Messens E, Van Montagu M, Schell J (1978) GC/MS determination of cytokinins in *Agrobacterium tumefaciens* cultures. Fresenius Z Anal Chem 290:125–126

Collins J, Brüning HJ (1978) Plasmids useable as gene-cloning vectors in an in vitro packaging by coliphage λ: "cosmids". Gene 4: 85–107

Currier TC, Nester EW (1976) Evidence for diverse types of large plasmids in tumor inducing strains of *Agrobacterium*. J Bacteriol 126:157–165

De Beuckeleer M, De Block M, De Greve H, Depicker A, De Vos R, De Vos G, De Wilde M, Dhaese P, Dobbelaere MR, Engler G, Genetello C, Hernalsteens JP, Holsters M, Jacobs A, Schell J, Seurinck J, Silva B, Van Haute E, Van Montagu M, Van Vliet F, Villarroel R, Zaenen I (1978) The use of the Ti-plasmid as a vector for the introduction of foreign DNA into plants. In: Proceedings IVth International Conference on Plant Pathogenic Bacteria, INRA, Angers, pp 115–126

De Beuckeleer M, Lemmers M, De Vos G, Willmitzer L, Van Montagu M, Schell J (1981) Further insight on the transferred-DNA of octopine crown gall. MGG 183:283–288

De Cleene M, De Ley J (1976) The host range of crown-gall. Botan Rev 42:389–466

De Greve H, Decraemer H, Seurinck J, Van Montagu M, Schell J (1981) The functional organization of the octopine *Agrobacterium tumefaciens* plasmid pTiB6S3. Plasmid 6:235–248

Depicker A, Van Montagu M, Schell J (1978) Homologous DNA sequences in different Ti-plasmids are essential for oncogenicity. Nature 275:150–153

Depicker A, De Wilde M, De Vos G, De Vos R, Van Montagu M, Schell J (1980) Molecular cloning of overlapping segments of the nopaline Ti-plasmid pTiC58 as a means to restriction endonuclease mapping. Plasmid 3:193–211

De Vos G, De Beuckeleer M, Van Montagu M, Schell J (1981) Restriction endonuclease mapping of the octopine tumor inducing pTiAch5 of *Agrobacterium tumefaciens*. Plasmid 6:249–253

Dhaese P, De Greve H, Decraemer H, Schell J, Van Montagu M (1979) Rapid mapping of transposon insertion and deletion mutations in the large Ti-plasmids of *Agrobacterium tumefaciens*. Nucleic Acids Res 7:1837–1849

Drummond MH, Gordon MP, Nester EW, Chilton M-D (1977) Foreign DNA of bacterial plasmid origin is transcribed in crown gall tumours. Nature 269:535–536

Eckhardt T (1978) A rapid method for the identification of plasmid desoxyribonucleic acid in bacteria. Plasmid 1:584–588

Ellis JG, Murphy PJ (1981) Four new opines from crown gall tumours – their detection and properties, MGG 181:36–43

Engler G, Depicker A, Maenhaut R, Villarroel-Mandiola R, Van Montagu M, Schell J (1981) Physical mapping of DNA base sequence homologies between an octopine and a nopaline Ti-plasmid of *Agrobacterium tumefaciens*. J Mol Biol 152:183–208

Firmin JL, Fenwick GR (1978) Agropine — a major new plasmid-determined metabolite in crown gall tumours. Nature 276:842–844

Garfinkel DJ, Nester EW (1980) *Agrobacterium tumefaciens* mutants affected in crown gall tumorigenesis and octopine catabolism. J Bacteriol 144:732–743

Genetello Ch, Van Larebeke N, Holsters M, Depicker A, Van Montagu M, Schell J (1977) Ti-plasmids of *Agrobacterium* as conjugative plasmids. Nature 265:561–563

Green C, Tibbetts C (1980) Targeted deletions of sequences from closed circular DNA. Proc Natl Acad Sci USA 77:2455–2459

Gurley WB, Kemp JD, Albert MJ, Sutton DW, Callis J (1979) Transcription of Ti plasmid-derived sequences in three octopine-type crown gall tumor lines. Proc Natl Acad Sci USA 76:2828–2832

Guyon P, Chilton M-D, Petit A, Tempé J (1980) Agropine in "null type" crown gall tumors: evidence for the generality of the opine concept. Proc Natl Acad Sci USA 77:2693–2697

Heffron F, So M, McCarthy BJ (1979) Insertion mutations affecting transposition of Tn3 and replication of a ColE1 derivative. Cold Spring Harbor Symp. Quant Biol 43:1279–1285

Hernalsteens JP, Van Vliet F, De Beuckeleer M, Depicker A, Engler G, Lemmers M, Holsters M, Van Montagu M, Schell J (1980) The *Agrobacterium tumefaciens* Ti plasmid as a host vector system introducing foreign DNA in plant cells. Nature 287:654–656

Hirsch PR, Van Montagu M, Johnston AWB, Brewin NJ, Schell J (1980) Physical identification of bacteriocinogenic, nodulation and other plasmids in strains of *Rhizobium leguminosarum*. J Gen Microbiol 120:403–412

Holloway BW (1979) Plasmids that mobilize bacterial chromosome. Plasmid 2:1–19

Holsters M, Silva B, Van Vliet F, Hernalsteens JP, Genetello C, Van Montagu M, Schell J (1978) In vivo transfer of the Ti-plasmid of *Agrobacterium tumefaciens* to *Escherichia coli*. MGG 163: 335–338

Holsters M, Silva B, Van Vliet F, Genetello C, De Block M, Dhaese P, Depicker A, Inzé D, Engler G, Villarroel R, Van Montagu M, Schell J (1980) The functional organization of the nopaline *A. tumefaciens* plasmid pTiC58. Plasmid 3:212–230

Hooykaas PJJ, Klapwijk PM, Nuti MP, Schilperoort RA, Rörsch A (1977). Transfer of the *Agrobacterium tumefaciens* Ti-plasmid to avirulent Agrobacteria and to *Rhizobium* ex planta. J Gen Microbiol 98:477–484

Hooykaas PJJ, Roobol C, Schilperoort RA (1979) Regulation of the transfer of Ti plasmids of *Agrobacterium tumefaciens*. J Gen Microbiol 110:99–109

Kaiss-Chapmann RW, Morris RO (1977) Trans-zeatin in culture filtrates of *Agrobacterium tumefaciens*. Biochem Biophys Res Commun 76:453–459

Kerr A (1969) Transfer of virulence between isolates of *Agrobacterium*. Nature 223:1175–1176

Kerr A (1971) Acquisition of virulence by non-pathogenic isolation of *Agrobacterium radiobacter*. Physiol Plant Pathol 1:241–246

Kerr A, Roberts WP (1976) *Agrobacterium:* correlation between and transfer of pathogenicity, octopine and nopaline metabolism and bacteriocin 84 sensitivity. Physiol Plant Pathol 9:205–211

Kerr A, Manigault P, Tempé J (1977) Transfer of virulence in vivo and in vitro in *Agrobacterium*. Nature 265:560–561

Klapwijk PM, Scheulderman T, Schilperoort RA (1978) Coordinated regulation of octopine degradation and conjugative transfer of Ti-plasmids in *Agrobacterium tumefaciens:* evidence for a common regulatory gene and separate operons. J Bacteriol 136:775–785

Leemans J, Shaw C, Deblaere R, De Greve H, Hernalsteens JP, Maes M, Van Montagu M, Schell J (1981a) Site-specific mutagenesis of *Agrobacterium* Ti plasmids and transfer of genes to plant cells. J Mol Appl Genet 1:149–164

Leemans J, Inzé D, Villarroel R, Engler G, Hernalsteens JP, De Block M, Van Montagu M (1981b) Plasmid mobilization as a tool for in vivo genetic engineering. In: Levy SB, Clowes RC, Koenig EL (eds) Molecular biology, pathogenicity and ecology of bacterial plasmids. Plenum Press, New York, pp 401–410

Lemmers M, De Beuckeleer M, Holsters M, Zambryski P, Depicker A, Hernalsteens JP, Van Montagu M, Schell J (1980) Internal organization, boundaries and integration of Ti-plasmid DNA in nopaline crown gall tumours. J Mol Biol 144:355–378

Loper JE, Kado CI (1979) Host-range conferred by the virulence-specifying plasmid of *Agrobacterium tumefaciens*. J Bacteriol 139:591–596

McCloskey JA, Hashizume T, Basile B, Ohno Y, Sonoki S (1980) Occurrence and levels of *cis*- and *trans*-zeatin ribosides in the culture medium of a virulent strain of *Agrobacterium tumefaciens*. FEBS Lett 111:181–183

Montoya A, Chilton M-D, Gordon MP, Sciaky D, Nester EW (1977) Octopine and nopaline metabolism in *Agrobacterium tumefaciens* and crown-gall tumor cells: role of plasmid genes. J Bacteriol 129:101–107

Nester EW, Garfinkel DJ, Gelvin SB, Montoya AL, Gordon MP (1981) A mutational and transcriptional analysis of a tumor inducing plasmid of *Agrobacterium tumefaciens*. In: Levy SB, Clowes RC, Koenig EL (eds) Molecular biology, pathogenicity and ecology of bacterial plasmids. Plenum Press, New York, pp 467–476

Ooms G, Klapwijk PM, Poulis JA, Schilperoort RA (1980) Characterization of Tn*904* insertions in octopine. Ti plasmid mutants of *Agrobacterium tumefaciens*. J Bacteriol 144:82–91

Ooms G, Hooykaas PJ, Moleman G, Schilperoort RA (1981) Crown gall plant tumors of abnormal morphology, induced by *Agrobacterium tumefaciens* carrying mutated octopine Ti plasmids; analysis of T-DNA functions. Gene 14:33–50

Otten LA, Schilperoort RA (1978) A rapid microscale method for the detection of lysopine- and nopaline dehydrogenase activities. Biochim Biophys Acta 527:497–500

Otten L, De Greve H, Hernalsteens JP, Van Montagu M, Schieder O, Straub J, Schell J (1981) Men-

delian transmission of genes introduced in plant by the Ti plasmids of *Agrobacterium tumefaciens*. MGG 183:209–213

Petit A, Delhaye S, Tempé J, Morel G (1970) Recherches sur les guanidines des tissus de crown gall. Mise en évidence d'une relation biochimique spécifique entre les souches d' *Agrobacterium* et les tumeurs qu'elles induisent. Physiol Vég 8:205–213

Petit A, Tempé J, Kerr A, Holsters M, Van Montagu M, Schell J (1978) Substrate induction of conjugative activity of *Agrobacterium tumefaciens*. Nature 271:570–572

Ruvkun GB, Ausubel FM (1981) A general method for site-directed mutagenesis in prokaryotes. Nature 289:85–88

Schell J, Van Montagu M (1977) On the transfer, maintenance and expression of bacterial Ti-plasmid DNA in plant cells transformed with *A. tumefaciens*. Brookhaven Symp Biol 29:36–49

Schell J, Van Montagu M, De Beuckeleer M, De Block M, Depicker A, De Wilde M, Engler G, Genetello C, Hernalsteens JP, Holsters M, Seurinck J, Silva B, Van Vliet F, Villarroel R (1979) Interactions and DNA transfer between *Agrobacterium tumefaciens,* the Ti-plasmid and the plant host. Proc R Soc Lond B 204:251–266

Schell J, Van Montagu M, Holsters M, Hernalsteens JP, Leemans J, De Greve H, Willmitzer L, Otten L, Schröder J, Shaw C (1981) The development of host vectors for directed gene-transfers in plants ICN-UCLA Symp Molec Cell Biol 23:557–575

Schröder J, Schröder G, Huisman H, Schilperoort RA, Schell J (1981) The mRNA for lysopine dehydrogenase in plant tumor cells is complementary to a Ti-plasmid fragment. FEBS Lett 129:166–168

Shortle D, Koshland D, Weinstock GM, Botstein D (1980) Segment-directed mutagenesis: construction in vitro of point mutations limited to a small predetermined region of a circular DNA molecule. Proc Natl Acad Sci USA 77:5375–5379

Smith EF, Brown NA, Townsend CO (1911) Crown gall of plants: its cause and remedy. US Dept Agric Bur Pl Ind Bull 213:1–200

Smith HH (1972) Plant genetic tumors. In: Braun AC (ed) Plant tumor research. S. Karger, Basel (Progress in experimental tumor research, vol. 15), pp 138–164

Southern EM (1975) Detection of specific sequences among DNA fragments separated by gel electrophoresis. J Mol Biol 98:503–518

Tempé J, Guyon P, Petit A, Ellis JG, Tate ME, Kerr A (1980) Préparation et propriétés de nouveaux substrats cataboliques pour deux types de plasmides oncogènes d'*Agrobacterium tumefaciens*. C R Acid Sci Paris 290 Série D:1173–1176

Tennhammer-Ekman B, Sköld O (1979) Trimethoprim resistance plasmids of different origin encode different drug-resistant dihydrofolate reductase. Plasmid 2:334–346

Thomashow MF, Nutter R, Montoya AL, Gordon MP, Nester EW (1980a) Integration and organisation of Ti-plasmid sequences in crown gall tumors. Cell 19:729–739

Thomashow MF, Nutter R, Postle K, Chilton M-D, Blattner FR, Powell A, Gordon MP, Nester EW (1980b) Recombination between higher plant DNA and the Ti plasmid of *Agrobacterium tumefaciens*. Proc Natl Acad Sci USA 77:6448–6452

Thomashow MF, Panagopoulos CG, Gordon MP, Nester EW (1980c) Host range of *Agrobacterium tumefaciens* is determined by the Ti plasmid. Nature 283:794–796

Van Larebeke N, Engler G, Holsters M, Van den Elsacker S, Zaenen I, Schilperoort RA, Schell J (1974) Large plasmid in *Agrobacterium tumefaciens* essential for crown gall-inducing ability. Nature 252:169–170

Van Larebeke N, Genetello C, Schell J, Schilperoort RA, Hermans AK, Hernalsteens JP, Van Montagu M (1975) Acquisition of tumour-inducing ability by non-oncogenic agrobacteria as a result of plasmid transfer. Nature 255:742–743

Van Larebeke N, Genetello C, Hernalsteens JP, Depicker A, Zaenen I, Messens E, Van Montagu M, Schell J (1977) Transfer of Ti-plasmids between *Agrobacterium* strains by mobilization with the conjugative plasmid RP4. MGG 152:119–124

Van Montagu M, Schell J (1979) The plasmids of *Agrobacterium tumefaciens*. In: Timmis K, Pühler A (eds) Plasmids of medical, environmental and commercial importance. Elsevier, Amsterdam, pp 71–96

Watson B, Currier TC, Gordon MP, Chilton M-D, Nester EW (1975) Plasmid required for virulence of *Agrobacterium tumefaciens*. J Bacteriol 123:255–264

Willmitzer L, De Beuckeleer M, Lemmers M, Van Montagu M, Schell J (1980) The Ti-plasmid

derived T-DNA is present in the nucleus and absent from plastids of plant crown-gall cells. Nature 287:359–361

Willmitzer L, Otten L, Simons G, Schmalenbach W, Schröder J, Schröder G, Van Montagu M, De Vos G, Schell J (1981) Nuclear and polysomal transcripts of T-DNA in octopine crown gall suspension and callus cultures. MGG 182:255–262

Wullems GJ, Molendijk L, Ooms G, Schilperoort RA (1981) Retention of tumor markers in F1 progeny plants from in vitro induced octopine and nopaline tumor tissues. Cell 24:719–727

Yadav NS, Postle K, Saiki RK, Thomashow MF, Chilton M-D (1980) T-DNA of a crown gall teratoma is covalently joined to host plant DNA. Nature 287:458–461

Yang F, McPherson JC, Gordon MP, Nester EW (1980) Extensive transcription of foreign DNA in a crown gall teratoma. Biochem Biophys Res Commun 92:1273–1277

Zaenen I, Van Larebeke N, Teuchy H, Van Montagu M, Schell J (1974) Supercoiled circular DNA in crown gall inducing *Agrobacterium* strains. J Mol Biol 86:109–127

Zambryski P, Holsters M, Kruger K, Depicker A, Schell J, Van Montagu M, Goodman HM (1980) Tumor DNA structure in plant cells transformed by *A. tumefaciens*. Science 209:1385–1391

Subject Index

Current Topics in Microbiology and Immunology

Editors: W. Henle, P. H. Hofschneider, H. Koprowski, O. Maaløe, F. Melchers, R. Rott, H. G. Schweiger, P. K. Vogt

Volume 91

1981. 70 figures. V, 284 pages
ISBN 3-540-10722-3

Contents:
B. N. Fields: Genetic of Reovirus. – *R. L. Erikson:* The Transforming Protein of Avian Sarcoma Viruses and Its Homologue in Normal Cells. – *D. H. Spector:* Gene-Specific Probes for Avian Retroviruses. – *T. Ben-Porat:* Replication of Herpesvirus DNA. – *G. Wick, R. Boyd, K. Hála, L. de Carvalho, R. Kofler, P. U. Müller, R. K. Cole:* The Obese Strain (OS) of Chicken with Spontaneous Autoimmune Thyroiditis: Review of the Recent Data. – *H. Kleinkauf, H. von Döhren:* Nucleic Acid-independent Synthesis of Peptides. – *P. H. Krammer:* The T-Cell Receptor Problem. – *W. S. Hayward, B. G. Neel:* Retroviral Gene Expression

Volume 92
Natural Resistance to Tumors and Viruses

Editor: O. Haller
1981. 22 figures. VI, 128 pages
ISBN 3-540-0732-0

Contents:
M. A. Brinton: Genetically Controlled Resistance to Flavivirus and Lactate-Dehydrogenase-Elevating Virus-Induced Disease. – *C. Lopez:* Resistance to Herpes Simplex Virus-Type 1 (HSV-1). – *O. Haller:* Inborn Resistance of Mice to Orthomyxoviruses. – *J.-L. Virelizier:* Role of Macrophages and Interferon in Natural Resistance to Mouse Hepatitis Virus Infection. – *V. Kumar and M. Bennett:* Genetic Resistance to Friend Virus-Induced Erythroleukemia and Immunosuppression. – *R. M. Welsh:* Natural Cell-Mediated Immunity During Viral Infections. – *R. Kiessling and H. Wigzell:* Surveillance of Primitive Cells by Natural Killer Cells

Volume 93
Initiation Signals in Viral Gene Expression

Editor: A. J. Shatkin
1981. 30 figures. V, 212 pages
ISBN 3-540-10804-1

Contents:
A. J. Shatkin: Introduction – Elucidating Mechanisms of Eukaryotic Genetic Expression by Studying Animal Viruses. – *R. Tjian:* Regulation of Viral Transcription and DNA Replication by the SV40 Large T Antigen. – *T. Shenk:* Transcriptional Control Regions: Nucleotide Sequence Requirements for Initiation by RNA Polymerase II and III. – *S. J. Flint:* Splicing and the Regulation of Viral Gene Expression. – *M. Kozak:* Mechanism of mRNA Recognition by Eukaryotic Ribosomes During Initiation of Protein Synthesis. – *R. M. Krug:* Priming of Influenza Viral RNA Transcription by Capped Heterologous RNAs. – *J. Perrault:* Origin and Replication of Defective Interfering Particles. – Subject Index

Volume 94/95

1981. 46 figures, 38 tables. Approx. 336 pages
ISBN 3-540-10803-3

Contents:
C. W. Ward: Structure of the Influenza Virus Hemagglutinin. – *H. G. Boman, H. Steiner:* Humoral Immunity in Cecropia Pupae. – *G. Hobom:* Replication Signals in Procaryotic DNA. – *W. Ostertag, I. B. Pragnell:* Differentiation and Viral Involvement in Differentiation of Transformed Mouse and Rat Erythroid Cells. – *J. Meyer:* Electron Microscopy of Viral DNA. – *J. Hochstadt, H. L. Ozer, C. Shopsis:* Genetic Alteration in Animal Cells in Culture

Springer-Verlag
Berlin
Heidelberg
New York

A. A. M. Gribnau, L. G. M. Geijsberts

Developmental Stages in the Rhesus Monkey (Macaca mulatta)

1981. 27 figures. VI, 84 pages
(Advances in Anatomy, Volume 68)
ISBN 3-540-10469-0

S. Ohno

Major Sex-Determining Genes

1979. 34 figures, 6 tables. XIII, 140 pages
(Monographs on Endocrinology, Volume 11)
ISBN 3-540-08965-9

T. Ohta

Evolution and Variation of Multigene Families

1980. 25 figures, 14 tables. VIII, 131 pages
(Lecture Notes Biomathematics, Volume 37)
ISBN 3-540-09998-0

Trisomy 21

An International Symposium. Convento delle
Clarisse, Rapallo, Italy, November 8–10, 1979.
Organizers: G. R. Burgio, M. Fraccaro, L. Tiepolo
Editors: G. R. Burgio, M. Fraccaro, L. Tiepolo,
U. Wolf
With contributions by numerous experts
1981. 55 figures, 55 tables. VI, 265 pages
(Human Genetics, Supplementum 2)
ISBN 3-540-10653-7

F. Vogel, A. G. Motulsky

Human Genetics

Problems and Approaches
1979. 420 figures, 210 tables. XXVIII, 700 pages
ISBN 3-540-09459-8

Springer-Verlag
Berlin
Heidelberg
NewYork